HANDBOOK OF MATERIAL FLOW ANALYSIS

For Environmental, Resource, and Waste Engineers

Second Edition

HANDBOOK OF MATERIAL FLOW ANALYSIS

For Environmental, Resource, and Waste Engineers

Second Edition

Paul H. Brunner • Helmut Rechberger

CRC Press
Taylor & Francis Group
Boca Raton London New York

CRC Press is an imprint of the
Taylor & Francis Group, an **informa** business

CRC Press
Taylor & Francis Group
6000 Broken Sound Parkway NW, Suite 300
Boca Raton, FL 33487-2742

First issued in paperback 2020

© 2017 by Taylor & Francis Group, LLC
CRC Press is an imprint of Taylor & Francis Group, an Informa business

No claim to original U.S. Government works

ISBN 13: 978-0-367-57409-3 (pbk)
ISBN 13: 978-1-4987-2134-9 (hbk)

Library of Congress Cataloging-in-Publication Data

Names: Brunner, Paul H., 1946- author. | Rechberger, Helmut, author.
Title: Handbook of material flow analysis : for environmental, resource, and waste engineers / Paul H. Brunner and Helmut Rechberger.
Other titles: Practical handbook of material flow analysis
Description: Boca Raton : Taylor & Francis, CRC Press, 2017. | Revised edition of: Practical handbook of material flow analysis. | Includes bibliographical references and index.
Identifiers: LCCN 2016027254 | ISBN 9781498721349 (hardback)
Subjects: LCSH: Materials management--Handbooks, manuals, etc. | Environmental engineering--Handbooks, manuals, etc.
Classification: LCC TS161 .B78 2017 | DDC 658.7--dc23
LC record available at https://lccn.loc.gov/2016027254

Visit the Taylor & Francis Web site at
http://www.taylorandfrancis.com

and the CRC Press Web site at
http://www.crcpress.com

To Sandra and Heidi

Contents

Preface to the Second Edition

A dozen years after the first edition was put on the market, the question of whether a second edition of the *Handbook of Material Flow Analysis* (MFA) is justified seems appropriate. From a pure numerical standpoint, the increase from a few hundred papers per year in 1975 to more than 5000 publications in 2015 proves that MFA is of growing interest to a rising number of researchers, readers, and writers. And from a content point of view, the range of important fields where MFA is applied and further developed is still expanding. More recently, MFA methodology has been extended and refined, particularly with respect to including dynamic MFA, uncertainty considerations, and supporting software. Hence, we hope that this second edition serves an even larger readership of engineers and natural scientists than the first one.

Originally, MFA was developed to describe and analyze single processes, e.g., in chemical production or in waste management. However, during the last two decades, there was rapid progress toward assessment of complex systems such as regional material balances, substance flows through large watersheds, and material flows and stocks in national and global economies. Today, the greatest challenge is to analyze the entire interaction of anthropogenic and natural systems, covering several scales and comprising many subsystems. Thus, MFA has become one of the core methodologies of scientific fields such as industrial ecology, resource management, and waste management. The second edition of the handbook takes into account all these new developments. In particular, newly added citations point to recent and relevant literature, additional case studies demonstrate the potential of MFA to solve industrial ecology challenges, and the new chapter on STAN software allows a uniform, standardized approach to establishing MFA. Compared to the first edition, the discussion of a few methods that are usually subsumed under MFA methodology such as materials accounting and input–output analysis has not been expanded. The reason is that, first, the MFA field, as mentioned previously, expanded rapidly, making it impossible to include all developments comprehensively in one handbook. Second, there are excellent books that cover materials accounting and input–output analysis that we recommend in the corresponding chapters. And third, the particular feature of this book is to combine the levels of material flow and substance flow analysis, to include uncertainties, and to present powerful software that is capable of combining goods, substances, and energy. In addition, we provide electronic materials to support readers, such as the homepage of the handbook containing the solutions to the problems at http://www.MFA-handbook.info and the STAN website that comprises additional case studies (http://www.stan2web.net/).

In spite of these changes, the book is still written for a general audience, guiding inexperienced as well as experienced engineers and scientists toward mastering the fine art of MFA from the very beginning to an advanced state of material balances of complex systems. The main purpose of the book is unchanged: to offer and promote a transparent and reproducible methodology as a basis for analysis, assessment, and improvement of anthropogenic systems. We hope that the methodology presented here will be widely applied, generating new knowledge, understanding, and data that can be used by the growing MFA community. After all, the task ahead of us, to utilize materials in a way that neither jeopardizes future generations nor puts at risk the environment and the availability of resources, requires a huge global effort.

We are particularly grateful for the support of our colleagues from Technische Universität Wien who dedicated their time and enthusiasm to the handbook: Oliver Cencic authored Sections 2.3 and 2.4, and David Laner contributed substantially to Chapters 2 and 3. The valuable backing of Rudolf Frühwirt in the statistical part of Chapter 2 is also acknowledged. We are thankful for contributions to the case studies by Dana Vyzinkarova. We appreciate the work on tables, graphs, bibliography, and copyright by Inge Hengl and Maria Gunesch. The initiative for this second edition came from Taylor & Francis senior editor Irma Shagla Britton, and we acknowledge her great support.

Paul H. Brunner and Helmut Rechberger
Andermatt and Vienna

MATLAB® is a registered trademark of The MathWorks, Inc. For product information, please contact:

The MathWorks, Inc.
3 Apple Hill Drive
Natick, MA 01760-2098 USA
Tel: 508 647 7000
Fax: 508-647-7001
E-mail: info@mathworks.com
Web: www.mathworks.com

Preface to the First Edition

About 40 years ago, Abel Wolman coined the term "metabolism of cities" in an article for the *Scientific American*. His pioneering view of a city as a living organism with inputs, stocks, and outputs of materials and energy has since inspired many others. Today, studies describing metabolic processes of companies, regions, cities, and nations are becoming more abundant. While the phenomenology of the anthroposphere is described in books and papers, the methodology lags behind. It is the purpose of this handbook to contribute to the establishment and dissemination of a rigid, transparent, and useful methodology to investigate the material metabolism of anthropogenic systems.

Both authors have been using and developing material flow analysis (MFA) for many years. We experienced the great value of this method in various fields such as environmental management, resources management, waste management, and water quality management. And we decided to fill the gap and to produce a book to share our experience with engineering students, professionals, and a wider audience. The aim is to promote MFA and to facilitate the use of MFA in a uniform way so that future engineers have a common method in their toolbox for solving resource-oriented problems.

The hidden agenda behind the handbook comprises two objectives: resource conservation and environmental protection, or "sustainable materials management." We believe that man-made activities should not destroy or damage natural resources and systems. Future generations must be able to enjoy resources and the environment as we do. We also believe that this goal can be achieved if technology and social sciences are developed further. The case studies presented in this book exemplify the potential of MFA to contribute to sustainable materials management.

This is a handbook. It is directed toward the practitioner. The 14 case studies demonstrate how to apply MFA in practice. The "Problems" sections following each chapter serve to exercise and deepen comprehension and expertise. The tool MFA is not perfect yet: it is still subject to further refinement. If the reader finds the handbook useful for better understanding and for designing anthropogenic systems, we have accomplished our goals. Since this book is not the endpoint yet, we appreciate any comments and suggestions you may have. We hope that the handbook encourages application of and discussion about MFA, and we are looking forward to your comments on the website http://iwr.tuwien.ac.at/ressourcen/mfa-handbook/home. You will also find the solutions to the problems on this site.

We are grateful to Oliver Cencic, who wrote Sections 2.3 and 2.4 about data uncertainty and MFA software. The support of the members of the Waste and Resources Management Group at the Vienna University of Technology in editing the final manuscript is greatly acknowledged. Demet Seyhan was instrumental in preparing the case studies on phosphorous. Bob Ayres and Michael Ritthoff supplied important comments on Section 2.5. We are indebted to Inge Hengl, who did all the artwork and managed the genesis and finalization of the manuscript in an excellent way. Helmut Rechberger thanks Peter Baccini for conceding the time to contribute to this handbook. Finally, we are particularly grateful for critical reviews by Bob Dean, Ulrik Lohm, Stephen Moore, and Jakov Vaisman, who evaluated a first draft of this handbook.

Paul H. Brunner and Helmut Rechberger
Vienna

Authors

Paul H. Brunner, recognized for his groundbreaking research work in the fields of waste and resource management, is professor emeritus at the Technische Universität Wien in Vienna, Austria. Together with Peter Baccini from ETH Zurich, he published the landmark book *Metabolism of the Anthroposphere*, presenting a new view of the interactions among human activities, resources, and the environment. His work focuses on advanced methods for waste treatment and on concepts to assess, evaluate, and design metabolic systems. For more than 40 years, Dr. Brunner has been engaged in research and teaching in Europe, Asia, and the Americas. As a member of many boards of journals and conferences, he successfully promoted the application of material flow analysis for improving decision making in resource and waste management.

Helmut Rechberger has been a professor for resource management since 2003 and heads the Research Center of Waste and Resource Management at Technische Universität Wien. He received his professional education in Vienna with Paul H. Brunner, at ETH Zurich with Peter Baccini, and at Yale with Tom Graedel. Since then, he has been engaged in research and teaching and has become internationally acknowledged for his research work in the fields of resource and waste management. Under his lead, the material flow analysis (MFA) software STAN has been developed. His current research interests are focused on the further development of methods to advance sustainable regional and urban resource management.

1

Introduction

1.1 Objectives and Scope

The meeting was long and intense. After all, more than $50 million had been invested in this waste treatment plant, and still, the objective of the treatment, namely, to produce recycling materials with a certain quality, could not be reached. Engineers, plant operators, waste management experts, financiers, and representatives from government were discussing means to improve the plant to reach the goals set. A chemical engineer took a piece of paper and asked about the content of mercury, cadmium, polybrominated diphenyl ethers (PBDEs), and some other hazardous substances in the incoming waste. The waste experts had no problem indicating a range of concentrations. The engineer then asked about the existing standards for the products, namely, cellulose fibers, plastic fraction, and compost. Again, he got the information needed. After a few calculations, he said, "If the plant is to produce a significant amount of recycling material at the desired specifications, it must be able to divert more than 80% of the hazardous substances from the waste received to the fraction that is incinerated. Does anybody know of a mechanical treatment process capable of such partitioning?" Since none of the experts present was aware of a technology to solve the problem at affordable costs, the financiers and government representatives started to question why such an expensive and state-of-the-art plant could not reach the objective. It was the old mayor of the local community, who experienced most problems with the plant because of citizens complaining about odors and product quality, who said, "It seems obvious: garbage in, garbage out; what else can you expect?"

The purpose of this book is to prevent such debacles. The methods presented here enable the reader to design processes and systems in view of careful resource management. *Resources* in this context stands for materials, energy, the environment, and wastes. Emphasis is laid on the linkage between sources, pathways, and sinks of materials, observing the law of conservation of matter. The book is a practical handbook; it is directed toward the practitioner. Hence, many case studies, examples, and problems are included. If the readers take advantage of these means for exercise, they will soon become well acquainted with the techniques needed for successful application of material flow analysis (MFA).

In addition to serving as a practical handbook, this volume also contains directions for those readers who are looking for sustainable resource management. The authors share the opinion that economic and technological progress will be most beneficial to mankind if long-term protection of the environment is assured, and if resources are used in a careful, little-dissipative way. In this book, we give evidence that current management of the anthroposphere may result in serious long-term burdens and that changes are needed to improve opportunities of today's and future generations. Such changes are already taking place, they are feasible, and they will eventually improve the quality of life. The authors are convinced that if decisions for changes are based, among others, on MFA, they will yield better results. They see the need for balance not only for technical systems but also for social and economic systems. However, this book has been written for a technical readership. Since the authors are engineers and chemists, social and economic science issues are sometimes mentioned in this volume but are never discussed to the necessary depth.

This book is directed toward engineers in the fields of resource management, environmental management, and waste management. Professionals active in the design of new goods, processes and systems will profit from MFA-based approaches because they facilitate incorporating environmental and resource consideration into the design process. The potential audience comprises private companies and consulting engineers operating in the aforementioned fields, government authorities, and educational institutions at the graduate and postgraduate level. In particular, the book is designed as a textbook for engineering students who are looking for a comprehensive and in-depth education in the field of MFA. It is strongly recommended to focus on the case studies and problems: they show best how MFA is applied in practice and how to interpret and use the results. At the end of each chapter, a problem section allows readers to exercise the newly acquired knowledge, to learn application of MFA, to gain experience, and to check one's ability to solve MFA problems.

The book is structured into four chapters: In the introductory chapter, a short overview of MFA is given in order to facilitate understanding of the main body of this chapter, the history, objectives, and application range of MFA. From a methodological point of view, Chapter 2 is the most important: it explains comprehensively the terms, definitions, and procedures of MFA, and discusses the software STAN best suited for MFA. Case studies are given in Chapter 3. They are essential in enabling the reader to experience the application range of MFA (STAN, 2016; Cencic, Rechberger, and Kovacs, 2006). The case studies make clear that the concept of MFA goes beyond simple input and output balances of single processes, and that analyzing flows and stocks of a complex real-world system is a challenging and often interdisciplinary task. The book ends in Chapter 4 with an outlook to potential future developments. Literature references are given at the end of the book. Problem sections accompany subchapters where appropriate.

1.2 What Are Material Flow Analysis and Substance Flow Analysis?

Material flow analysis (MFA), sometimes referred to as substance flow analysis (SFA) if a specific substance is the focus, is a systematic assessment of the state and changes of flows and stocks of materials within a system defined in space and time. MFA connects the sources, the pathways, and the intermediate and final sinks of a material. Because of the law of conservation of matter, the results of an MFA can be controlled by a simple mass balance comparing all inputs, stocks, and outputs of a process. It is this distinct characteristic of MFA that makes the method attractive as a decision-support tool in resource management, waste management, environmental management, and policy assessment.

An MFA delivers a complete and consistent set of information about all flows and stocks of a particular material over time within a spatially defined system. Through balancing inputs and outputs, the flows of wastes and environmental loadings become visible, and their sources can be identified. The depletion or accumulation of material stocks is recognized early enough, either to take countermeasures or to promote further buildup and future utilization (such as for urban mining). Moreover, if MFAs are performed for longer time periods, minor changes that are too small to be measured in short time scales but that could slowly lead to long-term damage also become evident.

Anthropogenic systems consist of more than material flows and stocks (Figure 1.1). Energy, space, information, and socioeconomic issues must also be included if the anthroposphere is to be managed in a responsible way. MFA can be performed without considering these aspects, but in most cases,

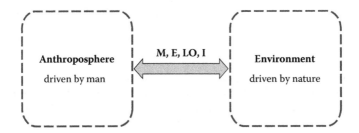

FIGURE 1.1
The two systems *anthroposphere* and *environment* exchange flows of materials (M), energy (E), living organisms (LO), and information (I).

these other factors are needed to interpret and make use of the MFA results. Thus, MFA is frequently coupled with the analysis of energy, economy, urban planning, and the like.

A common language is needed for the investigation into anthropogenic systems. Such commonality facilitates comparison of results from different MFAs in a transparent and reproducible way. In this handbook, terms and procedures to analyze, describe, and model material flow systems are defined, enabling a comprehensive, reproducible, and transparent account of all flows and stocks of materials within a system. The methodology, presented in greater detail in Chapter 2, is completely employed in the software STAN (see Chapter 2, Section 2.4).

The term *material* stands for both substances and goods. In chemistry, a *substance* is defined as a single type of matter consisting of uniform units (Atkins and Beran, 1992). If the units are atoms, the substance is called an element, such as carbon or iron; if they are molecules, it is called a chemical compound, such as carbon dioxide or iron chloride. *Goods* are substances or mixtures of substances that have economic values assigned by markets. The value can be positive (car, fuel, wood) or negative (municipal solid waste, sewage sludge). In economic terms, the word *goods* is more broadly defined to include *immaterial* goods such as energy (e.g., electricity), services, or information. In MFA terminology, however, the term *goods* stands for *material goods* only. Nevertheless, the link between goods as defined by MFA and other goods as used by economists can be important when MFA is applied, for example, for decisions regarding resource conservation.

A *process* is defined as a transport, transformation, or storage of materials. The transport process can be a natural process, such as the movement of dissolved phosphorous in a river, or it can be man-made, such as the flow of gas in a pipeline or waste collection. The same applies to transformations (e.g., oxidation of carbon to carbon dioxide by natural forest fires versus man-made heating systems) and storages (e.g., natural sedimentation versus man-made landfilling).

Stocks are defined as material reservoirs (mass) within the analyzed system, and they have the physical unit of kilograms. A stock is part of a process comprising the mass that is stored within the process. Stocks are essential characteristics of a system's metabolism. For steady-state conditions (input equals output), the mean residence time of a material in the stock can be calculated by dividing the material mass in the stock by the material flow in or out of the stock. Stocks can stay constant, or they can increase (accumulation of materials) or decrease (depletion of materials) in size.

Processes are linked by *flows* (mass per time) or fluxes (mass per time and cross section) of materials. Flows/fluxes across system boundaries are called *imports* or *exports*. Flows/fluxes of materials entering a process are named *inputs*, while those exiting are called *outputs*.

A *system* comprises a set of material flows, stocks, and processes within a defined boundary. The smallest possible system consists of just a single process. Examples of common systems for investigations by MFA are a region, a municipal incinerator, a private household, a factory, a farm, etc. The *system boundary* is defined in space and time. It can consist of geographical borders (region) or virtual limits (e.g., private households, including processes serving the private household such as transportation, waste collection, and sewer systems). When the system boundary in time is chosen, criteria such as objectives, data availability, appropriate balancing period, residence time of materials within stocks, and others have to be taken into account. This is discussed further in Chapter 2, Section 2.1.7.

If only one substance is the focus of an MFA study, such as phosphorus in Chapter 3, Section 3.5.2, such a study might be designated as SFA: SFA can be considered a special type of MFA with $n = 1$ (with n as the number of investigated substances). However, goods (e.g., mineral fertilizer) containing the substance (e.g., phosphorus) are instrumental for calculating flows. In most cases, the purpose of system design is to optimize substance flows, but this is usually done by changing flows of goods, because the substance is contained in a good. When focusing on the substance level with an SFA, one should keep in mind that both goods and substances (i.e., materials) are an integral part of MFA and SFA. MFA is the more universal term compared to SFA.

In addition to the basic terms necessary to analyze material flows and stocks, the notion of *activity* is useful when evaluating and designing new anthropogenic processes and systems. An activity comprises a set of systems consisting of flows, stocks, and processes of the many materials that are necessary to fulfill a particular basic human need, such as to nourish, to reside, or to transport and communicate (Baccini and Brunner, 2012). Analyzing material flows associated with a certain activity allows early recognition of problems such as future environmental loadings and resource depletions. One of the main questions for the future development of mankind will be, "Which sets of processes, flows, and stocks of goods, substances, and energy will enable long-term, efficient, and sustainable feeding of the increasing global population?" Equally important is the question, "How can the transportation needs of an advanced global population be satisfied without compromising the future resources of mankind?" When alternative scenarios are developed for an activity, MFA can help to identify major changes in material flows. Thus, MFA is a tool to evaluate existing systems for food production, transportation, and other basic human needs, as well as to support the design of new, more effective systems.

1.3 History of MFA

Long before MFA became a tool for the management of resources, wastes, and the environment, the mass-balance principle has been applied in such diverse fields as medicine, chemistry, economics, engineering, and life sciences. The basic principle of any MFA, the conservation of matter, or "input equals output," had already been postulated by Greek philosophers more than 2000 years ago. It was the French chemist Antoine Lavoisier (1743–1794) who provided experimental evidence that the total mass of matter cannot be changed by chemical processes: "neither man made experiments nor natural changes can create matter, thus it is a principle, that in every process the amount of matter does not change" (Vidal, 1985).

In the twentieth century, the field of MFA emerged in various areas at different times. Although the name of MFA had not been invented and a comprehensive methodology had not been established yet, many authors used the principle stated by the law of conservation of matter to balance processes. In process and chemical engineering, it was common practice to analyze and balance inputs and outputs of chemical reactions. In the economic field, Leontief introduced input–output tables in the 1930s (Leontief, 1977, 1966) and thus laid the basis for widespread application of input–output methods to solve economic problems. In the field of resource conservation and environmental management, the first studies appeared in the 1970s. Two areas of application were originally in the foreground: metabolism of cities and investigations into pollutants in regions such as watersheds or urban areas. In the following decades, MFA became a widespread tool in many fields, such as process control, waste and wastewater treatment, agricultural nutrient management, water quality management, resource conservation and recovery, product design, life cycle assessment (LCA), and others.

1.3.1 Santorio's Analysis of the Human Metabolism

One of the first reports about an analysis of material flows was prepared by Santorio Santorio (1561–1636), also called S. Sanctuarius, in the seventeenth century. The similarity of the conclusions regarding the analysis of the anthropogenic metabolism between Santorio and modern authors is astonishing. Santorio was a doctor of medicine practicing in Padua and President of the Venetian College of Physicians in Venice. His main interest was to understand the human metabolism. He developed a first method to balance inputs and outputs of a person (Figure 1.2a through c). Santorio measured the weight of the person, of the food and beverages he/she ate, and of the excretions he/she gave off. The result of his investigation was disappointing and surprising at the same time: he could not close the mass balance. However, he found that the visible material output of a person was less than half what the person actually takes in. He suspected that some yet-unknown insensible perspiration left the body at night. Thus, he wrapped the person in a hide during the night's sleep. But the little amount of sweat he collected by this procedure did not account for the large missing fraction. It was not known yet in the sixteenth century that the volume of air a person breathes

(a)

(b)

FIGURE 1.2
The experimental setup of Santorio Santorio (1561–1636) to analyze the material metabolism of a person. (a) The person is sitting on a chair attached to a scale. The weight of the food and the person are measured. (b) Despite the fact that all human excreta were collected and weighed, input A and output B do not balance. What is missing? *(Continued)*

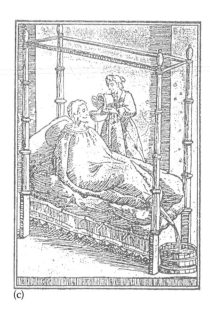

(c)

FIGURE 1.2 (CONTINUED)
The experimental setup of Santorio Santorio (1561–1636) to analyze the material metabolism of a person. (c) Santorio's measurements could not confirm his hypothesis that an unknown fluidum leaves the body during the night, but they proved that more than half of the input mass leaves the body by an unknown pathway.

has a certain mass. It should be remembered that Santorio lived a century before Lavoisier investigated oxidation processes and proved the existence of oxygen. Hence, Santorio did not account for the air a person was breathing, neither the intake of fresh air nor the emission of spent air. In 1614, Santorio published a book about his metabolic studies, *De Medicina Statica Aphorismi*, which made him well known as the "father of the science of human metabolism." He concluded that "a doctor, who carries the responsibility for the health of his patients, and who considers only the visible processes of eating and excreting, and not the invisible processes resulting in the loss of insensible perspiration, will only mislead his patients and never cure them from their disease." Considering that Santorio was certainly the first and probably the only one who performed such metabolic studies at that time, this statement may have secured him a visit from many patients.

The experiments of Santorio allow conclusions similar to recent MFA studies: First, it still is impossible to evaluate and optimize anthropogenic systems, i.e., to cure the patient, without knowing the material flows and stocks, that is, the metabolism of the system. Second, it is difficult to balance a process or system if basic information about this process or system such as major input and output goods is missing. Third, it happens quite often that inputs and outputs of a process or system do not balance, pointing to new research questions. And fourth, analytical tools are often not appropriate or precise enough to measure changes in material balances with the necessary accuracy.

1.3.2 Leontief's Economic Input–Output Methodology

Wassily W. Leontief (1906–1999) was an American economist of Russian origin. His research was focused on the interdependence of anthropogenic production systems. He searched for analytical tools to investigate the economic transactions between the various sectors of an economy. One of his major achievements, for which he was awarded the 1973 Nobel Prize in Economic Sciences, was the development of the input–output method in the 1930s (Major, 1938; Leontief, 1977). At the core of the method are so-called input–output tables: These tables allow systematic quantifying of the mutual interrelationship among the various sectors of a complex economic system. They connect goods, production processes, deliveries, and demand in a stationary as well as in a dynamic way. The production system is described as a network of flows of goods (provisions) between the various production sectors. Input–output analysis of economic sectors has become a widespread tool in economic policy making. It proved to be highly useful for forecasting and planning in market economies as well as in centrally planned economies, and was often applied to analyze the sudden and large changes in economies due to restructuring.

In order to investigate the effect of production systems on the environment, the original method was later expanded to include production emissions and wastes. More recently, the input–output method has been incorporated into LCAs to establish the Economic Input–Output LCA method (Hendrickson, Horvath, Joshi, and Lave, 1998). This expansion allows assessment of the relative emissions and resource consumption of different types of goods, services, and industries. The advantage of using input–output methodology for LCA is the vast amount of available information in the form of input–output tables for many economies. This information can be used for LCA as well.

1.3.3 Analysis of Urban Metabolism

Santorio has analyzed the physiological, "inner" metabolism of men. This is only a minor part of the modern anthropogenic turnover of materials. The "outer" metabolism, consisting of the use and consumption of goods not necessary from a physiological point of view, has grown much larger than the inner metabolism. Hence, in places with a high concentration of population and wealth such as modern urban areas, large amounts of materials, energy, and space are consumed. Today, most cities grow fast in population and size, and they comprise a large and growing stock of materials.

The first author to use the term *metabolism of cities* was Abel Wolman in 1965 (Wolman, 1965). He used available US data on consumption and production of goods to establish per capita input and output flows for a hypothetical American city of 1 million inhabitants. He linked the large amounts of wastes that are generated in a city to its inputs. Other authors have been fascinated, too, by the complex urban metabolism. They developed more specific methods to quantify the urban turnover of energy and materials, and investigated the effects of the large flows on resource depletion and the environment. Two prominent examples are the study of Brussels by Duvigneaud and Denayeyer-De Smet (1975) and of Hong Kong by Newcombe, Kalma, and Aston (1978).

In 1975, Duvigneaud and Denayeyer-De Smet analyzed the city of Brussels using natural ecosystems as an analogy (Duvigneaud and Denayeyer-De Smet, 1975). They assessed the total imports and exports of goods such as fuel, construction materials, food, water, wastes, sewage, emissions, etc., in and out of the city, and established an energy balance. The authors concluded that Brussels was highly dependent on its hinterland, the global economy surrounding the city of Brussels: all energy is imported. Since solar energy theoretically available within the city equaled all of Brussels's energy demand, this dependency could be reduced by shifting from fossil fuels to solar energy supply. Water produced within the city by precipitation is not utilized; all drinking water is imported. Materials such as construction materials and food are not recycled after their use; they are exported as wastes. The linear flows of energy and materials result in high pollution loads that deteriorate the quality of water, air, and soil of the city and its surroundings. The authors point out the necessity to change the structures of cities in a way that improves the utilization of energy and materials, creates material cycles, and reduces losses to the environment. As does Santorio, for the health of a person, they conclude that for continual welfare of a city, it is necessary to know the metabolism of the city. They recognize that only an interdisciplinary approach will succeed in analyzing, defining, and implementing the necessary measures of change.

Around the same time in the beginning of the 1970s, Newcombe and colleagues started their investigation into the metabolism of Hong Kong. This Asian city was experiencing a rapid transition period of high population growth and intense economic development due to its privileged position at the interface between Western trade and Eastern production and manufacturing. Hong Kong was an ideal case for metabolic studies, since city limits coincide more or less with state boundaries. Thus, in contrast to the Brussels case, economic data from state statistics were available for an accurate assessment of the import and export goods of Hong Kong. Also, Hong Kong differs from Brussels in its high population density, and lower per capita income and material throughput. The authors found that the material and energy used for Hong Kong's infrastructure was about one order of magnitude smaller than in more developed cities. They conclude that a worldwide increase in material consumption to the level of modern cities would require very large amounts of materials and energy and would have negative impacts on global resources and the environment. They also state that in order to find sustainable solutions for the future development of cities, it is necessary to know the urban metabolism, and hence, it is important to be able to measure the flows of goods, materials, and energy through urban systems. In 1997, König (1997) revisited the metabolism of Hong Kong, showing the impact of the large increase in material turnover on the city and its surroundings.

1.3.4 Regional Material Balances

At the end of the 1960s, the first studies on heavy metal accumulation in regions were initiated. In order to identify and quantify the sources of metals, methods such as pathway analysis and material balancing were developed and applied in regional studies. One of the groundbreaking regional material flow studies has been reported by Huntzicker, Friedlander, and Davidson (1975). In 1972, these authors established a balance for automobile-emitted lead in the Los Angeles basin, which was revisited by Lankey, Davidson, and McMichael (1998). The authors developed a material balance method that was based on the measurements of atmospheric particle size distributions, atmospheric lead concentrations, and surface deposition fluxes. The results of the studies show the important sources, pathways, and sinks for lead in the Los Angeles basin, and point to potential environmental problems and solutions. The authors state that their method did not allow them to elucidate every detail of substance dispersion in the environment. Yet the mass-balance approach brought forth all important pathways, their identification and quantification. They conclude that the "material balance-flow pathway approach" was, in general, a powerful tool for assessing the environmental impact of a pollutant. The tool is also well suited to evaluate environmental management decisions such as the reduction of lead in automobile fuels: the authors show that lead inputs and outputs were roughly in agreement, and that the reduction of lead in automobile fuel decreased the overall input of lead in the Los Angeles basin significantly. Their MFA method allowed identification of resuspended road dust as an important "secondary" source of lead that is expected to decrease slowly over time.

Another pioneering study was undertaken by Ayres et al. in the early 1980s (Ayres et al., 1989). These authors analyzed the sources, pathways, and sinks of major pollutants in the Hudson–Raritan basin for a period of 100 years from 1885 to 1985. They chose heavy metals (Ag, As, Cd, Cr, Cu, Hg, Pb, and Zn); pesticides (DDT, TDE, aldrin, BHC/lindane, chlordane, and others); and "other critical pollutants" [PCB, polycyclic aromatic hydrocarbon (PAH), N, P, total organic carbon (TOC), etc.] as objects of their investigation. One of the major motivations for this project was to explore the long-term effects of anthropogenic activities on aquatic environments, in particular, on fish populations in the Hudson River bay. The authors chose the following procedure:

- Definition of system boundaries
- Establishment of a model linking sources, pathways, and sinks for each of the selected pollutants
- Historical reconstruction of all major flows in the system
- Adoption of data either from other regions or deduced from environmental transport models, where major substance flows could not be reconstructed from historical records
- Validation of the model by comparing measured concentrations in the river basin with results calculated by the model

The authors used the same MFA methodology to balance single processes like coal combustion as well as complex combinations of processes such as consumption processes in the Hudson River basin. Despite the scarcity of data for the historical reconstruction, the authors were able to establish satisfying agreement between their model and values measured in the environment of the basin.

Ayres et al. used the results of their study to identify and discriminate the main sources and sinks for each pollutant. They were able to distinguish the importance of point sources and nonpoint sources, and that of production and consumption processes, which they identified as prevailing sources for many pollutants. Their model permitted them to predict changes in the environment due to alterations in population, land use, regulations, etc. In the following decade, Ayres expanded his studies from single substances to more comprehensive systems, looking finally at the entire "industrial metabolism" (Ayres, 1994). He used MFA methodology to study complex material flows and cycles in industrial systems. He aimed at designing a more efficient industrial metabolism by improving technological systems, by inducing long-term planning based on resource conservation and environmental protection, and by producing less waste and recycling more materials.

In the 1980s and 1990s, the papers by Huntzicker et al. and Ayres et al. were followed by many studies on substance flows in regions such as river basins, in nations, and on a global scale. (The term *region* is used in this chapter to describe a geographical area on the Earth's surface that can range from a small size of a few square kilometers to a large size such as a continent or even the globe. It is thus not used as defined in regional sciences, e.g., for urban planning.) Selected examples are as follows.

Rauhut and Balzer (1976) were among the first to publish national substance inventories similar to those of the US Bureau of Mines (1939–1969). Their studies were detailed enough to serve as a basis for policy decisions regarding heavy metals such as managing and regulating Cd in view of environmental protection.

Van der Voet, Klejin, Huppes, Udo de Haes, and others of the Centre of Environmental Science, Leiden University, in the Netherlands prepared several reports on the flows and stock of substances within the European economy and environment. They recognized the power of MFA as a tool to support policy decisions in the field of environmental management and waste management. Comprehensive Dutch MFA studies include chlorine (Kleijn, Van der Voet, and Udo de Haes, 1994), cadmium (Van der Voet, van Egmond, Kleijn, and Huppes, 1994), and other heavy metals (Gorter, 2000), as well as nutrients. Based on their experience with problems of transnational data acquisition, they advocated an internationally standardized use of MFA.

Stigliani and associates from the International Institute for Applied Systems Analysis (IIASA) in Laxenburg, Austria, assessed the flows of pollutants in the river Rhine basin using MFA methodology (Stigliani, Jaffé, and Anderberg, 1993). They point to the main sources and sinks of selected heavy metals and draw conclusions regarding the future management of pollutants in river basins. At the same institute, Ayres et al. were applying the material balance principle for selected chemicals. Their report (Ayres, Norberg-Bohm, Prince, Stigliani, and Yanowitz, 1989) concentrates on four widely used inorganic chemicals (Cl, Br, S, and N) and provides a better knowledge of their environmental implications. Also a new understanding on how societies produce, process, consume, and dispose of materials, and links of these activities to resource conservation and environmental change, is presented in this work. Ayres et al. achieved this by embedding MFA in the concept of industrial metabolism.

On a global scale, studies by geochemists such as Lantzy and Mackenzie (1979) become important for the understanding of large-scale geogenic and anthropogenic metal cycles of substances between lithosphere, hydrosphere, and atmosphere. In his books, Nriagu investigates the sources, fate, and behavior of substances such as arsenic, vanadium, mercury, cadmium and many others in the context of the protection of men and the environment (Nriagu, 1994). He presents a critical assessment of the chemistry and toxic effects as well as comprehensive information on local and global flows of these substances.

All the aforementioned regional and global studies are based on MFA. They have been selected to demonstrate the wide range of application of MFA and because they are recognized as pioneering studies in their research areas. The reader may certainly encounter more regional work of other authors active in the field that is not cited here.

1.3.5 Metabolism of the Anthroposphere

Baccini, Brunner, and Bader (Baccini and Brunner, 1991 and 2012; Baccini and Bader, 1996) extended MFA by defining a systematic and comprehensive methodology and by introducing the notions of "activity" and "metabolism of the anthroposphere" (cf. Chapter 2, Sections 2.1.8 and 2.1.9). Their main goal was to develop methods to analyze, evaluate, and control metabolic processes in man-made systems, and to apply methods to improve resource utilization and environmental protection on a regional level. Engaged in solving waste management problems, they recognized that the so-called filter strategy at the back end of the material chain is often of limited efficiency. It is more cost-effective to focus on the total substance flow and not just on the waste stream. Their integrated approach is directed toward the turnover of materials and energy, on activities and structures, and on the interdependency of these aspects in regions. In the project SYNOIKOS, Baccini, collaborating with a group of architects around Oswald, combined physiological approaches with structural approaches to analyze, redefine, and restructure urban regions (Baccini and Oswald, 1998). This project shows the full power of the combination of MFA with other disciplines to design new, more efficient, and sustainable anthropogenic systems.

Lohm, an entomologist studying the metabolism of ants, and Bergbäck (Lohm, Anderberg, and Bergbäck, 1994) were also among the first to use the notion of metabolism of the anthroposphere and to study metabolic processes by MFA. In their pioneering study of the metabolism of Stockholm, they focused on the stock of materials and substances in private households and the corresponding infrastructure. They identified the very large reservoir of potentially valuable substances such as copper and lead within the city. Lohm and Bergbäck drew attention to urban systems in order to prevent environmental pollution by the emissions of stocks, and to conserve and use the valuable substances hidden and, in part, hibernating in the city.

Fischer-Kowalsky et al. (1997) employed a somewhat similar set of tools and expanded the methodology by social science approaches. They coined the term "colonization" to describe the management of nature by human societies and investigated the transition from early agricultural societies to today's enhanced metabolism. Wackernagel, Monfreda, and Deumling (2002) developed a method to measure the ecological footprints of regions that is based in part on MFA. They conclude that regions in affluent societies use a very large hinterland for their supply and disposal, and suggest comparing and reducing ecological footprints of regions. They argue that progress had to be redefined: most of today's societies do not increase their welfare, if measured properly, and thus, they need to change the direction of their development (Wackernagel and Rees, 1996). Again based on MFA, Schmidt-Bleek (1997) and von Weizsäcker, Lovins, and Lovins (1997) from the Wuppertal Institute for Climate and Energy concluded the following: Considering environmental loadings and resource conservation, the turnover of materials in modern economies is much too high. In order to achieve sustainable development, they ask for a reduction of material flows by a factor of four to ten. Bringezu from the same group in Wuppertal established a platform for the discussion of materials accounting methodology called *Conaccount*, which was joined by many European research groups (Bringezu, Fischer-Kowalski, Klejin, and Palm, 1998). Also, the Wuppertal Institute started to collect and compare information about national material flows in several countries from Europe, Asia, and America. In *The Weight of Nations*, Matthews and colleagues document and compare the material outflows from five industrial economies (Austria, Germany, Japan, Netherlands, United States) (Matthews et al., 2000). They develop so-called physical indicators of material flows that complement national economic indicators such as gross domestic product (GDP). In 2000, a group of researchers from the Center for Industrial Ecology at Yale University headed by T. Graedel launched a several-year project to establish material balances for copper and zinc on a national, continental, and global level [the so-called stocks and flows project (STAF)] (Graedel, 2002).

1.3.6 Recent Developments in the Field of MFA

With respect to material flow accounting, that is, the reporting of economic data on flows and stocks of goods within a defined administrative unit, the number of regions and countries that participate in such has been increasing during the past 20 years (Dittrich, Bringezu, and Schütz 2012). Economy-wide material flow accounts aim at an aggregated overview about the physical flows of materials through economies as a basis for policy decisions regarding resources and environment. Major methodological efforts focus on (1) standardization and statistical recording and (2) the evaluation of resource efficiency based on indicators such as domestic material consumption (cf. Matthews et al., 2000; Bringezu, Schütz, and Moll, 2003; Science Communication Unit, 2013) or total material requirement (TMR). These indicators have been further developed, to provide a material input–oriented evaluation of products and services. Recent examples comprise the application of the TMR approach for assessing individual recycling processes like phosphorous recovery from steelmaking slag (Yamasue, Matsubae, Nakajima, Hashimoto, and Nagasaka, 2013) or for global material footprinting (Wiedmann, Schandl, Lenzen, Moran, Su, West, and Kanemoto, 2013): In analogy to carbon footprinting, the total amount of material resources that is required to sustain an economy is calculated, including domestic resources as well as resources exploited from the hinterland, the area beyond the system boundaries of a national economy.

In view of MFA, focusing on processes of material transport, transformation, and/or storage within an investigated system, the following general trends and developments can be discerned: While the first generation of MFA was directed toward the static description and analysis of a material flow and stock system, the need for dynamic MFA became apparent when the descriptive state was to be expanded to enable the investigation of stocks and flows as a consequence of specific developments and decisions. This requires a time-dependent model of a system. Due to the time reliance of most anthropogenic material flows and stocks, future developments of a material system can only be influenced in view of certain goals if the changes over time and the corresponding impact factors are known.

Hence, a key advancement in recent MFA is the widespread use of dynamic MFA, enabling the description of future states of a material system that follow from the current state by deterministic or stochastic functions. Since the first dynamic MFA studies appeared in the 1990s (e.g., Zeltner, Bader, Scheidegger, and Baccini, 1999; Kleijn, Huele, and van der Voet, 2000; D. B. Müller, Bader, and Baccini, 2004), the number of studies that investigate material flow systems over time using dynamic models increased considerably. A review of 60 recent studies on dynamic flow and stock modeling of metals is given by Müller, Hilty, Widmer, Schluep, and Faulstich (2014). Metals, in particular, are subject to dynamic MFA due to their widespread use and large accumulation in society and their potential value as secondary raw materials. But, also the management of hazardous organic materials has been investigated by dynamic MFA (e.g., Morf, Buser, Taverna, Bader, and Scheidegger, 2008). Knowledge about the evolution of material stocks is used to predict future material flows and thus to recover or exclude materials effectively as a function of the past and current stocks (Chen and Graedel, 2012; Müller, Hilty, Widmer, Schluep, and Faulstich, 2014). The following aspects proved to be important when establishing dynamic MFA:

- Extensive data sets are required to establish dynamic material flow models. This requires often considerable resources and use of imperfect data. Therefore, procedures for model calibration and data plausibility checks are required to evaluate the validity of model results (Buchner, Laner, Rechberger, and Fellner, 2015; Pauliuk, Wang, and Müller, 2013).

- Uncertainty and sensitivity analyses have become increasingly important for dynamic material flow modeling in order to evaluate the robustness of the results and identify critical parameters (e.g., Glöser, Soulier, and Tercero Espinoza, 2013; Melo, 1999).

- In dynamic MFA, material stocks are typically estimated using top-down approaches (based on an accounting of the net flows into or out of the stock over the past using product lifetime functions). Nevertheless, recently, efforts have been made to also produce independent bottom-up estimates for plausibility checks and model calibration (Buchner, Laner, Rechberger, and Fellner, 2015; Liu, Bangs, and Müller, 2011; Müller, Hilty, Widmer, Schluep, and Faulstich, 2014) or to investigate stock development over time completely based on bottom-up approaches (e.g., Tanikawa, Fishman, Okuoka, and Sugimoto, 2015). The latter is highly data intensive but allows for a potentially high resolution with respect to dimensions of interest such as space or institutional structures.

- An upcoming aspect of dynamic material flow modeling is the consideration of material qualities and contamination of recyclables to account for potential functionality limitations in recycling loops (in particular with respect to metal alloys and plastic mixtures). Indications that quality issues may, to some degree, constrain recycling have been observed and discussed for aluminum in several dynamic MFA studies (e.g., Hatayama, Daigo, Matsuno, and Adachi, 2012; Løvik, Modaresi, and Müller, 2014; Modaresi and Müller, 2012).

Despite the trend toward more dynamic MFA studies, static MFA will remain attractive and important. It provides a valuable snapshot of a system in time and is done at different levels of sophistication to investigate the patterns of material use and losses in a system. Reasons for preferring static MFA are as follows: (1) Often, it is appropriate and suffices for understanding of a material system. (2) To perform a static MFA is significantly less resource consuming than a dynamic MFA. (3) A dynamic MFA can often be more effectively performed if preceded by a static MFA.

The main fields of application of static and dynamic MFA are still environmental management and engineering, industrial ecology, resource and waste management, and anthropogenic metabolism (cf. the following Sections 1.4.1 to 1.4.5). In the environmental field, persistent organic pollutants (POPs) and nanomaterials have received considerable new attention from the MFA community (Morf et al., 2005; Vyzinkarova and Brunner, 2013; Gottschalk, Sonderer, Scholz, and Nowack, 2010). However, the latest interest of MFA authors appears to have shifted from environmental issues to resource issues: recent papers cover topics such as inventories of specific metals (rare-earth metals, platinum-group metals) or nutrients (mainly phosphorus); recovery of substances from individual waste streams like construction wastes or waste electric and electronic equipment (WEEE); and plastic recycling (Du and Graedel, 2011; Alonso et al., 2012; Habib, Schibye, Vestbø, Dall, and Wenzel, 2014; Kuriki, Daigo, Matsuno, and Adachi, 2010; Ott and Rechberger, 2012). MFA became instrumental for so-called criticality assessments, where the risks associated with (primary and secondary) raw material shortages are assessed from a company, country, or global perspective (cf. Graedel et al., 2011).

"Urban mining" is another area that has come into the focus of MFA research. Long-living stocks and their recycling potentials are of prime interest in this field. Hence, numerous MFAs have been performed to assess the abundance and value of anthropogenic resource stocks, e.g., to compare these stocks to geogenic resources (Johansson, Krook, Eklund, and Berglund, 2013; Lederer, Laner, and Fellner, 2014), to identify those urban stocks that are most effectively recoverable (Hashimoto, Tanikawa, and Moriguchi, 2007), or to model the depletion of hazardous metals (Cd, Hg) from these stocks (Månsson, Bergbäck, Hjortenkrans, Jamtrot, and Sörme, 2009).

Finally, *urban metabolism* as a part of the general topic *anthropogenic metabolism* became a recent research area with great potential for application of MFA. The fact that some of the latest waste- and resource-oriented research programs of the European Union (EU) in the framework of Horizon 2020 ask for an urban metabolism approach will boost the application of MFA research even further.

The EU expects urban planners and infrastructure developers to apply such approaches when aiming at increasing resource efficiency of materials and energy in cities. As a result, this should eventually (1) foster the implementation of the EU Thematic Strategies on the urban environment by reducing wastes and emissions and (2) promote the sustainable use of resources by saving primary resources as well as contributing to economic development and social welfare.

A few recent MFA studies might point to a new emerging field for MFA, the optimization of production processes in view of economic, resource, and environmental constraints, e.g., in manufacturing or in agricultural practice (Müller, 2013; Krolczyk 2015; Eisingerich 2015). It remains to be seen if these beginnings are followed by a new wave of MFAs that are performed primarily for economic reasons and focus on material transformation, logistics, and management in the business sector.

A new methodological development common for both static and dynamic MFA is the effort for creating consistent, reliable MFA databases that fulfill the mass-balance principle and include sufficient information about the uncertainty of results and underlying data. Beginning with early studies on concepts for the representation of uncertainty in MFA (Hedbrant and Sörme, 2001), the systematic evaluation of uncertainty during data acquisition and management as well as in model calculations has received increasing and widespread attention (e.g., Zoboli, Laner, Zessner, and Rechberger, 2015; Laner, Rechberger, and Astrup, 2014; Cencic and Frühwirth, 2015; Do, Trinh, and Nishida, 2014; Montangero and Belevi, 2007). Various approaches have been put forward to deal with uncertain quantities in material flow models ranging from qualitative descriptions of uncertainties to sophisticated mathematical models for uncertainty propagation and data reconciliation (e.g., Cencic and Frühwirth, 2015; Dubois, Fargier, Ababou, and Guyonnet, 2014). Due to the tremendous growth of computational power and options for default uncertainty analysis by MFA software (e.g., STAN) as well as other modeling tools, sensitivity and uncertainty analyses have become standard elements of state-of-the-art MFA (cf. Laner, Rechberger, and Astrup, 2014).

A more recent development in MFA is the enhanced combination of MFA and tools for environmental assessments such as LCA. Already two decades ago, it has been suggested to use MFA as a basis for life cycle inventories (LCIs) (Tukker and Kleijn, 1997). For improving data consistency, which is a critical issue for LCI and corresponding LCA (Ayres, 1995), MFA offers the benefit of fulfilling the mass-balance principle, a basic requirement for every MFA but not for LCA (Andersen, Boldrin, Christensen, and Scheutz, 2010; Tonini, Martinez-Sanchez, and Astrup, 2013). The use of MFA results in combination with LCI data and impact assessment methods to evaluate the environmental performance of material and energy flow systems has been put forward, particularly to assess waste treatment and recycling activities (e.g., Laner and Rechberger, 2007; Lederer and Rechberger, 2010; Brunner et al., 2016). Another step toward integrating MFA and LCA is the use of formal optimization to identify environmentally preferable setups of metabolic systems (e.g., with respect to eutrophication, global warming, resource depletion) (Höglmeier, Steubing, Weber-Blaschke, and Richter, 2015; Vadenbo, Hellweg, and Guillén-Gosálbez, 2014a; Vadenbo, Guillén-Gosálbez, Saner, and Hellweg, 2014b). Instead of working with predefined scenarios, the optimal use of a specific material is determined as a result of the optimization program, which is defined by given material flow networks, specific constraints, required functionalities (to be delivered), decision variables, and objective functions (see also Section 1.4.6). Such a combination allows for optimizing resource use from a systems perspective in consideration of physical constraints and interactions among different product systems and societal activities.

PROBLEMS—SECTIONS 1.1 THROUGH 1.3

Problem 1.1:

(a) MFA is based on a major principle of physics. Name it and describe its content. (b) What is the main benefit of the principle for MFA?

Problem 1.2:

Divide the following 14 materials into two categories of *substances* and *goods*: cadmium, polyvinyl chloride, molecular nitrogen (N_2), melamine, wood, drinking water, personal computer, steel, iron, copper, brass, separately collected wastepaper, glucose, and copper ore.

Problem 1.3:

Estimate roughly your total personal daily material turnover (not including the "ecological rucksack"). Which category is dominant on a mass basis: (a) solid materials and fossil fuels, (b) liquid materials (without fossil fuels), or (c) gaseous materials?

Problem 1.4:

The flux of zinc into the city of Stockholm is about 2.7 kg capita^{-1} year^{-1}; the output flux amounts to ca. 1.0 kg/capita/year; and the stock is about 40 kg/capita (Bergbäck, Johansson, and Mohlander, 2001). Calculate the time until the stock of zinc doubles, assuming stationary conditions.

The solutions to the problems are given on the website http://www.MFA-handbook.info.

1.4 Application of Material Flow Analysis

The history of material flow analysis (MFA) reviewed in Section 1.3 indicates the large variety of fields where MFA can be applied. In the following sections, these fields are expanded and elucidated further in a systematic way, discussing scale, scientific disciplines, and field of application. Individual case studies that exemplify the application of MFA in these fields are given in Chapter 3.

Scale: MFA is well suited to investigate physical and chemical systems on all scales, from individual processes to complex systems comprising many subsystems. If applied in a hierarchical way, beginning with single-unit processes and operations, progressing toward small systems comprising several processes and operations, and finally focusing on comprehensive systems providing important services to society, MFA can depict a complete picture of material use from sources to sinks. The general objective is the same for all levels, namely, to support the understanding of the functioning of the physical basis of society. Another common denominator is the mass-balance principle that has to be fulfilled by all MFA on all levels. It is important to note that for reaching the aforementioned objective of understanding a metabolic system, it is often necessary to include the substance level, too. While the level of goods yields information regarding economic aspects, the level of substances is necessary for conclusions regarding qualitative issues (e.g., product quality with respect to human health and the environment). Thus, many applications of MFA require investigation into both goods and substances. Because the mass-balance principle applies to both, the abundance of data about goods and substances enables a decrease in uncertainties (see Chapter 2, Section 2.3).

Single processes: MFA is often applied to assess and improve a single process because it links inputs (resources), stocks, and outputs (products, wastes, emissions) in a well-defined and transparent way. Knowing a mass balance forms the basis for evaluating and controlling resource consumption, environmental loadings, and costs of a production or a disposal process. Fields of application on this level are, for example, optimization of single industrial or disposal processes in view of economic, environmental, or resource constraints.

Small material systems, comprising a few processes, goods, and substances: The same applies as to single processes. In addition, the MFA yields the relationship between the individual processes and flows. Combined with other evaluation methods (cf. Chapter 2, Section 2.5), this allows identification of those processes, stocks, and flows that are most relevant to reach a certain objective. Thus, for operational as well as strategic decisions (e.g., of entrepreneurs), MFA yields an excellent physical basis.

Complex material systems, including a large number of processes, goods, and substances: On this largest scale, MFA is highly instrumental for supporting both strategic as well as policy decisions and evaluations. Particularly in the fields of waste, resource, and environmental management, MFA is often used to analyze large systems, such as for national economies (Matthews et al., 2000; OECD, 2008), global metal utilization (Glöser et al., 2013), or national waste management systems (Allesch et al., 2015). The main advantage is that

MFA can make use of the redundancy of a large-scale system: The law of conservation of mass requires that individual processes as well as the overall system and subsystems are balanced. Due to the many flows and stocks of goods and substances, such a large-scale system is often overdetermined, allowing for estimation of missing flows and for decreasing uncertainty (cf. Chapter 2, Section 2.3).

Scientific disciplines: MFA methodologies are tools for analyzing material flows from sources to sinks that have been applied in a large variety of scientific disciplines. The original use was mainly in chemistry (Lavoisier, see Vidal, 1985) and chemical engineering. After the introduction of input–output tables by Leontief (1966, 1977), economics soon became an important field for MFA. Ayres (1989) brought MFA back to natural and engineering sciences by applying it in the field of environmental protection, and later in resource management and industrial ecology (IE). In the early 1990s, Baccini and Brunner defined a new scientific discipline, "metabolism of the anthroposphere," comprising the analysis, evaluation, and design of anthropogenic systems and their interaction with environment and natural resources (Baccini and Brunner, 2012). Materials accounting was introduced by the Wuppertal Institute, bringing MFA to socioeconomic partners engaged in the so-called Conaccount conferences (see, e.g., Bringezu, 1997). Materials accounting and corresponding indicators were developed to follow and evaluate material flows and stocks within and between economies, among others, to measure resource efficiency. Thus, MFA is also applied in the field of policy science, where it has become an important tool for policy evaluation and policy decisions. Graedel (2004) widely applied MFA in IE, particularly metal flows on global scales. Today, MFA is a basic tool in engineering, management, economics, and policy science. This short and incomplete summary illustrates that MFA has been extensively applied by researchers from a wide array of scientific disciplines.

Field of application: The three main and specific areas for MFA application summarized in this section are resource management, environmental management, and waste management. In addition, IE and anthropogenic metabolism are overarching themes where MFA plays a key role. Today's emerging fields for MFA are manufacturing, process design, and engineering. In all of these fields, MFA has been used for analysis and optimization on all scales. The specific tasks tackled by MFA comprise—to mention a few—the improvement of resource efficiency; the design of measures for urban mining; pollution prevention and protection of, for example, surface water and groundwater; watershed and nutrient management; waste minimization; waste analysis; recycling; and the provision of final sinks.

Material flow accounts and related MFA-derived indicators are widely used for measuring material flows and resource productivity on national and international scales (OECD, 2008). Such accounts serve country governments for the development and implementation of national policies. A distinct feature of material flow accounts is that they consider the level of goods only; individual substances are not considered. Hence, this approach does not allow detailed impact assessments in view of environmental loadings or resource qualities, which require a focus on the substance level. On the other hand, material flow accounts are often established because they can be derived from national economic databases and thus are comparatively easily accessible, which is usually not the case for data about substance flows and stocks.

The potential uses and the limits of MFA in the main fields are further outlined in the following sections. In Section 1.4.5, some of the relevant results of the application of MFA in the field of anthropogenic metabolism are discussed in more detail. This facilitates the understanding and subsequent use of the term *anthropogenic metabolism* on one hand. On the other hand, it conveys a picture of the modern metabolic world and helps the reader to understand and appreciate the power of MFA to analyze, evaluate, and influence the future anthropogenic metabolism.

The methodological differences between MFA and other similar approaches such as pathway analysis, input–output analysis, and LCA are discussed in Chapter 2.

1.4.1 Environmental Management and Engineering

The environment is a complex system comprising living organisms, energy, matter, space, and information. The human species, like all other species, has used the environment for production and disposal. We produce food and shelter—drawing on soil, water, and air—and in return, we give back wastes such as feces, exhaust air, and debris. Environmental engineering has been described as (1) the study of the fate, transport, and effects of substances in the natural and engineered environments and (2) the design and realization of options for treatment and prevention of pollution (Valsaraj, 2000). The objectives of management and engineering measures are to ensure that (1) substance flows and concentrations in water, air, and soil are kept at a level that allows the genuine functioning of natural systems and (2) the associated costs can be carried by the actors involved.

MFA is used in a variety of environmental engineering and management applications, including environmental impact statements, remediation of hazardous-waste sites, design of air pollution control strategies, nutrient management in watersheds, planning of soil-monitoring programs, and sewage sludge management. All of these tasks require a thorough understanding of the flows and stocks of materials within and between the environment and the anthroposphere. Without such knowledge, the relevant measures might not be focused on priority sources and pathways, and thus, they could be inefficient and costly.

MFA is also important in management and engineering because it provides transparency. This is especially important for environmental impact statements. Emission values alone do not allow cross-checking if a change in boundary conditions (e.g., change in input or process design) still permits regulations to be met. However, if the material balances and transfer coefficients of the relevant processes are known, the results of varying conditions can be validated.

There are clear limits to the application of MFA in the fields of environmental engineering and management. MFA alone is not a sufficient tool to assess or support engineering or management measures. Nevertheless, MFA is an indispensable first step and a necessary basis for every such task, and it should be followed by an evaluation or design step, as described in Chapter 2, Section 2.5.

1.4.2 Industrial Ecology

Although earlier traces can be tracked down, the concept of IE evolved in the early 1990s (Frosch and Gallopoulos, 1989; Thomas and Spiro, 1994; Erkman, 1997; Desrochers, 2000; for a brief history, see also Erkman, 2002). Jelinski, Graedel, Laudise, McCall, and Patel (1992) provided a first definition and defined IE as a concept in which an industrial system is viewed not in isolation from its surrounding systems but in concert with them. The objective of IE is to optimize the total material cycle from virgin material to finished material, to component, to product, to waste product, and to ultimate disposal. A commonly agreed-upon description of IE can be retrieved from the website of the *Journal of Industrial Ecology* (www.wileyonlinelibrary.com/journal/jie), headquartered at Yale's School of Forestry and Environmental Studies. The journal designates IE as the rapidly growing field that systematically examines local, regional, and global materials and energy uses and flows in products, processes, industrial sectors, and economies. IE focuses on the potential role of industry in reducing environmental burdens throughout the product life cycle, from the extraction of raw materials to the production of goods, to the use of those goods, and to the management of the resulting wastes. While industrial metabolism, as defined, for example, by Ayres (1992) and Ayres and Simonis (1994), explores the material and energy flows through the industrial economy, IE goes farther. Similar to what is known about natural ecosystems, the approach strives to develop methods to restructure the economy into a sustainable system. The industrial system is seen as a kind of special biological ecosystem, or as an analogue of a natural system (Frosch and Gallopoulos, 1989). Other pioneers of IE even define it as the science of sustainability (Graedel and Allenby, 2002). In this context, it is worth mentioning that sustainability has not only an ecological but also social and economic dimensions, aspects that are addressed by Allenby (1999) and others, too.

Clearly, the variety of topics and approaches demonstrates the breadth of the field that IE attempts to span (Lifset and Graedel, 2002). This fact is also used to criticize the approach as being vague and mired in its own ambiguity and weakness. The legitimacy of the analogy between industrial and ecological ecosystems is also questioned (Commoner, 1997). However, there are several basic design principles in IE (adapted from Ehrenfeld, 1997) that suggest the utilization of MFA:

1. Controlling pathways for material use and industrial processes
2. Creating loop-closing industrial practices
3. Dematerializing industrial output
4. Systematizing patterns of energy use
5. Balancing industrial input and output to natural ecosystem capacity

The following applications illustrate the role of MFA in IE. First, a better understanding of industrial metabolism requires a description of the most relevant material flows through the industrial economy. This encompasses the selection of relevant materials on the economic *goods level* (e.g., energy carriers, mineral construction materials, steel, fertilizers) and on the chemical *substance level* (e.g., carbon, iron, aluminum, nitrogen, phosphorus, cadmium). The system boundaries must be defined in such a way that the pathways of materials are covered from cradle (exploitation) to grave (final sink for the material). The results of an MFA reveal the most important processes during the life cycle of a material, detect relevant stocks of the material in the economy and the environment, show the losses to the environment and the final sinks, and track down internal recycling loops. Additionally, MFA can be used to compare options on the process level and at the system level.

Second, the concept's imminent call for closed loops (realized, for example, in the form of a cluster of companies, a so-called industrial symbiosis) requires information about the composition of wastes to become feedstock again and about the characteristics of the technological processes involved. In particular, the implementation of industrial loops requires control by appropriate MFA, since loops have the potential of accumulating pollutants in goods and stocks. The fact that waste is recycled or reused is not yet a guarantee for a positive result. Three exemplary negative examples are the use of contaminated fly ash in cement production, the reuse of animal protein causing bovine spongiform encephalopathy (BSE), known as *mad cow* disease, and the unwanted transfer of PBDEs into products made of recycled plastics (cf. Chapter 3, Section 3.3.2.4).

A third objective in IE is dematerialization. This can be achieved by providing functions or services rather than products. Again, MFA can be used to check whether a dematerialization concept (e.g., the paperless office) succeeds in practice. Other ways of dematerialization are to prolong the lifetime of products (Truttmann and Rechberger, 2006) or to produce lighter goods.

The city of Kalundborg, Denmark, is frequently cited as an example of an industrial ecosystem in the IE literature. Materials (fly ash, sulfur, sludges, and yeast slurry) and energy (steam, heat) are exchanged between firms and factories within a radius of about 3 km (Chertow, 2000). Using waste heat for district heating and other purposes (e.g., cooling) has long been recognized as good industrial practice (known as power–heat coupling). The comparatively few material flow links between the actors in Kalundborg show that the concept of (apparently) closed loops is difficult to accomplish in reality. Material balancing is seen as a major tool to support sound industrial ecosystems.

1.4.3 Resource Management

There are two kinds of resources: first, natural resources such as minerals, water, air, soil, land, and biomass (including plants, animals, and humans), and second, human-induced resources such as the anthroposphere as a whole, including materials, energy, information (e.g., "cultural heritage," knowledge in science and technology, art, ways of life), and manpower. The human-induced or so-called anthropogenic resources are located in (1) private households, agriculture, industry, trade, commerce, administration, education, health care, defense, and security systems, and (2) infrastructure and networks for supply, transportation and communication, and disposal. Given the large-scale exploitation of mines and ores, many natural resources are massively transformed into anthropogenic resources (see Section 1.4.5). Thus, the growing stocks of the anthroposphere will become increasingly more important as a resource in the future.

A major task of resource management is the supply of an anthropogenic system with adequate resources. Resource availability and potential scarcity are therefore important aspects, and much research is being done to describe the supply situation for various commodities. It appears that there are different occurrences of resource constraints. First, reserves of a commodity may decline in quantity and quality after a certain time period of exploitation (geological scarcity). This phenomenon is often designated as peak X (X stands for oil, phosphorus, copper, etc.) (see Seppelt, Manceur, Liu, Fenichel, and Klotz, 2014). The peak theory is not undisputed, and real production declines cannot be observed so far. However, there is no doubt that the quality of many resource deposits is declining (lower ore grades, increasing contaminations). Second, scarcity can appear as a result of technological constraints (technological scarcity).

Zuser and Rechberger (2010) showed, for instance, that the realization of a political target (e.g., 25% of the world's electricity demand by photovoltaics in 2040) is technologically hardly feasible with the current and anticipated technology mix as, for example, the yearly production of tellurium has to be raised by a factor ranging from 30 to 180, etc., and the yearly production of gallium by factor 4 to 20. As both metals are by-products from zinc, lead, or aluminum mining, such an increase of the production rate within a few decades is a real economic challenge. Third, availability can be jeopardized if a dominating market supplier decides to reduce the output. Such situations could be observed in the past, e.g., for petroleum and phosphorus.

In the past decades, commodity markets for material resources have become more volatile. There are about 50 different commodity markets worldwide. The purely financial transactions are outnumbering the real physical trades of goods by far, and in part, this goes hand in hand with large price fluctuations (market scarcity).

One way to consider the different dimensions of resource constraints is criticality assessment of resources. Various approaches have been developed to quantify the supply risk. Such risks have geological, technological, economic, environmental, social, regulatory, and geopolitical components. In addition, they include vulnerability to supply restrictions on the corporate, national, and global levels (cf. Graedel et al. 2012). The result of such assessments is a list of commodities that are considered as critical for a specific economy. For instance, the European Union termed 20 commodities as being critical for its economy (cf. Figure 1.3).

Resource management comprises three steps: (1) analysis, planning, and allocation; (2) exploitation, upgrading, and utilization; and (3) recycling and final disposal of resources. MFA is of prime importance for analysis and planning. It is the basis for modeling resource consumption as well as changes in stocks, and therefore, it is important in forecasting potential scarcity of resources. MFA is helpful in identifying the accumulation and depletion of materials in natural and anthropogenic stocks, such as buildings, or soil and sediments. Without it, it is impossible to identify the shift of material stocks from *natural* reserves to *anthropogenic* accumulations. In addition, if MFA is performed in a uniform way at the front and back ends of the anthropogenic system, it is instrumental in linking resource management to environmental and waste management. It shows the need for final sinks and for recycling measures, and it is helpful in designing strategies for recycling and disposal.

Balancing all inflows and outflows of a given stock yields information on the time period until the stock reaches a critical state of depletion or accumulation. This could be the slow exhaustion of available phosphorus in agricultural soil due to the lack of appropriate fertilizer, or it could be the unnoticed buildup of valuable metals in a landfill of incinerator ash and electroplating sludge. It is difficult to estimate the change in the stock by direct measurement, especially for stocks with a high variability in composition and slow changes in time. In such cases, it is more accurate and cost-effective to calculate critical time scales (the time when a limiting or reference value is reached) by comparing the difference between input and output to the stock from its flow balance. Direct measurement requires extensive sampling programs with much analysis, and the heterogeneity of the flows produces large standard deviations of the mean values. Thus, it takes large differences between mean values until a change becomes statistically significant. A slow change in a heterogeneous material can be proved on statistical grounds only over long measuring periods. As a result, MFA is better suited and more cost-effective than continuous soil monitoring in early recognition of changes in resource quality, such as harmful accumulations in the soil.

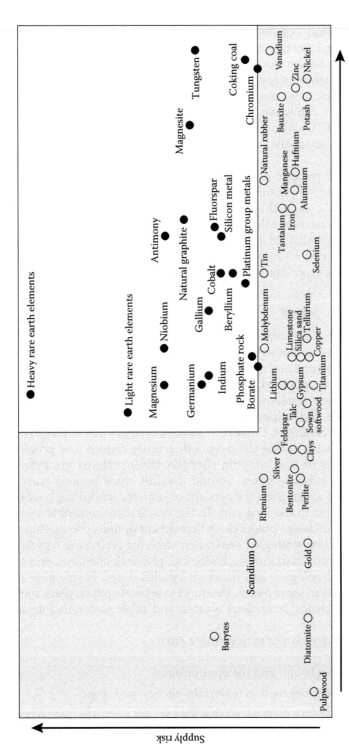

FIGURE 1.3

The Ad-hoc Working Group on defining critical raw materials identified 20 raw materials (upper-right corner) out of 54 candidates as critical with respect to economic importance and supply risk (EC, 2014).

For more information, see Obrist, von Steiger, Schulin, Schärer, and Baccini (1993) and Chapter 3, Section 3.1.1.

The use (and preferably conservation) of resources to manufacture a particular good or to render a specific service is often investigated by LCA. The result of an LCA includes the amount of emissions and the resources consumed. Since MFA can serve as the first step of LCA (contributing to the LCA inventory), MFA is also a basis for industrial resource conservation. For more information about the combination of MFA and LCA, see Laner and Rechberger (2016).

The quality and price of a resource usually depends on the concentration of valuable and/or hazardous substances. Thus, it is important to know whether a geogenic or anthropogenic process or system concentrates or dilutes a given substance. MFA is instrumental in the application of such evaluation tools as statistical entropy analysis (see Chapter 2, Section 2.5.1.8, and Chapter 3, Section 3.2.2.), which is used to compare the potential of processes and systems to accumulate or dilute valuable or hazardous substances.

1.4.4 Waste Management

Waste management takes place at the interface between the anthroposphere and the environment. The definition and objectives of waste management have changed over time and are still changing. The first signs of organized waste management appeared when people started to collect garbage and remove it from their immediate living areas. This was an important step regarding hygiene and helped to prevent epidemics. These practices were improved over the centuries. However, dramatic changes in the quantity and composition of wastes during the twentieth century caused new problems. First, the emissions of the dumping sites (landfills) polluted groundwater and produced greenhouse gases. Second, landfill space became scarce in densely populated areas. Even the concept of sanitary landfilling could not solve these problems in the long run. Today, waste management is an integrated concept of different practices and treatment options comprising prevention and collection strategies; separation steps for producing recyclables or for subsequent processing using biological, physical, chemical, and thermal treatment technologies; and different landfill types. People now have the opportunity (or, in some places, the duty) to separate paper, glass, metals, biodegradables, plastics, hazardous wastes, and other materials into individual fractions.

The goals of modern waste management are to

- Protect human health and the environment
- Conserve resources such as materials, energy, and space
- Treat wastes before disposal so that they do not need aftercare when finally stored in landfills

The last goal is also known as the final-storage concept and is part of the precautionary principle: the wastes of today's generation must not impose an economic or ecological burden to future generations. Similar goals can be found in many countries for modern waste management legislation (Resource Conservation and Recovery Act, 1976; BMUJF, 1990; BMUR, 1994; EKA, 1986; EC, 1975), which were written to comply with the requirements for sustainable development. The aforementioned goals make it clear that the focus in waste management is not necessarily on goods (paper, plastics, etc.) or on functions of materials (e.g., packaging). The focus is on the nature of the substances.

Hazardous substances endanger human health. The threat occurs when municipal solid waste (MSW) is burned in poorly equipped furnaces and volatile heavy metals escape into air. It is not the good leachate of a landfill that poses a danger to the groundwater. The danger resides in the cocktail of hazardous substances in the leachate of the landfill. The fact that a material has been used for packaging is irrelevant for recycling. What is important is its elemental composition, which determines whether it is appropriate for recycling. Hence, advanced waste management procedures are implemented to control and direct substances at the interface between anthroposphere and environment to achieve the following two goals.

1. Materials that can be recycled without inducing high costs or negative substance flows should be recycled and reused. Negative flows
can appear as emissions or by-products during the recycling process. The recycling process itself can also lead to enrichment of pollutants in goods and reservoirs. For example, the recycling process
can increase heavy metal contents in recycled plastics, or it can lead
to accumulation of metals in the soil when sewage sludge is applied
to agricultural fields.

2. Nonrecyclables should be treated to prevent the flow of hazardous substances to the environment. A tailor-made final sink—
defined as a reservoir where a substance resides for a long period
of time (>10,000 years) without having a negative impact on the
environment—should be assigned for each substance.

MFA is a valuable tool in substance management because it can cost-efficiently determine the elemental composition of wastes exactly (see Chapter 3, Section 3.3.1). This information is crucial if the goal is to assign a waste stream to the best-suited recycling/treatment technology and to plan and design new waste treatment facilities. For example, mixed plastic wastes that cannot be recycled for process reasons can be used as a secondary fuel in industrial boilers as long as their content of heavy metals and other contaminants is not too high (see Chapter 3, Sections 3.3.2.1 and 3.3.2.2).

MFA is also helpful in investigating the substance management of recycling/treatment facilities. For instance, substance control by an incinerator is different from substance control by a mechanical–biological treatment facility. Such information is a prerequisite for the design of a sustainable waste management system. North Rhine–Westphalia (Germany) is the first region that requires MFA by legislation as a standard tool in waste management planning (MUNLV, 2000).

Finally, MFA can contribute to the design of better products that are more easily recycled or treated once they become obsolete and turn into "waste." These practices are known as design for recycling, design for disposal, or design for environment.

An MFA-based total material balance shows whether given goals have been achieved. An MFA balance also identifies the processes and flows that have the highest potential for improvements.

Waste management is an integral part of the economy. Some experts who have experience with MFA suggest that waste management should be replaced by materials and resource management. They assert that controlling the material flows through the total economy is more efficient than the current practice of separating the management of wastes from the management of production supply and consumption.

1.4.5 Anthropogenic Metabolism

Baccini and Brunner (1991) applied MFA to analyze, evaluate, and optimize some of the key processes and goods of the metabolism of the anthroposphere. In a more recent study, Brunner and Rechberger (2001) systematically summarized relevant phenomena of the anthropogenic metabolism. The following examples illustrate the power of MFA to identify and reveal key issues of the anthropogenic metabolism such as resource management, environmental management, and waste management.

1.4.5.1 Unprecedented Growth

In prehistoric times, the total anthropogenic metabolism (input, output, and stock of materials and the energy needed to satisfy all human needs for provisions, housing, transportation, etc.) was nearly identical to human physiological metabolism. It was mainly determined by the need for food, for air to breathe, and for shelter. For modern man, the material turnover is 10 to 20 times greater (Figure 1.4). The fraction that is used today for food and breathing is comparatively small. More important is the turnover for other activities, such as to clean, to reside, and to transport and communicate (Table 1.1). These activities require thousands of goods and substances that were of no metabolic significance in prehistoric times.

The consumption of goods has increased over the past two centuries, and there are no clear signs yet that this will change. Figure 1.5 displays the growth of ordinary materials such as paper, plastic, and tires. Figure 1.6

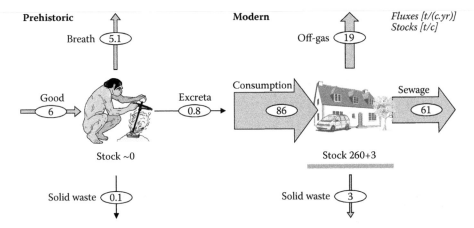

FIGURE 1.4

Material turnover of primitive man in his private household was about an order of magnitude smaller than today's consumption of goods. Note that the figure includes direct material flows only. Materials (and wastes) turned over outside of households to manufacture the goods consumed in households are larger than 100 t capita^{-1} year^{-1}. (Brunner, P.H. and Rechberger, H.: Anthropogenic metabolism and environmental legacies, in *Encyclopedia of Global Environmental Change*, Vol. 3, Munn, T., Ed. 2001. Copyright Wiley-VCH Verlag GmbH & Co. KGaA. Reproduced with permission.)

TABLE 1.1

Material Flows and Stocks for Selected Activities of Modern Man

Activity	Input, t Capita^{-1} Year^{-1}	Output, t Capita^{-1} Year^{-1}			Stock, t/Capita
		Sewage	Off-Gas	Solid Residues	
To nourish[a]	5.7	0.9	4.7	0.1	<0.1
To clean[b]	60	60	0	0.02	0.1
To reside[c]	10	0	7.6	1	100 + 1
To transport[d]	10	0	6	1.6	160 + 2
Total	86	61	19	2.7	260 + 3

Note: The most outstanding and unprecedented feature of today's economies is the very large stock of material that has accumulated in private households (Brunner and Rechberger, 2001).

[a] Includes all flows of goods associated with the consumption of food within private households, e.g., food, water for cooking, etc.

[b] Comprises water, chemicals, and equipment needed for laundry, dishwashing, personal hygiene, toilet, etc.

[c] Consists of the buildings, furniture, appliances, etc. needed for living.

[d] Accounts for materials (cars, trains, fuel, air, etc.) used for the transport of persons, goods, energy, and information to accommodate the needs of private households (including materials for road construction).

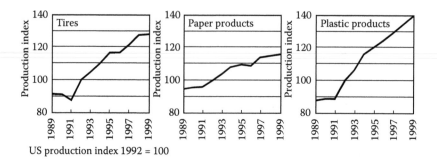

US production index 1992 = 100

FIGURE 1.5
Growth in consumer product consumption. The examples of paper, plastics, and tires consumed in the United States show the high growth rate of material flows from 1989 to 1999. (Anonymous, 2000). (Reprinted with permission from *Chem. Eng. News*, June 26, 2000, 78 (26), 49. Copyright 2000 American Chemical Society.)

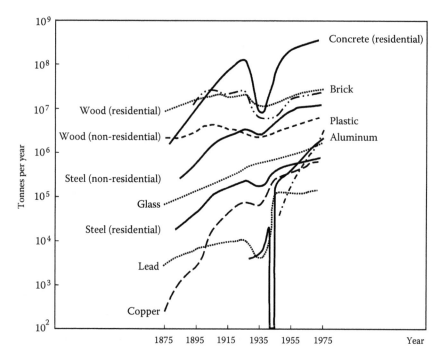

FIGURE 1.6
Growth of construction materials used in the United States from 1875 to 1975 (Wilson, 1990). Around 1900, wood was surpassed by concrete as the main construction material. Plastics and aluminum were among the fastest-growing construction materials in the 1970s. Changes in the consumption of construction materials alter the construction stock, too, and hence are important for future waste management; large amounts of plastics and metals such as aluminum and copper will have to be handled safely and efficiently. (Reprinted from Recycling of demolition wastes. Wilson, D., in *Concise Encyclopedia of Building and Construction Materials*, pp. 517–518. Copyright 1990, with permission from Elsevier.)

shows the increase in construction materials in the United States for a period of 100 years.

Growth of material flows is closely associated with economic growth. Economic progress is defined in a way that causes an increase in material turnover. It is important to develop new economic models that decouple economic growth from material growth, thus promoting long-term welfare without a constant increase in resource consumption.

The need for a new economic model becomes even more evident when one considers the growth rates of potentially hazardous substances such as heavy metals or persistent and toxic organic substances. On the level of substances, the increase in consumption is by far greater than one order of magnitude. For instance, the global anthropogenic lead flow (Figure 1.7) increased in the last few thousand years by about 10^6, i.e., by six orders of magnitude.

Material growth is not just an issue from a quantitative point of view; it is also of qualitative concern. If total material flows have increased by one to two orders of magnitude and certain hazardous substances by more than five orders of magnitude, then the stock as well as the output of the anthropogenic system will be highly enriched with such substances. Hence, in the future, stocks and wastes will have to be managed with greater care to avoid harmful accumulations of these substances. It becomes clear that, from an environmental point of view, the growth of substance flows is more important than the growth rate of mass goods.

1.4.5.2 Anthropogenic Flows Surpass Geogenic Flows

A consequence of the constant increase in material exploitation due to the large per capita consumption is that the anthropogenic flow of certain substances already surpasses the geogenic (or "natural") flow of such substances. While this is quite clear for those organic chemicals that are exclusively man-made (such as polyvinylchloride [PVC]), it is less obvious for metals like Cd (Figure 1.8). For this metal, the man-made flow from the lithosphere to the anthroposphere is about three to four times larger than the geogenic flows by erosion, weathering, leaching, and volcanoes. As a consequence, the Cd concentrations will increase in some environmental compartments such as the soil. The Cd stock in the anthroposphere represents an unwanted legacy for future generations, which will have to bear the high cost of managing this stock very carefully to avoid hazards associated with this toxic metal.

The global phenomenon of anthropogenic flows surpassing natural flows has been demonstrated on a national scale, too. Bergbäck (1992) compared the rates for weathering and erosion with those of anthropogenic emissions in Sweden. He found that emission rates of lead, chromium, and cadmium are exceeding natural rates. Similar results have been observed in other

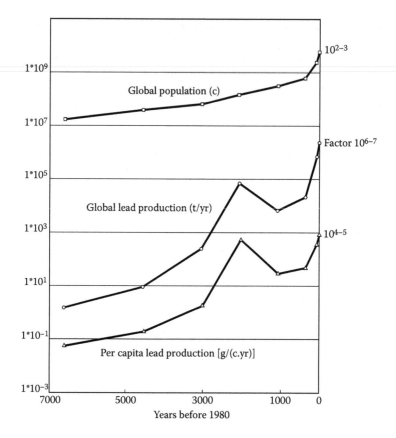

FIGURE 1.7
Increase in global production of lead. Due to tremendous progress in mining technology and economic development, lead mining increased from 0.1 g capita^{-1} year^{-1} to 1000 g capita^{-1} year^{-1}. Taking into account population growth, which increased at a smaller rate than the per capita lead production, the total amount of lead produced from mining increased from 1 t/year to more than 3 million tons per year (lead data from Settle and Patterson, 1980). (Brunner, P.H. and Rechberger, H.: Anthropogenic metabolism and environmental legacies, in *Encyclopedia of Global Environmental Change*, Vol. 3, Munn, T., Ed. 2001. Copyright Wiley-VCH Verlag GmbH & Co. KGaA. Reproduced with permission.)

countries with advanced economies. This demonstrates that modern economies with a large material turnover slowly change the concentrations of substances in environmental compartments (Figure 1.9).

1.4.5.3 Linear Urban Material Flows

Figure 1.10 displays the material flows through Vienna. Like in any other city, the flows are mainly linear. To change from a linear flow system to a cycling system requires more than new technologies. It means a complete change in

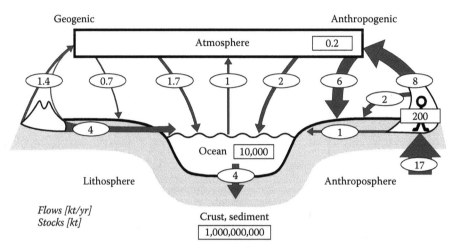

FIGURE 1.8
Global flows (kt/year) and stocks (kt) of cadmium in the 1980s. Man-made flows (right-hand side) surpass natural flows (left-hand side). Comparatively high atmospheric emissions lead to a significant accumulation of anthropogenic cadmium in the soil. The stock of cadmium in the anthroposphere grows by 3% per year. In the future, this stock needs to be managed carefully if negative effects on the environment are to be avoided. (From Brunner, P.H. and Baccini, P., *Neue Zürcher Zeitung, Beilage Forsch. Technik, 70, 65, 1981.* With permission.)

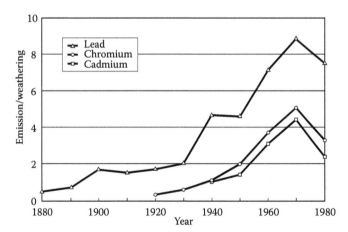

FIGURE 1.9
Rates of anthropogenic emissions of lead, chromium, and cadmium compared with weathering rates in Sweden. (From Bergbäck, B., Industrial Metabolism: The Emerging Landscape of Heavy Metal Emission in Sweden, PhD thesis, Linköping University, Sweden, 1992. With permission.)

lifestyle, value system, and priorities, as well as large-scale changes in technology and the economy. At present, such a radical change is not in sight, nor is there convincing evidence that full material cycles are feasible and needed (Becker-Boost and Fiala, 2001). While a gradual shift to a cycling economy taking place in several countries has the advantage of less demand for primary

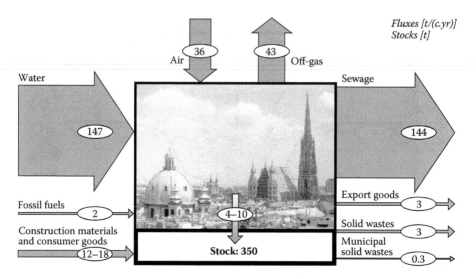

FIGURE 1.10
In the 1990s, about 200 t of materials were consumed per capita per year in the city of Vienna, and somewhat less than this amount left the city in a linear way. The small difference of 4 to 10 t capita^{-1} year $^{-1}$ accumulates, resulting in a continuous growth of the stock of 350 t/capita, which doubles in 50 to 100 years. Basically, the modern anthropogenic metabolism is a linear-throughput reactor, with less than 5% of materials being involved in regional cycles. Note the differences from Figure 1.3 and Table 1.1 (material turnover in private households): the numbers for Vienna also include material turnovers outside private households in industry, the service sector, public service and administration, etc. (Brunner, P.H. and Rechberger, H.: Anthropogenic metabolism and environmental legacies, in *Encyclopedia of Global Environmental Change*, Vol. 3, Munn, T., Ed. 2001. Copyright Wiley-VCH Verlag GmbH & Co. KGaA. Reproduced with permission.)

materials and of increased resource efficiency, it is not yet a guarantee for less environmental pollution.

1.4.5.4 Material Stocks Grow Fast

Because material inputs in growing regional economies are, as a rule, larger than outputs, most modern regions accumulate a large material stock within their boundaries. Exceptions are regions that exploit and export resources on a large scale, such as coal, metal ores, sand, gravel and stones, or timber. There are many stocks in the anthroposphere: mining residues left behind as tailings; material stocks of industry, trade, and agriculture; urban stock of private households and of the infrastructure for transport, communication, administration, education, etc., and the comparatively small but growing stock of wastes in landfills. The difference in stock accumulation between prehistoric (<1 t) and modern humans is striking. Present-day material stocks amount to about 200 to 400 t per urban citizen. This stock has to be managed and maintained. Today's decisions about the renewal and maintenance of the urban stocks, such as buildings and networks for energy supply, transport, and communication, have far-reaching consequences. The residence time of materials in the stock can range up to more than 100 years. This means that once a material is built into the stock, it will not show up quickly in the output of the stock, namely in waste management.

For the management of resources and the environment, the urban stock is important for several reasons:

1. It is a growing reservoir of valuable resources and holds tremendous potential for future recycling.
2. It represents a mostly unknown source, whose importance is not yet in focus, which awaits assessment with respect to its significance as an economic resource and as a threat to the environment.
3. It is a potential long-term source of large flows of pollutants to the environment. Urban stocks contain more hazardous materials than so-called hazardous-waste landfills, which are, at present, a focus of environmental protection.

Planners and engineers are asked to design new urban systems. In the future, the location and amount of materials in a city's stock should be known. Materials should be incorporated into the stock in a way that allows easy reuse and environmental control. Economies are challenged to maintain high growth rates, building up ever-larger stocks, and setting aside sufficient resources to maintain and renew this stock properly over long periods of time.

1.4.5.5 Consumption Emissions Surpass Production Emissions

In service-oriented societies, most production emissions are decreasing, while some consumer emissions, for example, CO_2 and heavy metals, are increasing. This is due to the tremendous efforts that have been undertaken in many countries to reduce industrial emissions. On the other hand, inputs into the consumption process still grow, and thus, consumer outputs such as emission loads and waste amounts are rising, too (Ivanova et al., 2016). Anderberg, Bergbäck, and Lohm (1989) indicate that for heavy metals in Sweden, industrial activities were the most important emission sources up to the 1970s. After that period, nonpoint consumer-oriented emissions became dominant (Figure 1.11).

An actual example, which is based on MFA, is given in Figure 1.12. A galvanizing company optimized its production process so that most of its zinc residues and wastes were either internally or externally reused. From the point of view of the production process, the company has protected the environment in the best possible way. An MFA reveals that now the largest zinc flow to the environment is caused by the corrosion of the zinc coating on the

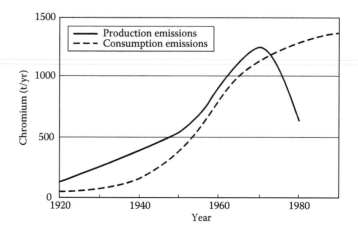

FIGURE 1.11
Consumption emissions versus production emissions for chromium in Sweden. Production emissions decreased due to pollution abatement measures as well as decreasing activities of heavy industry, and consumer emissions increased due to rising demand in consumer goods and dissipative applications of materials. (From Anderberg, S. et al., *Ambio*, 18, 4, 216, 1989. With permission from the Royal Swedish Academy of Sciences.)

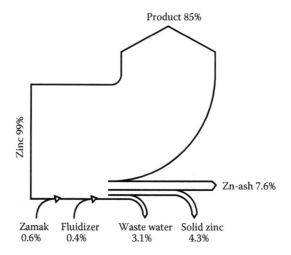

FIGURE 1.12
Zinc flow through an advanced electroplating plant: 85% of the zinc used as surface coating for corrosion protection is sold with the product. Nearly all of the 15% zinc not sold is recycled. The emissions to the environment during production are very small. During the lifetime use of the Zn-coated product, most of the zinc is likely to be dispersed in the environment due to corrosion. Hence, despite the effort to protect the environment by pollution prevention measures in the factory (cleaner production), most zinc ends up in the environment. (Brunner, P.H. and Rechberger, H.: Anthropogenic metabolism and environmental legacies, in *Encyclopedia of Global Environmental Change*, Vol. 3, Munn, T., Ed. 2001. Copyright Wiley-VCH Verlag GmbH & Co. KGaA. Reproduced with permission.)

product: during the lifetime of the galvanized surface, much of the coating (zinc) is lost to the environment. Thus, despite pollution prevention measures during the electroplating process, as much as 85% of the zinc may end up in the environment. Since the corrosion process is slow and the residence time of metals on surfaces can be rather long, the metal flows to the environment last for years to decades. The legacies of today's protected surfaces are tomorrow's environmental loadings. At present, the significance of such emissions has not been assessed and evaluated. There are no methods available yet to measure and appraise these sources of pollutants to the environment. For state-of-the-art surface coating, it is not the production waste that causes the largest losses to the environment; it is the product itself that is the important source of environmental loadings.

This new situation is due to, on the one hand, advanced technology and legislation in the field of industrial environmental protection. On the other hand, the high and still growing rate of consumption has led to large stocks discharging significant amounts of materials in a hardly noticeable way into the water, air, and soil. Examples of such consumer-related emission sources are the weathering of surfaces of buildings (zinc, copper, iron, etc.), the corrosion of and erosion from vehicles and infrastructure for traffic (chassis, tires, brakes, guardrail, light post), the emission of carbon dioxide and other greenhouse gases due to space heating or cooling and transportation, and the growing problem of nitrogen overload. Cities are hot spots of such non-point sources. The new situation requires a shift in pollution abatement. In order to be efficient, environmental management must concentrate on the priority sources. Thus, in the future, the output of the consumer has to be in the focus of the product designer and environmental regulator. Since consumer behavior and lifestyle determine the material throughput through the anthroposphere, it is important to explore means to influence this behavior.

It is more difficult to prevent consumer-related emissions than it is to stop industrial pollution:

1. It is a much more demanding challenge for a company to change or discontinue a product than to add an air pollution–reducing device to its production facility.

2. The number of consumer sources is many orders of magnitude larger than that of production facilities.

3. The individual emission may be very small, and only the multiplication by the large number of sources may cause a hazard for the environment. Thus, consumer awareness of the overall relevance of his/her small contribution must be created first in order to be able to tackle the problem.

4. In general, emissions of industrial sources can be reduced with a much higher efficiency than widespread small-scale emissions from households or cars.

1.4.5.6 Changes in Amount and Composition of Wastes

Due to the large growth in consumption, the amount and composition of wastes are changing and will continue to vary in the future, too. For materials with long residence times, such as construction materials, a huge stock is being built up before wastes become noticeable at the back end of the life cycle. Waste management inherits most of the materials in the stock. Due to the increasing complexity of the goods produced, wastes comprise rising amounts of new materials, many of them being composed of mixtures that cannot be taken apart by physical methods. The long-lived stock needs to be protected from degradation by microorganisms, ultraviolet light, temperature, weathering, erosion, etc. Thus, it contains more hazardous substances (heavy metals and persistent organic materials as stabilizers and additives) than the short-lived consumer products that are recycled today (packaging materials, newsprint, glass). Therefore, in the future, a large amount of hazardous substances will have to be removed from the wastes of the long-lived stock if materials of the stock are to be recycled safely.

1.4.6 Combination of MFA with Other Approaches

Optimization of processes and systems must be based on sound analysis and evaluation, and requires appropriate instruments. The idea of a toolbox and the role of MFA, input–output analysis, LCA, risk assessment, and other methods of systems analysis for environmental decision making has been discussed before (e.g., Wrisberg, Udo de Haes, Triebswetter, Eder, and Clift, 2002; Finnveden, Björklund, Ekvall, and Mosberg, 2006). However, the potential of combinations of MFA with other methodologies reaches well beyond environmental decision making and embraces all fields described in this section (Section 1.4). Of particular value is the combination with MFA when it comes to optimizing anthropogenic systems as a whole in view of multiple criteria. The reason is as follows:

The value of an MFA is twofold: First, it enables the expert to better understand an investigated system. In order to describe the material flows and processes and their changes in sufficient detail, systems boundaries must be drawn, and the crucial elements of the system have to be identified, linked, and quantified according to the mass-balance principle. This usually results in new knowledge about the system that was not available before the investigation. Second, the results of the MFA form a transparent and reproducible basis for subsequent evaluation. This allows checking for the significance of the results in view of relevant aspects such as resource management, waste management, environmental impacts, or economic issues. Based on such evaluations, systems or relevant processes can be changed and optimized with respect to given goals.

Usually, this second aspect is the true motivation for an MFA: A material flow (e.g., the metal copper [Graedel et al., 2004] or the nutrient phosphorus [Zoboli, Laner, Zessner, and Rechberger, 2015]) through an economy is examined in view of the resource potentials and the environmental impacts. Or a waste management system is compared to the statutory objectives, such as protection of human health and the environment, and conservation of resources (e.g., Allesch and Brunner, 2015).

An MFA points out flows, stocks, and changes in stocks of materials. Per se, it is not yet a value judgment. An MFA alone cannot discriminate between beneficial and harmful flows and stocks. It takes additional information and criteria for evaluation: (1) Assessment methodologies like LCA offer such value-laden criteria (ozone depletion, acidification, human toxicity, ecotoxicity, and the like). (2) Comparing MFA results of anthropogenic substance flows with geogenic flows and stocks yields information about accumulation or depletion of substances. If paired with limiting values for geogenic stocks (e.g., standards for soil and water), MFA results become highly valuable for environmental assessments. (3) The combination of MFA and entropy analysis allows for estimating if a metabolic system of processes increases or decreases dissipation of a substance. (4) Economic analysis can assign monetary values to the flows and stocks, and thus allows evaluation of a system in view of economic aspects.

MFA is instrumental for all subsequent evaluation steps such as LCA or economic analysis. It ensures that the data used for the evaluation step are consistent, robust, and transparent. The advantage of applying MFA is that all materials entering a system must be taken into account in order to fulfill the mass-balance condition. Some evaluation systems do not require mass balances. This may eventuate in neglecting "hidden" material flows that come to light only when all inputs have to be matched by outputs. Particularly if an MFA is performed on the level of substances (substance flow analysis [SFA]), the full mass balance of the substance forces the evaluator to take all hazardous flows into account, which is not necessarily the case for an evaluation method that is not based on the mass-balance principle.

In practice, MFA and evaluation methods are connected by, first, establishing a quantitative MFA and, second, using the results of this MFA as input for the evaluation step by LCA or other evaluation methods.

Examples for the combination of MFA with LCA-based evaluation approaches are the comparison of technologies for the treatment of sewage sludge (Lederer and Rechberger, 2010), a comparative analysis of optimal treatment technologies for end-of-life cooling appliances (Laner and Rechberger, 2007), or the evaluation of national waste management systems (Brunner et al., 2016). Further integration of MFA and LCA has been achieved within formal optimization frameworks, which allow for developing optimal resource use strategies from a system's perspective. Based on an MFA, the resource flows are mapped, and transfer coefficients for different processes are determined. On this basis, life cycle assessment is used to identify the most environmentally beneficial solution fulfilling the required functionality (e.g., satisfying the resource demand) and respecting given constraints (e.g., plant capacities, qualities of inputs to various utilization pathways). Such an optimization framework has been applied to sewage sludge management in the Swiss region of Zürich (Vadenbo, Hellweg, and Guillén-Gosálbez, 2014a; Vadenbo, Guillén-Gosálbez, Saner, and Hellweg, 2014b). The proposed modeling approach allowed for identifying environmentally preferable scenarios of sewage sludge management compared to the status quo and for exploring trade-offs between different environmental objectives in view of alternative sewage sludge management strategies. Apart from the optimization of waste resource utilization, formal optimization has also been used to assess wood utilization on a regional level (Höglmeier, Steubing, Weber-Blaschke, and Richter, 2015). The wide range of potential applications for wood and their associated environmental impacts make it an interesting subject of optimization modeling. In particular, the effect of wood utilization on other resource systems (e.g., wood as a fuel replaces other fuels in energy conversion processes) is important for the overall environmental performance of a specific wood utilization strategy.

From an anthropogenic resource perspective, MFA has been put forward as the major tool for prospecting and exploring anthropogenic material stocks and flows (Lederer, Laner, and Fellner, 2014). In such a setting, MFA is combined with environmental and economic assessments on a technology (e.g., Fellner et al., 2015) or project level (e.g., Winterstetter, Laner, Rechberger, and Fellner, 2015), to distinguish anthropogenic materials that can be (potentially) exploited in an economic way from others that cannot. Another use of MFA in the context of resource use and raw material supply is within criticality assessments (Graedel et al., 2011). Here, MFA is typically used as a basis for determining the current recycling level to assess the degree of material recirculation for a given raw material (cf. Nassar et al., 2011). Along this line, MFA has been combined with statistical entropy analysis to evaluate the dissipative character of a material flow system (Rechberger and Brunner, 2002). The basic idea is that the more dissipative the use of a substance (resource), the harder it is to recycle and, therefore, the less sustainable is this material use pattern (Georgescu-Roegen, 1971; Rechberger and Graedel, 2002; Yue, Lu, and Zhi, 2009).

From an environmental risk perspective, MFA has been applied as a basis to investigate the pathways of hazardous substances with respect to clean cycles and appropriate sinks (Brunner, 2010). The objective is to investigate if a particular strategy to manage a hazardous substance leads to accumulation or depletion in the anthroposphere or environment. Such enrichments or depletions are, per se, not yet a strong criterion for policy decisions, but they point out the need for further assessments and possibly action (Vyzinkarova and Brunner, 2013). If paired with toxicity data, they allow the drawing of more specific conclusions. MFA-based indicators assessing the allocation of substances into suitable sinks have been put forward as a tool for "early warning" (Kral, Brunner, Chen, and Chen, 2014). In this context, MFA was also combined with risk assessment to identify risks associated with the use patterns of a substance (Chen, Crawford-Brown, Chang, and Ma, 2014). Such MFA-based screening can be particularly valuable if direct measurements for potential exposure routes are not available and/or in case of emerging pollutants, which are currently accumulating in anthropogenic stocks. For an example, see Gottschalk, Sonderer, Scholz, and Nowack (2010).

For a more detailed discussion of evaluation methods for MFA results, see Chapter 2, Section 2.5.

1.4.7 Policy Evaluations and Decisions

To evaluate and design policies in the fields of resource management, waste management, and environmental management is a challenging field, requiring reliable knowledge bases. In the case of existing policies, the evaluation step is quite straightforward because of available information, data, and methodologies. The design step for new policies is more demanding and requires anticipating forthcoming changes and progress based on past developments. Often, scenario analysis is used to test the outcome of different policies (e.g., Brunner et al., 2015). MFA is an excellent tool for such analysis. It forms a common basis for describing the status quo and is instrumental for building scenarios. MFA allows simple changes of relevant system elements, such as the introduction of new processes (change in transfer coefficients), new products (addition and depletion of goods), or new substances (ditto). Also, external impacts can be considered, such as the demand for a particular degree of resource efficiency, or new regulations phasing out hazardous substances. Modeling metabolic systems by MFA makes it easy to recalculate scenarios if new transfer coefficients or changing flows and stocks are introduced.

In the field of the metabolism of the anthroposphere, anthropogenic material flows are responsible for the benefit of utilization, for resource depletion/accumulation, and for environmental pollution. In order to predict the effect of changes and to design policies to prevent depletion and pollution, knowledge about stocks and flows of goods and substances is indispensable (Baccini and Brunner, 2012; Baccini and Bader, 1996). Theories and empirical work from evaluation research in resource and environmental policies have been used as foundations for a paradigm shift in environmental policy (Pesendorfer, 2002): While during the first decades of environmental protection, the focus was on the output of the industrial and consumption process, namely, emissions, the following policy effort was on reducing the material input. An example for such input-oriented and MFA-based policy is the so-called materials accounting approach based on economic input–output analysis (Miller and Blair, 1985), allowing, for example, prediction of the impact of investments and of technological changes.

MFA is well suited to support policy makers, serving a wide range of groups such as ministries, government agencies, companies, and international organizations. It is particularly useful for development and evaluation of national and regional policies in the fields of environmental protection, and resource and waste management, and is also applied in policies regarding economic trade and technology development. MFA facilitates also policy decisions in industry, trade, and nongovernmental organizations regarding new technologies and evolving management schemes.

As shown in Chapter 1, Section 1.4.6, MFA alone does not suffice as a decision- and policy-supporting tool. It requires combination with other methods that take into account criteria such as economic values, product qualities, and environmental loadings to enable sound judgments.

Why is MFA particularly well suited for policy decisions and evaluations?

- Mass-balance principle: Decision makers and stakeholders need reliable, reproducible, and transparent information for their rulings. The mass-balance principle offers this. Particularly if underpinned with pictures such as STAN graphs or Sankey diagrams (see Chapter 2, Section 2.4), the significant results of MFA can be displayed straightforwardly, and the message relevant for decision making can be conveyed in a clear and comprehensible manner. MFA is instrumental for policy makers to understand: "What goes in must come out."

- Goods and substances: Policy decisions are most often multidimensional, covering several fields of interest. MFA forms a common basis to combine different fields, such as economy, environment, and resources (primary as well as secondary resources): the reason is that MFA comprises goods as well as substances. While goods, as economic entities, cover the economic aspects, the various substances determine the qualities, for example, of goods, resources, and the environmental impact. Hence, MFA results yield relevant information for both levels, and since goods and substances are also mathematically linked in an MFA (e.g., by the software STAN), policy makers get a full and reliable picture of the comprehensive material system. This is in contrast to materials accounting as defined by the Wuppertal Institute observing the economic level of goods only (Matthews et al., 2000). On this level, one cannot assess and identify effects on the impact level, e.g., the environment. MFA based on goods and substances is designed to tackle specific problems close to the impact and to link the place of impacts and sources in a rigid way. This ensures high effectiveness of policy decisions.

- System understanding: In contrast to evaluation methods such as LCA, MFA depicts a physical system from sources to sinks (or from imports to internal stocks and exports) in a comprehensive way. Systems boundaries are coherent and apparent, each process within the system is characterized by specific transfer coefficients, and each flow and stock of goods and substances has a numerical value, facilitating the understanding of the functioning of the material system. This is also of importance for dissemination of existing experience and knowledge.

- Uncertainty: Policy decisions have an inherent risk because of uncertainties of the underlying knowledge base (Morgan, Henrion, and Small, 1990). Often, decision makers have the choice between several variants that have different costs and benefits in view of, for example, product quality, environmental loadings, and resource conservation. If the uncertainties between the variants are the same, decisions are comparatively easy. If each variant has a different uncertainty, deciding may become a highly challenging task. MFA allows taking into account uncertainty in a systematic and transparent way (see Chapter 2, Section 2.3) and thus prepares a numerical, mass balance–approved basis for decision making under uncertainty.

- Uniform metric: MFA provides a simple, uniform metric that is based on common knowledge and scientific traditions. It represents a guide for measuring material flows and stocks that—if followed— allows comparing of data and results (e.g., resource productivities) from different regions (Klinglmair et al., 2016), countries, companies, and others. It enables those involved to take identical approaches when collecting statistical data and to compare their data and results with other policy analysts and researchers. It is important to note that at present, international harmonization of MFA methodology has not yet taken place. In the future, common MFA definitions and indicators for general comparison and standardization are needed.

- Monitoring instrument: Monitoring material flows from source to sink is required for many policy fields, e.g., for linking environmental loadings with the corresponding emissions, for predicting resource accumulations in urban mining, or to indicate opportunities for increasing performance and resource efficiency. MFA serves as an excellent monitoring tool. It is often much more economical than traditional monitoring systems that are based on sampling and chemical analysis. Monitoring by MFA is a highly effective support for policy evaluations and suited for both an input-oriented approach such as in materials accounting (Pesendorfer, 2002) and an output-oriented approach focusing on emissions and accumulations.

- Early recognition: If (static or dynamic) MFA systems take several time periods into account, changes in anthropogenic and environmental stocks can be observed. This allows for early recognition of beneficial or harmful accumulations and depletions, e.g., in the urban material stock, or in soil, water, and air. The benefit for industry, trade, commerce, and public services can be quite substantial because, for example, future constraints can be assessed, and necessary prevention or plant capacities can be provided for.

- Setting priorities: Because MFA draws a complete picture of a material system, the importance of the magnitude of flows and stocks of individual processes can be discerned. Hence, if for economic motives, or for reasons of environmental protection or resource conservation, material reductions or increases are intended, an MFA allows the identification of those flows and stocks that are most relevant.

- Interdisciplinarity: Policy decisions are most often based on a range of knowledge bases from various fields, such as economics, engineering sciences, and natural sciences. MFA is useful for analyzing issues that cut across different policy areas and support decisions that have implications in several fields, such as socioeconomic, environmental, and engineering topics. MFA is a tool that can act as a mutual denominator for all disciplines involved. It often serves as a common backbone for all stakeholders engaged in a particular decision. If a sound physical system is defined, other disciplines may find it easier to link their aspects to the shared objectives.

1.5 Objectives of Material Flow Analysis

While in the earlier sections, the fields of application of material flow analysis (MFA) have been discussed, this final part of Chapter 1 focuses on the specific goals of MFA.

In general, MFA is the method of choice when it comes to investigating the flows and stocks of any material-based system. MFA gives insight into the behavior of a material system, and when combined with other tools such as life cycle assessment, energy flow analysis, economic analysis, and consumer-oriented analysis, it facilitates the understanding and control of an anthropogenic system. The specific objectives of MFA are listed here. They are

- To understand a metabolic system qualitatively by selecting the relevant processes, material flows, and stocks in well-defined, uniform terms, and by creating a physical model selecting system boundaries and linking processes and material flows
- To find the key quantitative flows, processes, and stocks in this metabolic system, linking inputs (sources), stocks, and outputs (sinks), enabling the tracking of, for example, wastes and emissions to the original material input or process of origin
- To follow the material system over time, (1) with a focus on past developments or (2) to forecast the future based on past and present trends, or based on assumptions about progress such as new technologies or changing drivers like consumer behavior
- To reduce complexity of the system as far as possible while still guaranteeing a sound and robust basis for decision making
- To apply a mass-balance approach for cross-checking and identifying deficits, which is a specific feature distinguishing MFA from

other accounting techniques that are based on economic data and not mass flows

- To present results about flows and stocks of a system in a concentrated, comprehensive, and understandable way, thus facilitating transparency, universal understanding, and communication of methodology and results to a wider audience of stakeholders by appropriate visualization of the procedures and results, e.g., by Sankey-style diagrams
- To form a basis for sensitivity and uncertainty analysis to reveal key sensitivities and uncertainties of flows and stocks
- To serve as a basis and support for assessment tools such as LCA, entropy and economic analysis, input–output studies, etc.
- To serve the management of environment, resources, and wastes in the following:
  - Early recognition of potentially harmful or beneficial accumulations and depletions of stocks, as well as timely prediction of future environmental loadings.
  - Setting priorities regarding measures for environmental protection, resource conservation, and waste management: Which measure has potentially the largest effect? What comes first?
  - Design of products, processes, and systems that promote environmental protection, resource conservation, and waste management (green design, ecodesign, design for recycling, design for disposal, etc.).

As mentioned before, MFA is one of the major tools in resource management, environmental protection, and other fields. In contrast to assessment methods such as LCA, risk assessments, or economic analysis, evaluation of material flows and stocks is not an objective of MFA. The objective of MFA is to serve as a physical backbone for such evaluations. An MFA without support from other methodologies cannot discern impacts; hence, this is not a goal for MFA. This may be different for materials accounting, where the paradigm is set as "the less flows, the better," and a goal of general material flow reduction by a factor 4 or 10 is proposed (Schmidt-Bleek, 1998). For more information about the objectives of the tools available for environmental management, see the work of Udo de Haes et al. (2000).

PROBLEMS—SECTIONS 1.4 AND 1.5

Problem 1.5:

Assuming you want to repeat Santorio's experiment, which of the three approaches do you prefer for measuring the input of the test person? (a) Weigh the mass of food the person eats, (b) determine the difference in the mass of the person before and after eating the meal by weighing the person twice, or (c) both. Take into consideration that different scales are usually used to weigh 1 kg and 100 kg.

Problem 1.6:

In order to conserve resources, factor 4 and factor 10 have been proposed (Schmidt-Bleek, 1997; von Weizsäcker, Lovins, and Lovins, 1997). The total turnover of materials should be decreased by 75% and 90%, respectively. (a) How much can you reduce the flow of materials through your household by existing technologies and at affordable costs? Take Table 1.1 as a reference and discuss for each activity the potential to reduce material flows by Factor 4. (b) What do you recognize as the main advantage and disadvantage of the factor X philosophy?

Problem 1.7:

The global flows and stocks of cadmium in the 1980s are presented in Figure 1.7. Two decades later, an (unpublished) review in 2003 reveals the following changes in the flows and stocks of the anthroposphere: cadmium exploitation has slightly increased and amounts to 18 to 20 kt/year. The anthropogenic stock has more than doubled (+300 kt). Due to environmental protection, emissions to air and water have both decreased by about 80%. Data about the residence time of goods containing cadmium suggest that about 20 kt/year of cadmium are changing into goods that are no longer in use; they are either discarded as wastes or "hibernating" in the anthroposphere. The amount of cadmium transferred to landfills is not known; it may range between 2 and 20 kt/year. (a) Discuss the effect of reduced atmospheric cadmium emissions on the soil. (b) In view of the increase in anthropogenic stock from 200 to 500 kt of cadmium during the last 20 years, assess the cadmium flow to landfills. (c) Around the year 2000, NiCd batteries were responsible for about 75% of all cadmium used. What measures do you suggest to control problems that may arise from the use and disposal of cadmium batteries?

Problem 1.8:

Prepare a quantitative list of solid wastes that you produce in your household (in kg/capita/year). Start with an inventory of major inputs of goods in your household. Use Table 1.1, your solution to Problem 1.3, and your own observations. Include all wastes from materials such as your home, car, water, food, etc., and arrange them in groups according to their collection (e.g., MSW, construction wastes, hazardous wastes).

Problem 1.9:

Read the latest two issues of the *Journal of Industrial Ecology*, and look for recent applications of MFA. Identify the goals of the MFA, the procedures, and the results, and discuss whether the conclusions and implications have been appropriately visualized by MFA. Evaluate whether MFA was the only possible approach to reach the objectives or whether other methods would have allowed the same conclusions.

The solutions to the problems are given on the website http://www.MFA -handbook.info.

2

Methodology of Material Flow Analysis

This chapter introduces and describes the terminology of material flow analysis (MFA). The exact definition of *terms* and *procedures* is a prerequisite for generating reproducible and transparent results and for facilitating communication among users of MFA. The definitions as developed by Baccini and Brunner (1991) during the 1980s are employed. As described in Chapter 1, Section 1.1.3, there have been several parallel developments of MFA, resulting in diverse usage and meanings for single terms. For example, the terms *substance* and *materials* have been applied to substances as well as goods, and sometimes several names are used for the same object. Needless to say, the presence of such a babel within the global MFA community complicates the application and usage of MFA, especially for the new user. Differences appear not only in the employment of single terms but also on a higher conceptual and methodological level. In this chapter, these aspects are addressed and discussed whenever appropriate. The authors' intention is to contribute to a generalized terminology in MFA.

2.1 MFA Terms and Definitions

2.1.1 Substance

Looking up *substance* in a dictionary may yield the following definitions relevant to the MFA field: the physical matter of which a thing consists; material (Guralnik and Friend, 1968); matter of particular or definite chemical constitution (Gove, 1972), or: matter; or material of specified, especially complex, constitution (Morris, 1982). For *material*, one may find the following: of matter; of substance (Guralnik and Friend, 1968); the elements, constituents, or substances of which something is composed or can be made (Gove, 1972); and the substance or substances out of which a thing is or may be constructed (Morris, 1982). All of these definitions seem convincing, but they are not precise enough for our purposes. As a matter of fact, it would be difficult to decide based on these definitions whether *wood* should be addressed as a *substance* or *material*. In everyday language both terms are used more or less synonymously. Therefore, MFA relies on the term *substance* as defined by chemical science. *A substance is any (chemical) element or compound composed of uniform units. All substances are characterized by a unique and identical*

constitution and are thus homogeneous (Sax and Lewis, 1987). Using this definition makes clear that wood is not a substance. It is composed of many different substances such as cellulose, hydrogen, oxygen, and many others. Thus, a
substance consists of identical units only. If these units are atoms, one speaks
about a (chemical) element such as carbon (C), nitrogen (N), or cadmium
(Cd). In cases where the units represent molecules such as carbon dioxide
(CO_2), ammonium (NH_3), or cadmium chloride ($CdCl_2$), they are designated
as (chemical) compounds. *In MFA, chemical elements and compounds both are
correctly addressed as substances.*

Note that some authors use an additional and complementary definition
for substance as follows: substance is the umbrella term for matter that differs only in size and shape but not in other specific properties such as color,
density, electric conductivity, solubility, etc. (Holleman and Wiberg, 1995).
This means that a substance has to have a shape. According to this definition, it is not correct to identify Cd in a piece of polyvinyl chloride (PVC)
as a substance. As soon as the substance Cd is incorporated and dispersed
in the PVC matrix, forming a new heterogeneous mixture of homogeneous
substances, Cd loses its shape and, therefore, the status of a substance. In
MFA, this definition is not useful, since MFA tracks the fate of substances
through a system regardless of the physical and chemical form. Hence, the
MFA definition of substances excludes shape. Thus, it is correct to refer to Cd
and other substances in a window frame made of stabilized PVC.

For MFA, substances are designated as *conservative* when they are not
destroyed or transformed in a process. This applies to all elements and certain compounds that remain stable under process conditions. For example,
cadmium can be transformed to cadmium chloride in a combustion process.
Hence, it is not possible to establish a closed balance for cadmium chloride
for the process. Cadmium chloride has to be addressed as a nonconservative
substance, while the element cadmium can be designated as conservative
because it does not disappear. In chemistry, the definition of conservative is
somewhat different and is attributed to substances that are nonreactive.
Also, in the environmental sciences, the term *conservative* characterizes substances that are persistent in the environment, having characteristics that are
resistant to ordinary biological or biochemical degradation.

Substances are important for environmental and resource management.
They determine the value or the hazardousness of a good. Many MFAs are
carried out to determine flows of potentially hazardous substances to the
environment, and to find out more about the fate of these substances in environmental compartments such as water bodies and soils. Other MFA studies are commenced to understand better the accumulation or depletion of
resources in an anthropogenic system. Often these are substances such as
heavy metals (Cu, Zn) or nutrients (N, P). Material flows through society
can pose a problem because of both their composition (substances) and their
quantity. The importance of discussing substances *and* materials is further
explained in Section 2.1.10.

2.1.2 Good

The term *good* describes merchandise and wares. It is mostly used as a plural (*goods*). The opposite of a *good* is not a *bad*, meaning that waste materials are also goods (Brunner, 2002). The term is not used as an adjective but as a noun. *Goods are defined as economic entities of matter with a positive or negative economic value. Goods are made up of one or several substances.* Hence, a correct designation for wood is found: it is a good. Examples for goods are drinking water (includes, besides H_2O, calcium and other trace elements and is therefore not a substance), mineral ores, concrete, TV sets, automobiles, garbage, and sewage sludge. All of these materials are valued and financially rated by the economy. Their turnover in terms of economic value and, less often, mass is usually reported in all kinds of statistics (by governments, national or multinational organizations and associations, companies, etc.). Hence, production figures (e.g., t/year) of most goods are more or less well known. Such information is a prerequisite for establishing material balances.

Sometimes, the words *product, merchandise,* or *commodity* are used synonymously with *goods*. The advantage of these synonyms is that they are common words. Their drawback is that *product* usually designates the output but not the input of a process or reaction. *Merchandise* and *commodity* are usually used to describe goods with a positive economic value and are rarely applied to negatively valued goods like garbage and sewage sludge. There are only a few goods that have no economic value, such as air or fog.

2.1.3 Material

The term *material* has already been explained in Section 2.1.1. While daily language does not distinguish between *substance* and *material*, in MFA, *material* serves as an umbrella term for *both* substances and goods. So carbon as well as wood can be addressed as a material.

2.1.4 Process and Stock

A *process* is defined as the *transformation, transport, or storage* of materials (Baccini and Brunner, 1991). Note that some authors prefer the term *reservoir* instead of *process. Reservoir* is a narrower term and does not consider all characteristics that can be attributed to a process. While the term *process* implies that something is changed in its quality (feedstock becomes product and waste), a reservoir mainly changes its quantity (mineral deposit, groundwater, landfill). So *reservoir* is merely a synonym for *stock*, and *process* is the more extensive term comprising change in quality and quantity (location being part of quality).

Materials are transformed throughout the entire economy on different levels. Transformation takes place in primary production processes such as in the mining and metal industry, where metals are extracted from mineral ores. Consumption processes such as private households transform goods into wastes and emissions. Examples of transformation processes are

- The human body, where food, water, and air are transformed mainly into CO_2, urine, feces, and human biomass
- The entirety of private households in a defined region, where countless inputs are converted to sewage, wastes, emissions, and some useful outputs
- Wastewater treatment plants, where sewage is transformed to purified wastewater and sewage sludge
- A car, where fossil fuels are oxidized to CO_2 at a rate of 0.003/s
- A power station with a fuel transformation rate of 60/s (turnover, 20,000 times that of a car)
- The metabolism of a city, where many inputs are converted to corresponding outputs
- The total agriculture of a collective of states, such as the US or all countries of a river catchment area, transforming inorganic nitrogen to organic nitrogen, etc.

Transformation of materials is not restricted to anthropogenic processes but is even more relevant for natural systems. Examples are forests that transform carbon dioxide into oxygen and biomass, and soils where biomass residues are transformed into humus and CO_2.

The *transport* of goods, persons, energy, and information can be described as a process, too. During a transportation process, the goods moved are not transformed but relocated over a certain distance. The transportation process covers all material flows that are needed to carry out transport and all wastes and emissions that result from the transport. Transformation and transport processes are both symbolized by rectangular boxes (see Figure 2.1). Usually, processes are defined as black-box processes, meaning that processes within the box are not taken into account. Only the inputs and the outputs are of interest. If this is not the case, the process must be divided into two or more subprocesses.

This facilitates investigation and analysis of the functions of the overall process in greater detail (see Figure 2.2). If a process is subdivided into several subprocesses, the ensemble of subprocesses, flows, and stocks is called a subsystem. In MFA modeling, it is sometimes convenient to define several layers of subsystems. The advantage is that on the top layer, there still exists a comparatively simple system allowing easy visualization and comprehension of the main flows and stocks. If a user wants to understand more details, he/she can penetrate step by step deeper into the system, thus facilitating complete transparency of the underlying processes, flows, and stocks. Also, if a process has to be improved with respect to its input and/or output, a better understanding of the inner mechanisms of the process is needed. This

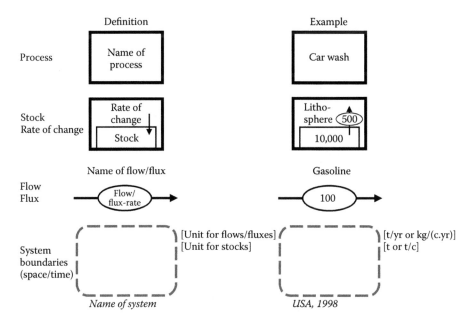

FIGURE 2.1
Main symbols used in MFA diagrams. If space is limited, ovals around flow/flux rates may be omitted.

can only be achieved if the subprocesses and their connections by flows are investigated. Another case that requires disintegration of a process is when data for single input or output flows are missing or not plausible. Then the determination of the inner flows can help to solve such a data problem.

There is one exception to the black-box approach: The third type of process is used to describe the quantity of materials within a process. The total amount of materials stored in a process is designated as the *stock of materials* in the process. Both the mass of the stock as well as the rate of change of the stock per unit time (accumulation or depletion of materials) are important parameters for describing a process. A *final sink* is a process where materials have very long residence time (t_R >1000 year).

Examples of storage processes are private households storing goods like concrete, furniture, electrical appliances, etc., for years to decades, or landfills where human society stores most of its nonrecyclables for centuries to millennia. Natural storage processes are the atmosphere and the oceans for carbon dioxide, or soils and sediments for heavy metals. A stock within a process is symbolized by a smaller box within the "process" box (see Figure 2.1).

Stocks have become more important in industrial ecology research of the past years as MFA studies showed that advanced economies have built up significant material stocks. During the last five decades, billions of tons of materials have been accumulated in buildings, settlements, and

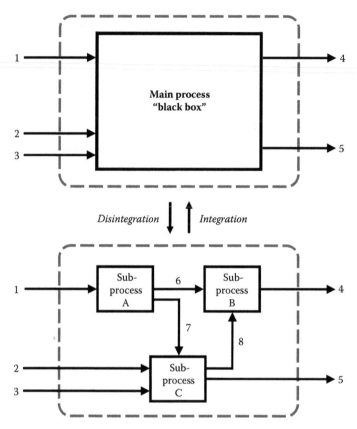

FIGURE 2.2
Opening up a black box by subdividing a single process into several subprocesses provides
additional information about the black box.

infrastructure networks. These are durable goods with long residence times
(several decades). In a certain sense, such stocks act as a buffer, meaning that
the stored materials become outputs ("secondary resources") only after a
long residence time. When the buffer function of the stock decreases due to,
for example, economic or technological changes, outputs of storage processes
may become important feedstocks for inputs into the economy. Therefore, it
is important to understand the stock dynamics such as the composition of
the stock, the average residence times of materials, and its changes and drivers. These stocks are usually determined by a dynamic MFA (top-down) or a
bottom-up approach (cf. Chapter 3, Section 3.2.3).

2.1.5 Flow and Flux

For MFA, the terms *flows* and *fluxes* have been often used in a random way.
A *flow* is defined as a "mass flow rate." This is the ratio of mass per time

that flows through a conductor, e.g., a water pipe. The physical unit of a flow might be given in units of kg/s or t/year. A *flux* is defined as a flow per "cross section." For the water pipe, this means that the flow is related to the cross section of the pipe. The flux might then be given in units of kg s^{-1} m^{-2}. Note that flux and flow for a system (the water pipe) have different units. The flux can be considered as a *specific* flow. Integration of the flux over the cross section yields the total flow (see Figure 2.3).

In MFA, commonly used cross sections are a person, the surface area of the system, or an entity such as a private household or an enterprise. Some authors also consider the system itself as a cross section and prefer the term *fluxes* for all material exchanges between processes. For instance, the total flow of oil into a region is then designated as a flux. The cross section is the *region* of study. The same flow per capita is also designated as a flux. The cross section in this case is one inhabitant of the region.

In this handbook, a slightly different definition is used. The actual exchange of materials determined for a system is designated as a flow. The system itself is *not* considered to be a cross section. Only specific flows that are related to a cross section are designated as fluxes (see Figure 2.4).

Table 2.1 gives some examples for flows, cross sections, and associated fluxes.

The advantage of using fluxes is that they can be easily compared among different processes and systems, since fluxes are specific values (such as density and heat capacity). Usually, fluxes are kept in mind as reference values for a quick and dirty assessment of anthropogenic systems, processes, or material flows: an affluent society generates 500 kg capita^{-1} year^{-1} of municipal solid waste (MSW); an economical car consumes no more than 5 L/100 km of gasoline.

An *arrow* symbolizes flows and fluxes (see Figure 2.1). For each flow and flux, a *process of origin* and a *process of destination* have to be defined. The

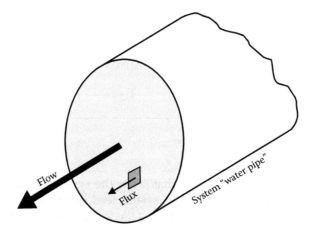

FIGURE 2.3
Total flow and specific flow (flux) of water through a pipe.

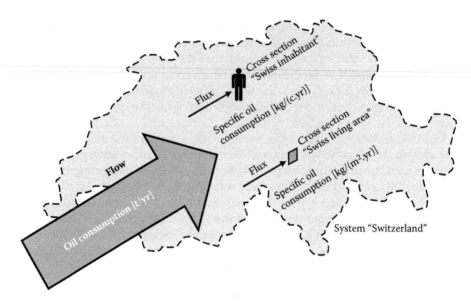

FIGURE 2.4
Flows and fluxes for the system *Switzerland*. Fluxes are specific flows and are related to a cross section.

TABLE 2.1

Examples of Flows and Fluxes

	System	Cross Section	Numerical Value of Cross Section	Flow	Flux
Paper consumption	Switzerland	Swiss population	7.3 million	1.8 million t/year[a]	246 kg capita^{-1} year^{-1}
Waste treatment	Municipal solid waste (MSW) incinerator	Grate	50 m²	15 t/h	300 kg m^{-2} h^{-1}
Emission of SO$_2$	Switzerland	Area of state	42,000 km²	30,000 t/year[a]	0.7 g m^{-2} year^{-1}
Total deposition of nitrogen[b]	Vienna	Area of city	415 km²	1400 t/year	3.4 g m^{-2} year^{-1}

[a] In 2000.
[b] Paumann, Obernosterer, and Brunner, 1997.

process of origin of a flow or flux that *enters* the system is called an *import process*. Flows and fluxes that *leave* the system lead to *export processes*. By definition, both import and export processes are located outside of the system. Therefore, they are not balanced, and stocks in export or import processes are not taken into account (see Figure 2.7). The benefit of designating import and export processes becomes evident when the design of an MFA system becomes difficult because of data scarcity: since export and import processes must not be balanced, it may be attractive to position a process where too little information is available outside of the system boundaries. The flow of a good is indicated with $\dot{m}$ and the flow of a substance with $\dot{x}$.

2.1.6 Transfer Coefficient

Transfer coefficients (TCs) describe the partitioning of a good or substance in a process, either for a specific single input or for the sum of all inputs. They are defined as follows (Equations 2.1 through 2.4) and illustrated in Figure 2.5.
Inputs with different TCs:

$$\dot{m}_{output,j} = \sum_{i=1}^{k_I} TC_{i,j} \cdot \dot{m}_{input,i} \tag{2.1}$$

$$\sum_{j=1}^{k_O} TC_{i,j} = 1 \tag{2.2}$$

Inputs with uniform TCs:

$$\dot{m}_{output,j} = TC_j \cdot \sum_{i=1}^{k_I} \dot{m}_{input,i} \tag{2.3}$$

$$\sum_{j=1}^{k_O} TC_j = 1 \tag{2.4}$$

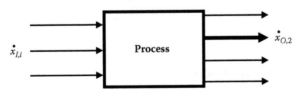

FIGURE 2.5
The transfer of substance x into output flow 2 is given as $TC_2 = \dot{x}_{O,2} / \sum_i \dot{x}_{I,i}$.

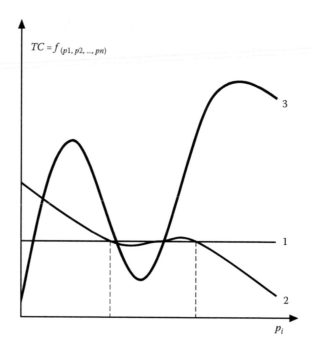

FIGURE 2.6

Different kinds of transfer coefficients: TC_1 is independent of parameter p_i; TC_2 can be treated as a constant within a certain range of p_i; TC_3 is highly sensitive to variation in parameter p_i.

where

    k_I = number of input flows

    k_O = number of output flows

TCs can be defined for each output good or substance of a process. Multiplied by 100, the TC gives the percentage of the total throughput of a substance that is transferred into a specific output good (designated as partitioning). It is a material-specific value. Furthermore, TCs are technology-specific values and stand for the characteristics of a process. TCs are not necessarily constant. They can depend on many variables such as process conditions (e.g., temperature, pressure) as well as input composition. Sometimes, TCs can be regarded as constant within a certain range (see Figure 2.6). This makes them useful for sensitivity analysis of the investigated system and for scenario analysis.

2.1.7 System and System Boundaries

The *system* is the actual object of an MFA investigation. A system is defined by a group of MFA elements, the interaction between these elements, and the boundaries between these and other elements in space and time. It is a group of physical components connected or related in such a manner as to form and/or act as an entire unit (Patten, 1971; Ayres, 1994). An open system interacts with its surroundings; it has either material or energy imports and exports, or both. A closed system is conceived as a system with complete isolation, preventing material and energy flows across the system boundary. In MFA, the physical components are processes, and the connections/relations are given by the flows that link the processes. A single process or a combination of several processes can represent a system. The actual definition of the system is a decisive and demanding task. Poor results of MFA can often be traced back to an unsuitable system definition. The system boundaries are defined in time and space (temporal and spatial system boundary).

The temporal boundary depends on the kind of system inspected and the given problem. It is the time span over which the system is investigated and balanced. Theoretically, it can range from 1 s for a combustion process to 1000 years for landfills. Commonly applied temporal boundaries are 1 h, 1 day, or 1 year. For anthropogenic systems such as an enterprise, a city, or a nation, periods of 1 year are chosen for reasons of data availability. Because financial accounting and other reporting is typically done on an annual basis, information that covers periods of 1 year is usually easier to retrieve than data for shorter or longer time periods.

The spatial system boundary is usually fixed by the geographical area in which the processes are located. This can be the premises of a company, a town or city, a region such as the Long Island Sound or the Danube River watershed, a country such as the United States, a continent such as Europe, or the whole planet. Abstract areas also can serve as a boundary definition when MFA is applied to a specific part of the economy, e.g., the waste management system of a county, the average private household, or a virtually assembled region within a defined economy. For most studies, it is necessary to define the spatial system boundary also for the third dimension (vertically). Above the Earth's surface, usually the first 500 m of the atmosphere is considered. Within this layer (so-called planetary boundary layer), the main exchange of air (and pollutants) between regions takes place, and there is negligible material exchange with the air compartment above. Under the Earth's surface, the system boundary has to be extended far enough that the groundwater flows are included. A rectangle with broken lines and rounded corners symbolizes the spatial system boundary in Figures 2.1 and 2.7).

For material balances of larger systems such as nations or continents, the term *cycle* is frequently used (Ayres, 1994; Van der Voet, 1996; Van der Voet, Guinée, and Udo de Haes, 2000; Graedel and Allenby, 2002). This term stems from the grand biochemical cycles of carbon, oxygen, nitrogen, hydrogen, sulfur, and phosphorus that drive the biosphere. While these natural cycles indeed show cyclic behavior, anthropogenic systems hardly do (e.g., see Chapter 1, Section 1.4.5.3 on urban material flows, and Chapter 3, Section 3.2.2.1

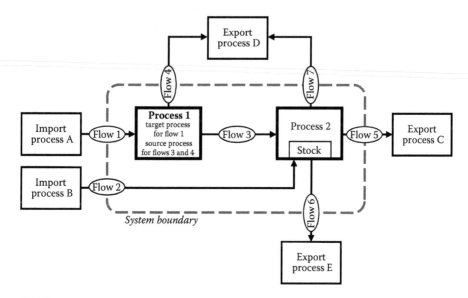

FIGURE 2.7
Exemplary MFA system illustrating selected terms. Ovals of import and export flows are placed directly on the system boundary to facilitate visual understanding of the system. Note that processes outside the system boundary must not be balanced.

on copper flows). Therefore, applying the term *cycle* to anthropogenic systems such as urban areas or national economies could be misleading.

2.1.8 Activities

Regardless of a community's social, cultural, technical, or economic development, there is a set of basic human needs such as to eat, to breath, to reside, to communicate, to transport, and others (Baccini and Brunner, 1991). The main goal of a sustainable economy is to satisfy these needs best at the least cost. An activity is defined as comprising all relevant processes, flows, and stocks of goods and substances that are necessary to satisfy a certain human need. Human beings also have nonmaterial needs, such as love, security, intellectual exchange, social recognition, etc., that cannot be quantified and are therefore not tangible for an MFA approach. These nonmaterial needs are important drivers for the material needs, but they are not discussed here in a quantitative way.

The purpose of defining activities is to facilitate the analysis of a given way of fulfilling human needs, to evaluate the constraints and optimization potentials, and to suggest strategies and measures for optimizing the way the needs are satisfied.

The most important activities can be defined as follows. (Note that this list is not complete. Additional activities such as *leisure, health and sports*, etc., can be defined, if necessary, to analyze and solve a particular resource-oriented problem (Baccini and Brunner, 1991).

2.1.8.1 To Nourish

This activity comprises all processes, goods, and substances used to produce, process, distribute, and consume solid and liquid food. *To nourish* starts with agricultural production (e.g., the goods *seed corn, water, air, soil,* and *fertilizer,* and the process *crop raising,* e.g., beans); food production (e.g., the process *cannery,* the good *canned beans*); distribution (e.g., the process *grocery*); and consumption (e.g., the processes *storage, preparation,* and *consumption* of canned beans in private households); it ends with the release of off-gases (breath), feces and urine, and solid wastes (can, leftover beans) to the atmosphere and the waste and wastewater treatment systems. These systems already belong to the activity *to clean,* as discussed in the following paragraph. Figure 2.8 shows the main processes for the activity.

2.1.8.2 To Clean

In anthropogenic processes, *wanted* materials are often separated from *unwanted* materials. When sugar is produced from sugar cane, sucrose is separated from cellulose and impurities. In dry cleaning, dirt is removed from the surface of clothes by organic solvents such as perchloroethylene. People need to remove dirt and sweat from their body surfaces. Also, they need to remove materials not useful for their metabolism and wastes from their body, such as carbon dioxide in breath, salts in urine, or undigested biomass in feces. Since many of these processes are called *cleaning,* the separation of valuable from useless materials has been defined as the activity *to clean.* It is an essential physiological activity, since it is a necessity for human beings to keep material input (food and beverages) and output (feces, urine, and off-gas) in a balance. If humans do not separate CO_2 and excretion products from their body, they cannot survive. Other examples for cleaning processes on the individual level are *private and commercial laundry* (see Figure 2.9), *dishwashing,* and house cleaning; on the industrial level, there is *refinery, metal purification,* and *flue gas treatment*; and on the community level, there is *sewage and waste treatment* and *public cleaning. To clean* is also a very important activity for individual and public health.

2.1.8.3 To Reside and Work

This activity comprises all processes that are necessary to build, operate, and maintain residential units and working facilities. Important processes are *building construction, operation and maintenance of buildings, machine construction, operation and maintenance of machinery, manufacture of furniture and household appliances, manufacture of clothing and leisure appliances,* and *consumption.*

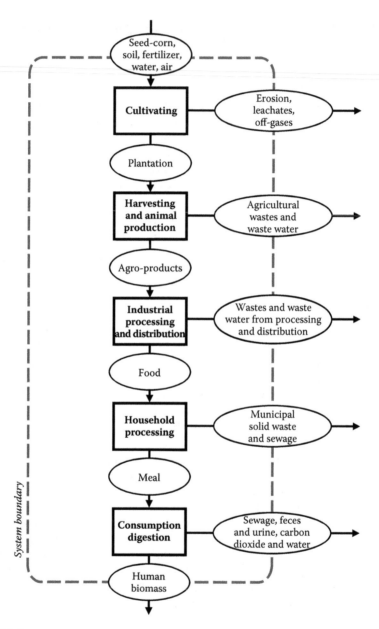

FIGURE 2.8
Process chain of main processes associated with the activity *to nourish*. Required goods to operate the processes (energy, machinery, tools, etc.) are omitted. (With kind permission from Springer Science+Business Media: *Metabolism of the Anthroposphere*, 1991, Baccini, P. and Brunner, P. H.; Baccini, P. and Brunner, P. H. (2012). *Metabolism of the Anthroposphere—Analysis, Evaluation, Design (2nd Edition)*. Cambridge, MA: MIT Press.)

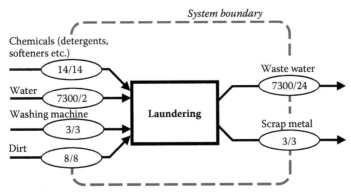

Fluxes [kg/(c.yr)]
Total mass/dry matter

FIGURE 2.9

Mass flows of goods for the activity *to clean* through the process *laundering*, which is a subprocess of the process *household*. About 100 kg of water is needed to separate 1 kg of dirt from textiles (in kg/capita/year; left number, total mass flow, right number, flow of dry matter). (With kind permission from Springer Science+Business Media: *Metabolism of the Anthroposphere*, 1991, Baccini, P. and Brunner, P. H.; Baccini, P. and Brunner, P. H. (2012). *Metabolism of the Anthroposphere—Analysis, Evaluation, Design (2nd Edition)*. Cambridge, MA: MIT Press.)

TABLE 2.2

Subprocesses and Their Input and Output Goods for the Process *Building Construction*

Process	Subprocess	Inputs	Outputs
Building construction	Concrete production, steel and metal production, quarry, lumber mill, energy supply	Gravel, sand, stone, limestone, marl, metal ores, wood, fuels, water, air	Buildings, construction and demolition waste, wastewater, off-gas

Table 2.2 gives an example for subprocesses and related goods for the process *building construction*. Figure 2.10 outlines a flowchart for the activity *to reside and work*.

The functions and services that are expected from a building are manifold: A house should provide an agreeable temperature and humidity to its user. This can be realized by different heating and cooling systems, different types of wall construction, and the use of different materials for better insulation. But other approaches are also possible to fulfill the service *agreeable temperature* during the cool season. Besides measures for the outer skin (insulation of the walls of a building), a combination of reduced heating and wearing a pullover (insulation of the human skin) can also fulfill the task. All three approaches (heating, insulation, and clothes) result in different material and energy consumption.

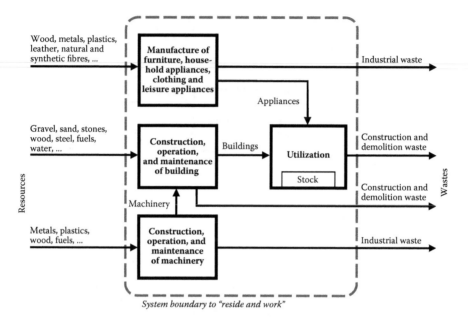

FIGURE 2.10
Relevant processes and goods for the activity *to reside and work*. Only solid wastes are indicated. (With kind permission from Springer Science+Business Media: *Metabolism of the Anthroposphere,* 1991, Baccini, P. and Brunner, P. H.)

2.1.8.4 To Transport and Communicate

This activity comprises all processes that have been developed to transport energy, materials, persons, and information. The processes range from *road construction, operation, and maintenance of networks and transport vehicles* to *administration* and *production and transport of printed and electronic media*. Rapid technological progress causes swift changes in processes and goods that are associated with this activity. Today's networks are established by roads, railway tracks, air tracks, cables, satellites, radio, etc. Hence, there are many ways of transmitting information over longer distances. This can be done by the transport of persons; the transport of information carriers (printed paper, compact disc, magnetic recording, etc.); or transmission via cable, fiber optics, radio, etc. This activity can be used to analyze which way of transmitting information is less resource consuming. Transport systems and strategies for goods and persons can also be compared by applying the activity approach to evaluate different transportation modes (e.g., gasoline versus hydrogen- or battery-powered engines) as well as different management modes (individual car versus using the car pool of a service provider). The effect of self-driving vehicles on material and energy demand of the activity *to transport and communicate* is also of high interest.

Goods and processes can be part of several activities. For instance, an automobile is used to drive to work and to do business, such as a taxi. It belongs to the activity *to transport and communicate* as well as to the activity *to reside and work* because it is an essential part of both. It is important to note that there are no strict rules regarding the allocation of processes to activities. It is done according to the utilitarian principle: if it serves the purpose to analyze and optimize a certain activity, it should be allocated to this activity.

2.1.9 Anthroposphere and Metabolism

The human sphere of life—a complex system of energy, material, and information flows in space—is called the *anthroposphere* (Baccini and Brunner, 1991). It is part of planet Earth and contains all processes that are driven by mankind. The anthroposphere can be seen as a living organism. In analogy to the physical processes in plants, animals, lakes, or forests, the metabolism of the anthroposphere includes the uptake, transport, and storage of all substances; the total biochemical transformation within the organism; and the quantity and quality of all off-products such as flue gas, sewage, and wastes (Baccini and Brunner, 2012). The complementary part to the anthroposphere is designated as the environment. The anthroposphere interacts with the environment via extraction of resources (air, water, and minerals) and emission of off-products and wastes. The anthroposphere can be defined as part of any region where human activities take place.

As a first step, the anthroposphere can be divided into the four compartments (main processes in Figure 2.11). (Note that, for simplicity, Figure 2.11 does not comply strictly with MFA syntax.)

1. Agriculture
2. Industry, trade, and commerce
3. Private households (consumption)
4. Waste management

All along the anthropogenic process chain, there is exchange of materials and energy with the environment, which comprises four compartments: atmosphere, hydrosphere, pedosphere, and lithosphere. Often, terms such as *water*, *air*, and *soil* (for pedosphere) are used synonymously. Some authors use the terms *technosphere*, *biosphere*, or *socioeconomic sphere* instead of anthroposphere.

Sometimes, the interface between anthroposphere and environment is not clear. For example, a soil that is used by humans can be regarded as a part of the anthroposphere as well as of the environment. Hence, the definition of the anthroposphere is certainly subjective. Some authors claim that all soils belong to the anthroposphere because anthropogenic trace substances have been detected in all soils on Earth; there are no soils anymore that are in a

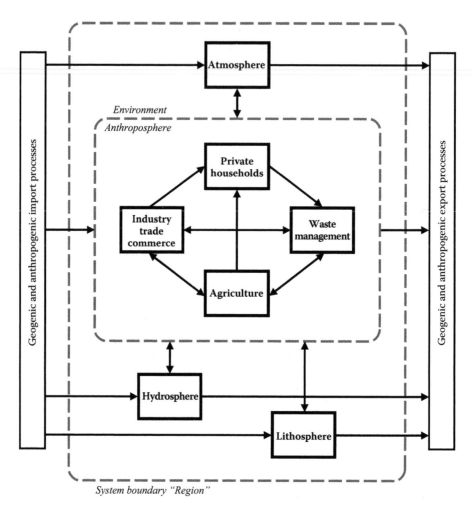

FIGURE 2.11
A region where anthropogenic activities take place can be divided into the subsystems *anthroposphere* and *environment*. Exchange of materials takes place within the subsystems as well as between the subsystems and processes outside the region.

natural, uninfluenced state, and thus, they do not belong to the environment anymore. Other authors include only those soils in the anthroposphere that are actively managed by mankind. In actual MFA practice, such allocation problems are usually of little relevance. They should be acknowledged and solved when encountered but not overemphasized.

The term *metabolism* is used in many different fields and contexts. Initially, it was used to name the turnover, which is the uptake, internal transformation, and emission of energy and matter within living bodies (organisms). Later on, due to the compelling analogy between man-made systems and biological organisms, the term *metabolism* was applied to anthropogenic as well as geogenic (natural) processes and systems (Baccini and Brunner, 1991; Ayres, 1994). In MFA, metabolism of a system stands for the transfer, storage, and transformation of materials within the system and the exchange of materials with its environment. Metabolism is applied to anthropogenic as well as geogenic (natural) processes and systems. A similar term is *physiology*. While at present, metabolism prevails in the literature, it remains to be seen if physiology will be used more often in future.

2.1.10 Material Flow Analysis

MFA is a method to describe, investigate, and evaluate the metabolism of anthropogenic and geogenic systems. MFA defines terms and procedures to establish *material balances* of systems. A complete description of how to perform an MFA is given in Section 2.2. As described before, MFA includes balances of both goods and substances. In Chapter 1, Section 1.1.4, and in Section 2.1.1, it has been explained why it is often important to base decisions on substance balances and not on balances of goods alone.

One of the first steps in MFA is to establish a mass balance of goods for the system of study. This is the basis for the determination of balances of selected substances. The reason for making the couple *goods/substances* so central in MFA is that a substance balance without the underlying mass balance of goods does not provide the full information that is required for policy analysis or decision making.

First, for each substance flow in a system, one has to know the flow of goods on which it is based. This is important because often, the flow of goods can be controlled directly, but the flows of substances can only be influenced indirectly. For example, application of sewage sludge can cause accumulation of heavy metals in agricultural soil. The flow of heavy metals to soil can be indirectly controlled through an action directed toward *sewage sludge*, which is a good.

Second, it is important to know the concentration of a substance in a good. For example, if a hazardous-substance flow has to be reduced, one must know the concentration of the substance in the good. The higher the concentration, the smaller the quantity of the good must be in order to achieve an acceptable flow to the environment. This quantity is decisive when developing an appropriate strategy for reduction or prevention.

Third, establishing mass balances for goods *and* substances helps to detect sources of error that otherwise could not be found with mass balances for either goods or substances alone. Figure 2.12 shows Sankey diagrams of mass balances of goods and cadmium for an MSW incinerator. The information given by the two graphs combined is more comprehensive and allows one to draw more conclusions than would be possible using only a single figure on goods or substances.

In addition to MFA, the term *substance flow analysis* (SFA) is also occasionally used. Some authors do not distinguish between goods, materials, and

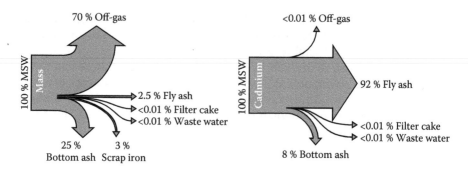

FIGURE 2.12
Mass balances for goods and cadmium of an MSW incinerator (Schachermayer, Bauer, Ritter, and Brunner, 1994). Each balance alone provides limited information. The ensemble of all balances yields a comprehensive understanding of the material flows in an MSW incinerator. Hence, both balances of goods and substances are necessary.

substances in a rigid way as described in this chapter. Therefore, SFA is sometimes used as another name for what is here defined as MFA. At other times, it refers to mass balances of one or several substances (Van der Voet, 1996; Van der Voet, Guinée, and Udo de Haes, 2000). Graedel and Allenby (2002) differentiate between elemental flow analysis (EFA), molecular flow analysis (MoFA), SFA, and MFA. In cases where stocks are considered, EFA mutates to elemental stock and flow analysis (ESFA), and so on. The differentiation is based not only on the element, molecule, substance, or material of study but, also, on the objective and scope of the analysis. A zinc balance for a system can be the result of an EFA, SFA, or MFA, depending on whether the emphasis is on the atom (EFA), on the different speciation of zinc as compounds (SFA), or on the reservoirs (i.e., processes) among which zinc is exchanged (MFA). Other schools use the term *MFA* to designate exclusively mass balances of goods without considering substances. These studies are frequently termed as *bulk MFA*, and the term *material* is used synonymously for goods and not for substances. Such studies at the level of goods do not include environmental impacts, in contrast to SFA, which does. Bulk MFA studies have been shown to provide valuable information and to create awareness about the turnover of goods in developed economies.

In the following sections, the term *MFA* stands for the investigation and quantification of the metabolism of anthropogenic and geogenic systems. This includes the definition and selection of system boundaries as well as relevant goods, substances, and processes. In MFA, mass balances of both goods and substances are essential.

2.1.11 Materials Accounting

Materials accounting (sometimes referred to as materials bookkeeping) is the routine updating of the results of an MFA by a few measurements of the key flows and stocks only. Materials accounting yields time series and enables detection of trends. It is instrumental for the control of measures in resource management, environmental management, and waste management. The basis for materials accounting is an MFA, which serves to aid in understanding the system of study and in identifying the appropriate parameters (flows, stocks, concentrations) for the routine analysis. The art of materials accounting is to find those key parameters that yield maximum information at minimum cost. Materials accounting can be applied to all sizes of systems, from single companies to national economies. *Materials accounting* is used in a more restricted way by the proponents of the Wuppertal Institute (Bringezu, Fischer-Kowalski, Kleijn, and Palm, 1997). They use the term exclusively for investigations on the level of goods and do not include substances in their deliberations.

PROBLEMS—SECTION 2.1

Problem 2.1:

Consider the MFA system in Figure 2.13 and detect nine errors.

Problem 2.2:

The following list comprises physical units for flows, fluxes, and stocks. Assign the units to the three categories and define some

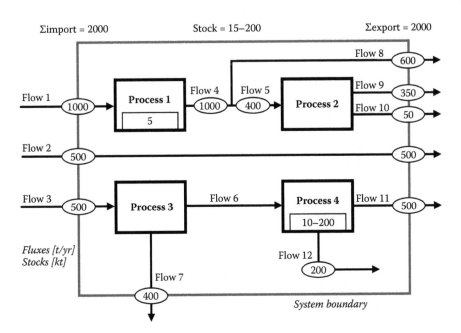

FIGURE 2.13
MFA flowchart with multiple errors.

more units: t/year, L m^{-2} s^{-1}, kg/h, mg/capita, m/s, Tg, kg capita^{-1} year^{-1}, μg m^{-2} year^{-1}, kg/m^3, L/day.

Problem 2.3:

(a) Assess your average daily water consumption. First, list the processes of your daily life that consume water. Second, try to assess the specific water use of these processes. Third, add up your estimates to a final result in L capita^{-1} day^{-1}. Use ranges when estimates are uncertain. (b) Some data can be easily found on the Internet. Compare your estimates with the result of an Internet search.

Problem 2.4:

MSW is separated by a mechanical sorting plant into a combustible fraction (30%) and another fraction (70%) that is digested by a subsequent biochemical process. Approximately half of the latter fraction is transformed to biogas during the anaerobic treatment. (a) Draw the system as a quantitative flowchart. (b) The concentrations of Hg are as follows: MSW input, 1.5 mg/kg; combustible fraction, 1.0 mg/kg; biogas, 0.005 mg/kg; digestion product, 3.4 mg/kg. Calculate all mass flows and transfer coefficients for Hg.

Problem 2.5:

Allocate the following substances to one of three categories of element concentrations in the Earth's crust: matrix elements (>10%), trace elements (<1%), or elements in between (1–10%): Au, Cr, O, Na, Cu, Ca, Hg, Sb, Ag, Mg, Mn, Zn, Ti, V, Fe, Si, Rb, Cl, Al, As, K, Pb, Se, H, C, Tl, and Cd.

Problem 2.6:

Assign the following goods and processes to activities (to nourish, to clean, to reside and work, to transport and communicate): brick, cement kiln, electrostatic precipitator, chicken farm, fertilizer, fuel oil, gasoline, jeans, mobile phone, paper, shovel excavator, sal ammoniac, truck, vacuum cleaner. Note that multiple assignments of terms to activities are possible.

Problem 2.7:

What types of processes (transformation, transport, storage) make up the metabolic system shown in Figure 2.14?

Problem 2.8:

Which of the following processes are anthropogenic? Which are geogenic? Agriculture, agricultural soil, atmosphere, compost production, crop production, forest management, landfill, pedosphere, lithosphere, planetary boundary layer, river, volcano, zoo.

The solutions to the problems are given on the website http://www.MFA-handbook.info.

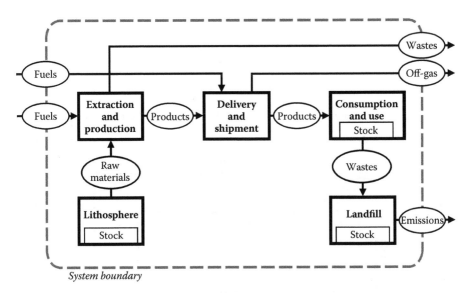

FIGURE 2.14
Flowchart comprising different types of processes.

2.2 MFA Procedures

An MFA consists of several steps that are discussed in stages in the following sections. In general, an MFA begins with the definition of the problem and of adequate goals. Then, relevant substances, goods, processes, and appropriate system boundaries are selected. These are the steps that comprise the system definition and lead to a *qualitative* model. Next, mass flows and stocks of goods and substance concentrations in these flows are determined by measurements, literature data, or estimations. Flows and stocks are then balanced based on the principle of mass conservation, which has to be fulfilled for each process as well as for the entire system. If possible, uncertainties are considered. The result of this second step is a *quantitative* model, which is usually generated using MFA software. Such software supports calculation and presentation of results in an appropriate way to visualize conclusions and to facilitate implementation of goal-oriented decisions. It is important to note that these procedures must not be executed in a strictly consecutive way. The procedures have to be optimized iteratively. The selections and provisions that are taken during the course of the MFA have to be checked continuously. If necessary, they must be adapted to accommodate the objectives of the project. In general, it is best to start with rough estimations and provisional data, and then to constantly refine and improve the system and data until the predefined certainty of data quality has been achieved (see Figure 2.15).

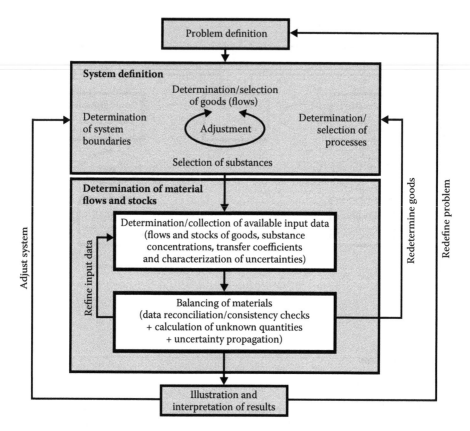

FIGURE 2.15
Course of action to perform an MFA. The procedure consists of an iterative process, taking into account objectives, means, and uncertainties, which are often quite far apart. (With kind permission from Springer Science+Business Media: *Special Types of Life Cycle Assesment*, M. Finkbeiner (Ed.), Material flow analysis, 2016, Laner, D., and Rechberger, H., based on Brunner and Rechberger, 2004.)

2.2.1 Selection of Substances

There are various approaches to choosing substances relevant for an MFA. On the one hand, they depend on the purpose of the MFA, and on the other hand, they depend on the kind of system on which the MFA is based.

First, legislation such as the Clean Air Act or standards such as building, material quality, and safety codes provide listings of relevant substances that are regulated. This approach benefits from existing knowledge and ensures that all substances that have already been selected by the respective authorities are considered for inclusion in the MFA.

Second, the relevance of substances in the important flows of goods has to be evaluated. At the beginning of a study, a practicable rule to determine those flows is as follows. Group all import and export flows of goods of the system into solids, liquids, and gases. For each group, select as many flows as needed to cover at least 90% of the total mass flow of the group. This yields a set of important flows of goods (i) for the system. An elegant way to determine indicator substances (j) is to establish the ratio of substance concentrations in the selected flows of goods (c_{ij}) to geogenic reference materials (c_{geog}). For solid material flows, the average concentrations in the Earth's crust ($c_{geog} = c_{EC}$) can be taken as a reference. Similarly, for liquid material flows, the average concentration in geogenic water bodies ($c_{geog} \equiv c_{hydro}$) can be taken as a reference. For gaseous material flows, the average concentration of the atmosphere ($c_{geog} = c_{atmo}$) can be taken as a reference. Substances with a ratio of $c_{ij}/c_{geog} > 10$ are candidates to be selected for the study (see Table 2.3). The proportions among the substance-specific ratios for a good also have to be considered. If all or most of the ratios are <10, just select those substances with the highest ratios. Note that this rule is only a tool that will assist in the selection process. The result has to be checked for consistency and reasonability during the course of the MFA. These two approaches can be used

TABLE 2.3

Selection of Substances for MFA by Goods/Earth's Crust Ratios

Substance	Earth's Crust (EC)[b] mg/kg	Printed Circuit Boards (PCB)[e] mg/kg	Ratio[a] PCB/EC	Coal (C)[d] mg/kg	Ratio[a] C/EC	Mixed Plastic Waste (MPW)[c] mg/kg	Ratio[a] MPW/EC
Ag	0.07	640	**9100**	–	–	–	–
As	1.8	–	–	10	**5.6**	1.3	0.7
Au	0.003	570	**190,000**	0.01	3.3	–	–
Cd	0.15	395	**2600**	1	7	73	**490**
Cl	130	–	–	1000	8	–	–
Cr	100	–	–	20	0.2	48	0.5
Cu	50	143,000	**2900**	20	0.4	220	4
Hg	0.02	9	450	0.15	8	1.3	**65**
Mn	1000	–	–	40	0.04	17	0.02
Ni	75	11,000	150	22	0.3	10	0.1
Pb	13	22,000	**1700**	20	1.1	390	**30**
Sb	0.2	4500	**22,500**	2	**10**	21	**110**
Se	0.05	–	–	3	**60**	6.7	**130**
Sn	2.5	20,000	**8000**	8	3	4	2
Zn	70	4000	57	35	0.5	550	8

[a] The ratio of element concentration in selected goods to element concentration in the Earth's crust helps to identify substances that may be of environmental relevance or of importance as a resource. Candidates for MFA study are in boldface.

[b] Krauskopf, 1979.

[c] Heyde and Kremer, 1999.

[d] Tauber, 1988.

[e] Legarth et al., 1995.

when the system of study is determined already by the scope of the study. For example, these approaches could be used when the task of the project is to determine the metabolism of a manufacturing firm or a special wastewater treatment technology in order to find potentials for optimization.

The third case stems from the fact that many MFA studies are carried out to determine a system's metabolism of one or several substances for resource and/or environmental impact aspects. In such cases, the selection of substances is part of the project definition (e.g., project "copper household of Europe"). Also, when activities such as *to nourish* are investigated, some substances such as C, N, and P are selected for investigation by MFA.

As mentioned before, a main objective of MFA is to reduce as much as possible the number of parameters that have to be taken into account. To achieve this goal, so-called indicator elements are selected. This is an important means of achieving maximum information with minimum effort. An indicator element represents a group of similar substances. It shows a characteristic physical, biochemical, and/or chemical behavior that is a specific property of all members of the group. Indicator substances can therefore be used to predict the behavior of other substances. For combustion and high-temperature processes, for instance, substances can be separated into *atmophile* and *lithophile* elements. In geochemistry, elements are grouped as follows (Krauskopf, 1979): siderophile elements are those elements that are concentrated in the Earth's iron core; chalcophile elements tend to combine with sulfur in sulfide minerals; lithophile elements generally occur in or with silicates; atmophile elements are relatively prominent in air and other natural gases. Atmophile elements have a lower boiling point and thus are transferred to the off-gas. Examples of atmophile elements are Cd, Zn, Sb, Tl, and Pb. Thus, Cd serves well as an indicator for this group. Lithophile elements and their compounds have a high boiling point and therefore tend to remain in the bottom ash or slag. Lithophile elements include Ti, V, Cr, Fe, Co, and Ni. Fe is representative for this group. Note that the boiling point of the element is not the only decisive factor in whether a substance tends to behave as an atmophile or lithophile in a process. The boiling points of the occurring species (chlorides, oxides, silicates, etc.) are also important. It is obvious that indicator elements have to be chosen in accordance with the system observed and processes analyzed. The aforementioned indicator metals are appropriate when thermal processing of MSW is investigated. In biochemical processing, where evaporation is not a path to the environment for most metals except mercury, other indicators are likely to be more appropriate.

The selection of substances depends on the scope, the grade of precision, and the resources (financial and human) that are available for the MFA study. Experience with MFA shows that many anthropogenic and natural systems can be roughly characterized by a comparatively small number of substances, such as about 5 to 10 elements. Table 2.4 gives a list of frequently used indicator elements in MFA.

TABLE 2.4

Examples of Indicator Elements in MFA

Indicator Element	Symbol	Relevant Activity	Properties
Carbon (organic)	C_{org}	N, R&W, C, T&C	Carrier of chemical energy
			Carrier of nutrients
			Main matrix element in many toxic compounds
Nitrogen	N	N, C, T&C	As NO_3^-, an essential nutrient
			As NO_x, a potential air pollutant
			As NH_4^+ and NO_2^-, toxic for fish
			Eutrophication of aquatic ecosystems
Fluorine	F	C	As F^-, a strong inorganic ligand
			In incineration, HF is formed, a strong acid
Phosphorous	P	N, C	As PO_4^{3-}, an essential nutrient
			Eutrophication of aquatic ecosystems
Chlorine	Cl	N, C	Forms as Cl^-, soluble salts
			Forms stable, sometimes toxic chloro-organic compounds (e.g., PCB and dioxins)
Iron	Fe	R&W, T&C	Forms as Fe^{3+}, poorly soluble oxides, and hydroxides
			As metal, easily oxidized under atmospheric conditions (H_2O and CO_2)
			Recycling of metallic iron is economical
			Resource in construction
Copper	Cu	R&W, C, T&C	Forms as Cu^{2+}, stable complexes with organic ligands
			As metal, an important electrical conductor
			Even in small concentrations, toxic for unicellular organisms
			Resource in construction
Zinc	Zn	R&W, C	Forms as Zn^{2+}, soluble salts
			Important anticorrosive and rubber additive
			Atmophile
			Resource in construction
Cadmium	Cd	C, T&C	Stabilizer for PVC, pigment, anticorrosive
			Constituent of rechargeable batteries
			Toxic for humans and animals, less so for plants
			Atmophile
Mercury	Hg	C	Forms metal–organic and toxic compounds under reduced conditions
			Atmophile and liquid metal
Lead	Pb	R&W, C, T&C	Forms as Pb^{2+}, stable complexes with natural organic ligands
			Lithophile element
			Was and still is an important additive in gasoline

Source: With kind permission from Springer Science+Business Media: *Metabolism of the Anthroposphere*, 1991, Baccini, P. and Brunner, P. H.; Baccini, P. and Brunner, P. H. (2012). Metabolism of the *Anthroposphere—Analysis, Evaluation, Design (2nd Edition)*. Cambridge, MA: MIT Press.

Note: C, to clean; N, to nourish; R&W, to reside and work; T&C, to transport and communicate.

2.2.2 System Definition in Space and Time

The spatial system boundary is usually determined by the scope of the project (carbon balance of a community, MFA of a power plant, etc.). It coincides often with the politically defined region, the estate of a company, or a hydrologically defined region such as the catchment area of a river. Often, the only possibility is to define system boundaries as administrative regions, such as nations, states, or cities, because information is systematically collected on these levels. One advantage of choosing these kinds of system boundaries is that the political and administrative stakeholders are within the regional boundary. Thus, measures based on the results of MFA can be implemented more easily in such administratively defined regions.

In general, any system should be chosen to be as small and consistent as possible while still being broad enough to include all necessary processes and material flows. Consider the following example. From the point of view of resource conservation as well as environmental protection, nitrogen flows and stocks in a city are important. The people in the city must be fed, and the water and air leaving the city must not be polluted. The production and processing of food to nourish the people in the city results in emissions in the agricultural region that supplies the food, the so-called hinterland. Agricultural losses from farming as well as production wastes from sugar refineries, canneries, frozen food processing, etc., yield nutrient flows that are either directly or indirectly led to the river surface waters. Obviously, the choice of the system boundary is important. If the system includes the city only, the major emission flow is not observed (see Figure 2.16).

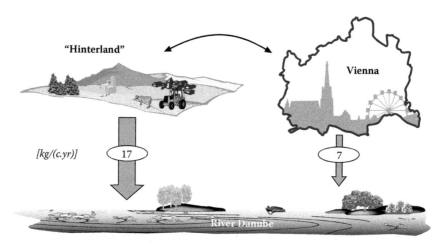

FIGURE 2.16
To nourish related nitrogen fluxes (in kg capita⁻¹ year⁻¹) to the river Danube from the city of Vienna and the corresponding hinterland. (From Obernosterer, R. et al., *Materials Accounting as a Tool for Decision Making in Environment Policy—Mac TEmPo Case Study Report—Urban Metabolism*, The City of Vienna, 1998.)

If both the hinterland and the city are included within the system boundary, it becomes apparent that the flow of nutrients to the river is more than twice as large as measured in the city alone, and that the city is responsible for large emissions in the hinterland, too. Hence, it is crucial to include the hinterland within the scope of study. The most appropriate solution for the definition of the system boundary has to be individually established for each case study and objective.

Temporal system boundaries are comparatively easier to determine. This is especially the case when average flows and stocks over a longer period of time are of interest. In this case, the time span of investigation, which is identical with the temporal system boundary, has to be extended long enough to outweigh the momentary unsteadiness of the system. For many anthropogenic systems, this is the case for time periods of investigation of 1 year. If higher resolutions in time are preferred instead of averages, shorter time periods may be more appropriate for system boundaries in time. Short time periods allow detection of short-term anomalies and nonlinear flows.

2.2.3 Identification of Relevant Flows, Stocks, and Processes

After selecting substances and defining system boundaries, a first rough balance of goods is carried out for the system. Information about material flows is taken from the literature or other sources such as company and national reports. Sometimes, data have to be assessed, e.g., by contacting experts or visiting government agencies. At this stage, material flows <1% of the total system throughput are neglected. Nevertheless, these small flows can make a significant contribution to one or more substance balances that are established later. Therefore, when in a subsequent step, the substance flows are investigated in greater detail, it is important to check whether these small neglected flows are relevant in view of the objectives of the investigation.

The number of processes necessary to describe the system depends on the objectives of the study and on the complexity of the system. Generally, processes can be subdivided into subprocesses, and vice versa, any number of processes can be merged into a single process (see Figure 2.2). In most cases, systems comprising more than 15 processes (exclusive of import and export processes) turn out to be clumsy and unnecessarily complex. Remember that one of the main purposes of MFA is to develop simple and reliable models to picture reality. For mathematical formulations, let the number of flows in a system be k, the number of processes be p, and the number of substances be n. According to the mass-balance principle, the mass of all inputs into a process equals the mass of all outputs of this process plus a storage term that considers accumulation or depletion of materials in the process (Equation 2.5 and Figure 2.17; the symbol ˙ designates a flow or flux).

$$\sum_{k_I} \dot{m}_{input} = \sum_{k_O} \dot{m}_{output} + \dot{m}_{stock} \tag{2.5}$$

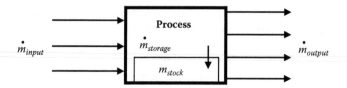

FIGURE 2.17
Consideration of a stock within a process.

TABLE 2.5

Development of a Data Spreadsheet in the Course of an MFA[a]

Goods	Flow Rate, t/Year	Concentration of Substance $\underline{S_1}$, $\underline{S_2}$, S_3,..., S_n, Unit					Substance Flow Rate $\underline{S_1}$, $\underline{S_2}$, S_3,..., S_n, Unit					
$\underline{G_1}$	$\underline{\dot{m}_1}$	•		•	...	•		•	•	•	...	•
$\underline{G_2}$	$\underline{\dot{m}_2}$	•	•	•	...	•		•	•	•	...	•
$\underline{G_3}$	$\underline{\dot{m}_3}$	•	•	•	...	•		•	•	•	...	•
⋮	⋮	⋮	⋮	⋮	...	⋮		⋮	⋮	⋮	...	⋮
$\underline{G_k}$	$\underline{\dot{m}_k}$	•	•	•	...	•		•	•	•	...	•

Note: Entries are underlined (selected goods and substances, flow rates of goods). Additionally, required data are marked with •. G = name of good; S = name of substance.
[a] See also Tables 2.6 and 2.7.

If inputs and outputs do not balance, one or several flows are either missing or have been determined erroneously. The mass-balance principle applies to systems as well as processes. A true material balance of a process or system is only achieved if all input and output flows are known, and if either $\dot{m}_{stock} = 0$ or $\dot{m}_{stock}$ can be measured. In practice, $\dot{m}_{stock}$ is often calculated by the difference between inputs and outputs. Table 2.5 shows how data can be managed during the course of an MFA.

2.2.4 Determination of Mass Flows, Stocks, and Concentrations

Information about mass flows is usually taken from databases or measured directly or indirectly on site. Regional, national, and international administrative bodies such as bureaus of statistics, industrial associations, professional societies, and consumer organizations can be good sources of specific data, sometimes including time series. Such information often comprises figures on production, consumption, and sales of commodities and goods (Kelly and Matos, 2014). Data on waste flows, emissions of pollutants, and substance concentrations in air, water, and soil are systematically collected by national and international environmental protection agencies. Papers in scientific journals, proceedings, and books are also rich sources of information. The search for and the collection, evaluation, and handling of data are core tasks in MFA. The more experience MFA experts have, the less time they need and the better results they get when acquiring data for an MFA.

Some material flows are assessed based on assumptions, cross-comparisons between similar systems, or so-called proxy data. *Proxies* are figures that help in approximating or estimating the actual data of interest. An example of proxy data is given as follows. The objective of a study is to determine the total loss of zinc by the wear of tires from passenger cars in the United States. In Sweden, Landner and Lindeström (1998) assessed the loss per car as 0.032 kg Zn/year. Assuming that this figure is also valid for the United States, the total loss of zinc in the United States can be determined by multiplying the Swedish figure for Zn loss by the active US vehicle fleet of 140 million passenger cars, yielding approximately 4500 t/year. This is probably an underestimation because driving distances in the United States are longer than in Sweden. It depends on the objective of the study whether this approximation is sufficiently exact. The proxy datum in this case is the specific loss per car of 0.032 kg Zn/year determined for Sweden. It is always necessary to check whether a proxy is eligible to be transferred from one system to another.

Depending on the financial resources available for an MFA, mass flows of goods and substance concentrations actually can be measured. If large systems or extended time periods are to be balanced, this can be quite costly. Therefore, flows, stocks, and concentrations are preferentially measured in smaller systems such as a wastewater treatment plant, a company, a farm, or a single private household. Such field studies require intensive and timely planning of the measuring procedure and campaign. Brunner and colleagues (Schachermayer, Bauer, Ritter, and Brunner, 1995; Morf, Ritter, and Brunner, 1997; Morf and Brunner, 1998) carried out such measurements for MSW incinerators. They developed a plan to minimize sampling locations, the number and quantities of samples, and time of sampling. Such elaborate sampling plans are instrumental in keeping project costs low while producing rigorous and reproducible results. Nevertheless, it is very rare that a balance between inputs and outputs of a measured system yields an error less than 10% of the total flow. In principle, flows and concentrations can also be measured directly in larger systems, such as regions or entire watersheds. In any case, it is necessary to assess whether the required accuracy can be achieved by using available data or whether additional data have to be collected. Table 2.6 shows the progress for the data table after mass flows and concentrations have been measured or determined.

TABLE 2.6

Data Spreadsheet after Determination of Substance Concentrations[a]

Goods	Flow Rate, t/Year	Concentration of Substance $\underline{S}_1, \underline{S}_2, \underline{S}_3,..., \underline{S}_n,$ mg/kg					Substance Flow Rate $\underline{S}_1, \underline{S}_2, \underline{S}_3,...,$ $\underline{S}_n,$ Unit				
G_1	$\dot{m}_1$	$\underline{c}_{11}$	$\underline{c}_{12}$	$\underline{c}_{13}$	$\cdots$	$\underline{c}_{1n}$	•	•	•	$\cdots$	•
G_2	$\dot{m}_2$	$\underline{c}_{21}$	$\underline{c}_{22}$	$\underline{c}_{23}$	$\cdots$	$\underline{c}_{2n}$	•	•	•	$\cdots$	•
G_3	$\dot{m}_3$	$\underline{c}_{31}$	$\underline{c}_{32}$	$\underline{c}_{33}$	$\cdots$	$\underline{c}_{3n}$	•	•	•	$\cdots$	•
$\vdots$	$\vdots$	$\vdots$	$\vdots$	$\vdots$	$\cdots$	$\vdots$	$\vdots$	$\vdots$	$\vdots$	$\cdots$	$\vdots$
G_k	$\dot{m}_k$	$\underline{c}_{k1}$	$\underline{c}_{k2}$	$\underline{c}_{k3}$	$\cdots$	$\underline{c}_{kn}$	•	•	•	$\cdots$	•

Note: New entries are underlined. Additionally required data are marked with •. G = name of good; S = name of substance.

[a] See also Tables 2.5 and 2.7.

2.2.5 Assessment of Total Material Flows and Stocks

The substance flows ($\dot{x}$) that are induced by the flows of goods can be directly calculated from the mass flows of the goods ($\dot{m}$) and the substance concentrations (c) in these goods (Equation 2.6).

$$\dot{x}_{ij} = \dot{m}_i \cdot c_{ij} \qquad (2.6)$$

where
  $i = 1,..., k$ as the index for goods
  $j = 1,..., n$ as the index for substances

Again, the mass-balance principle applies for each substance in each process in the entire system. As for goods, the reason for a failure to balance can be missing flows; this source of error is eliminated when establishing the balance at the level of goods (see Section 2.2.3). Another reason could be errors in substance concentrations. Balance differences between input and output of 10% are common and are usually not significant for the conclusions. Section 2.3 shows how balances can be optimized and how data gaps and uncertainty can be handled using mathematical and statistical tools. Table 2.7 shows the complete data table after the final step of adding substance flow rates.

There are two ways of assessing the amount of materials in stocks. First, the total mass of the stock can be determined either by direct measurement of the mass or by assessing the volume and the density of the stock. This approach is often used when stocks do not change significantly for long time periods ($\dot{m}_{stock}/m_{stock} < 0.01$). This is the case for stocks, for example, in natural processes such as soils or large lakes. The second approach is applied to fast-changing stocks ($\dot{m}_{stock}/m_{stock} > 0.05$). Then the magnitude of the stock can be calculated by the difference between inputs and outputs over an appropriate time span ($t_0 - t$). The size of the stock at t_0 has to be known (see Figure 2.18).

TABLE 2.7

Completed Data Spreadsheet[a]

Goods	Flow Rate, t/Year	Concentration of Substance $\underline{S}_1, \underline{S}_2, \underline{S}_3,..., \underline{S}_n$, mg/kg					Substance Flow Rate $\underline{S}_1, \underline{S}_2, \underline{S}_3,...,$ $\underline{S}_n$, kg/Year				
G_1	$\dot{m}_1$	c_{11}	c_{12}	c_{13}	...	c_{1n}	$\underline{\dot{X}}_{11}$	$\underline{\dot{X}}_{12}$	$\underline{\dot{X}}_{13}$	...	$\underline{\dot{X}}_{1n}$
G_2	$\dot{m}_2$	c_{21}	c_{22}	c_{23}	...	c_{2n}	$\underline{\dot{X}}_{21}$	$\underline{\dot{X}}_{22}$	$\underline{\dot{X}}_{23}$	...	$\underline{\dot{X}}_{2n}$
G_3	$\dot{m}_3$	c_{31}	c_{32}	c_{33}	...	c_{3n}	$\underline{\dot{X}}_{31}$	$\underline{\dot{X}}_{32}$	$\underline{\dot{X}}_{33}$	...	$\underline{\dot{X}}_{3n}$
⋮	⋮	⋮	⋮	⋮	...	⋮	⋮	⋮	⋮	...	⋮
G_k	$\dot{m}_k$	c_{k1}	c_{k2}	c_{k3}	...	c_{kn}	$\underline{\dot{X}}_{k1}$	$\underline{\dot{X}}_{k2}$	$\underline{\dot{X}}_{k3}$	...	$\underline{\dot{X}}_{k\ln}$

Note: Final entries (substance flow rates) are underlined. G = name of good; S = name of substance.

[a] See also Tables 2.5 and 2.6.

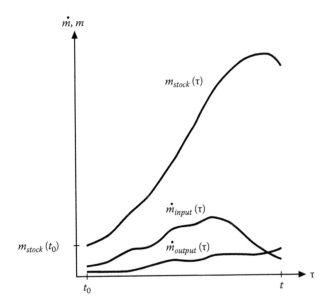

FIGURE 2.18

The stock of a nonsteady-state process. Functions $\dot{m}_{input}$ and $\dot{m}_{output}$ have to be known to calculate m_{stock}.

Usually, $\dot{m}_{input}$ and $\dot{m}_{output}$ are functions of time. For the range $0.01 >= r <= 0.05$, an expert judgement is required to decide which way to assess the stock is more appropriate.

Applying Equation 2.7, the stock (m_{stock}) can be calculated for any time t.

$$m_{stock}(t) = \int_{t_0}^{t} \dot{m}_{input}(\tau)\,d\tau - \int_{t_0}^{t} \dot{m}_{output}(\tau)\,d\tau + m_{stock}(t_0) \qquad (2.7)$$

For rough assessments, $\dot{m}_{input}$ and $\dot{m}_{output}$ are often assumed to be time independent. Fast-changing stocks are characteristic of anthropogenic activities. Examples are landfills, metals in urban settlements, plastic materials and electric appliances in private households, and nutrients in agricultural soils.

2.2.6 Calculation of Results: Static and Dynamic MFA

Calculation of results for the case of static MFA using MFA software such as STAN (cf. Section 2.4) is comparatively simple. Once the qualitative model has been designed and entered into the software, data on mass flows and substance concentrations added, and uncertainties assessed (cf. Section 2.3), the mass balance can be calculated. If sufficient data are available, unknown flows including their uncertainties can be determined by error propagation. If the system is overdetermined, even data reconciliation can be performed. In any case, the software user has to check the calculated results with respect to their consistency and reasonability. Typical checks are as follows: (1) Control the uncertainties of calculated flows: These uncertainties are the results from uncertainties of the input data. If they appear rather large (>25%), then one should check the applied uncertainties for each single input data. Maybe some uncertainties were assumed too broad and can be reasonably narrowed down. If this is not possible, then large uncertainties for calculated flows have to be accepted, and this outcome has to be recognized when discussing the results. (2) Control the extent of the reconciliation step: This gives insight about the structural quality of the model and the consistency of the input data. Large data reconciliation (e.g., large difference between input and calculated data) is an indication either that one or several flows are missing (structural deficiency) or that data are in conflict (bad data quality). The latter can be detected by software (e.g., STAN), and serves as a starting point for a data consistency check. Data for flows that are strongly reconciled (even when the reconciliation stayed within the allowed range) should be checked for misprints or conversion errors and plausibility, and if possible, confirmed by additional data. Generally, marginal data reconciliation is a strong indication for a consistent model and good data. The extent of the reconciliation process can be quantified (cf. Section 2.3.4) by the average of all reconciliations divided by the average of all maximally allowed reconciliations. In typical outcomes for well-described MFA studies on a national level, the reconciliation process demands less than 20% of the allowed range.

The case of dynamic MFA is more complex. While a static MFA is rather concerned with generating a better understanding of a material system based on simple accounting principles (i.e., mass-balance equations), a dynamic MFA is primarily used to investigate the stock buildup of materials in society (i.e., secondary resources) and in the environment (i.e., dissipative losses) based on the investigation of material flows over time.

The mathematically simpler form of static MFA models puts more focus on the underlying data and can therefore serve as a basis for improving material flow databases. Static analyses are well suited to identify general patterns of material use and losses for a specific time increment (typically 1 year) to characterize sources, pathways, and sinks of materials (Laner and Rechberger, 2016). Dynamic models are more complex and have a higher data demand than static models, which typically poses a challenge with respect to checking the plausibility of results. In this respect, the combination of static and dynamic MFA offers the chance of checking results of the dynamic model for specific points in time, for which detailed snapshots of the material flows and stocks have been established (cf. Chapter 3, Section 3.2.3). The advantage of dynamic MFA is that it offers more explanatory power by identifying trends over time and enabling extrapolation of system behavior into the future. Thus, dynamic MFA is used to understand the consequences of changes in a time-dependent system over time and to analyze alternative scenarios with respect to future materials management. Overall, whether to use a dynamic or static model depends on the goal and scope of the MFA, the system under investigation, and the data availability. In the past, most MFA studies used static models to investigate material flow systems, but since the late 1990s, dynamic models have become increasingly popular, with the primary focus on the investigation of material stocks in society (e.g., Zeltner, Bader, Scheidegger, and Baccini, 1999). Metals, in particular, have been subject to dynamic MFA due to the large accumulated metal stocks and their potential value as secondary raw materials (cf. Chen and Graedel, 2012; Müller, Hilty, Widmer, Schluep, and Faulstich, 2014). Two different approaches can be used to estimate material stocks (cf. Figure 2.19). The top-down approach is used in dynamic MFA and is based on accounting of the net flows into or out of a stock over the past. The bottom-up approach consists of summing up the amount of material contained in all relevant goods of the various sectors. While the latter approach is preferred in static MFA, due to the lack of time series data in the model, it can also be used in dynamic MFA to directly investigate the material stock over time (cf. Müller, Hilty, Widmer, Schluep, and Faulstich, 2014).

The two different approaches are schematically illustrated in Figure 2.19 for determining the stock in buildings of a specific material at time T. For the top-down estimate a time series of input–output balances is used to calculate the total stock, while the bottom up-estimate is based on deriving the total stock from the material intensities in all relevant products (buildings in the case of Figure 2.19).

The formula for estimating the material stock $S(T)$ at time T using the bottom-up approach is shown in Equation 2.8. The stock is the sum of the contents of the material of interest c_i for all relevant products or product groups P_i at the time T. This approach appears to be particularly useful, if the material is mainly used for a few applications, for which sufficient data are available (e.g., platinum-group metals in vehicle catalytic converters).

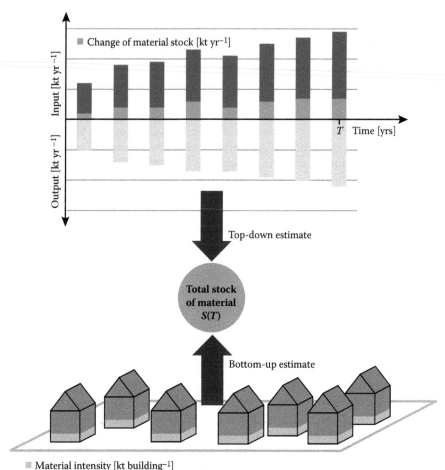

FIGURE 2.19
Schematic illustration of estimating the total stock of a specific material in buildings at time T using either (i) a top-down approach based on time series data for the material inputs, outputs, and stock changes until T (applied in dynamic MFA), or (ii) a bottom-up approach based on the material intensity of all relevant buildings at time T (typically applied in static MFA). (With kind permission from Springer Science+Business Media: *Special Types of Life Cycle Assessment*, M. Finkbeiner (Ed.), Material flow analysis, 2016, Laner, D., and Rechberger, H.)

For materials with diverse fields of application, stock estimates using the bottom-up approach are often incomplete and therefore tend to underestimate the actual anthropogenic stock, as discussed for aluminum in Chapter 3, Section 3.2.3.

$$S(T) = \sum_{i=1}^{I} P_i(T) \cdot c_i(T)$$

(2.8)

$$S(T) = S(0) + \sum_{t=1}^{T} \dot{m}_I(t) - \dot{m}_O(t) \qquad (2.9)$$

The basic formula for the top-down approach is described in Equation 2.9 for the stock of a material $S(T)$ at the time T as the sum of the net additions to stock in previous years. The net addition to stock is calculated by subtracting the output mass in a specific year $t\left(\dot{m}_O(t)\right)$ from the input mass in that year $\left(\dot{m}_I(t)\right)$. To calculate the total material stock, the initial stock at the beginning of the time series $S(0)$ is included, too (see Equation 2.9). This approach has been popular for deriving stock estimates of current in-use metal stocks based on historic production and consumption data, and also for making projections on the availability of future secondary resources (Müller, 2006; Glöser, Soulier, and Tercero Espinoza, 2013; Pauliuk, Wang, and Müller, 2013). However, historical data on the outputs from use sectors are rarely available; therefore, the output of obsolete products is often calculated based on lifetime functions. Such functions are defined for specific products and end-use sectors, with outputs being calculated by accumulation of the fraction of all former inputs becoming obsolete in the respective year (Müller, Hilty, Widmer, Schluep, and Faulstich, 2014). Lifetime distribution functions frequently used for MFA and in the field of system reliability are the Dirac delta distribution or the Weibull distribution. Furthermore normal, lognormal, beta, and gamma distributions are used in dynamic MFA to derive output flows based on the residence time of products in the stock. Though the lifetime function method is most widely applied to calculate outflows from different stocks, for some stocks, an alternative method of using leaching coefficients may be better suited. The coefficients determine the fraction of stock leaving the stock in a specific year, which means that all the material in the stock has an equal chance of exiting the stock (cf. van der Voet, 2002). This method can also be used to describe the leaching of metals from a landfill or metal emissions due to the weathering of surfaces.

2.2.7 Presentation of Results

One of the main advantages of MFA in policy assessment and decision making is the fact that the result of an MFA can easily be summarized in a figure such as a Sankey diagram or a picture of all relevant processes and stocks and flows of a system. Such figures facilitate comprehension, transparency, and ultimately quick understanding of the issues at stake, which is highly relevant for today's busy decision makers. Hence, it is important to present the results of an MFA in an appropriate way. The relevant results of the study have to be condensed into a clear message that can be presented in an easily comprehensible manner. The main goal of the presentation is to stage this message to make it clear, understandable, reproducible, and trustworthy. It is important to keep in mind that there are two crucial audiences for an

MFA study. On the one hand, there are the technical experts in MFA, LCA, and environmental and resource management; and on the other hand, there are the stakeholders, who are not familiar with MFA terminology and procedures and who probably have no technical or scientific background. It is often this second, decision-oriented audience that is more important because it controls the policy- and decision-making processes. Hence, the results of an MFA have to be presented first in a comprehensive technical report and then in an executive summary.

For the technical report, the following content is suggested. The first page consists of an abstract that does not contain any jargon or technical terms and that can be understood by the general public. The abstract is followed by a table of contents. Next, objectives of the study and questions to be addressed are given in detail. The state of knowledge is summarized, including the relevant literature and previous studies by other authors. In the following chapter, a full account of the procedures is given: which methodology is applied, how is the system defined, where do the data come from, how are uncertainties treated, and how are the results calculated or determined? It is necessary to state clearly the source of all data used. Reproducibility and transparency of the procedures are of utmost importance. In the chapter on results and discussions, the numerical results are presented as well as the conclusions and consequences for the field of study, such as additional research needs or actions to be taken by the stakeholders. A summary completes the presentation. The summary is a full account of the total study and is distinctly different from the abstract. Emphasis must be laid on a discussion about goals and questions: what are the answers to the initial (research) questions, and have the goals been reached? While the abstract contains only the main message, the summary recapitulates the full content with objectives, procedures, results, and conclusions in a concise and concentrated form not exceeding a few pages. The abstract is well suited as reference material for literature databases. The summary serves as an overview for the swift reader who does not want to spend much time reading the full account of the work but still wants to know how the authors proceeded. Literature references are listed at the end of the report. In order to fulfill the requirements of reproducibility, it may be appropriate to attach an annex or link to a webpage containing supplementary data, databases, and calculations.

The summary also serves as the foundation for the executive summary, even though there are important differences between the two. The executive summary is written for an audience that is not familiar with MFA. Thus, no technical jargon may be used. For example, it cannot be assumed that the reader of an executive summary knows and appreciates the differences between goods and substances. Also, the executive summary is not the place to explain this difference. Hence, the art of writing an executive summary is to choose the words and language the audience understands while conveying the MFA message in a clear and transparent way to the customer.

To maximize the impact of the MFA findings, figures that visualize the results and conclusions are of fundamental importance. Several standard graphs—flowcharts, Sankey diagrams, partitioning diagrams, etc.—are proven means of illustrating the results. A flowchart is a drawing that comprises all processes, stocks, material flows, and imports and exports entering and leaving the system. All processes, stocks, and flows must be named and quantified. To help the reader rapidly assess the importance of a flow, the width of all flows should be drawn proportionally to the numerical value of the flow. This type of illustration, known as a Sankey diagram, is used to display material, energy, and cash flows. The figure must include system boundaries and the units of flows and stocks. The figure also provides an overview of the system's behavior by showing the sum of all imports at the top left-hand corner (above the system boundary), the sum of all exports at the top right-hand corner (above the system boundary), the total stock, and the changes in stock between these two numbers (e.g., see Figure 2.20). This method of presentation allows the reader to check at a glance whether materials are accumulated or depleted in the system, and to identify which sources, pathways, and sinks are most important. In the summary, $n + 1$ such flowcharts can be drawn, 1 for the flow of goods and n for the various substance flows.

Transfer coefficients of single processes can be visualized by partitioning diagrams (see Figure 2.12). Such diagrams supply easily understandable information about the different behaviors of substances in a process or system, especially when many substances are investigated.

2.2.8 Materials Accounting

As defined in Section 2.1.11, materials accounting stands for the routine, usually long-term analysis of material flows and stocks in a defined system by measuring as few key flows and stocks as possible over time. Such materials accounting allows early recognition of harmful loadings and depletions of substances in stocks. Examples include regional accounting of heavy metals in order to assess long-term soil pollution, accounting for plastic flows to detect future recycling potentials, accounting for carbon to support decisions regarding climate change, accounting for metals (iron, copper, lead) in order to identify future resource potentials, and others.

2.2.8.1 Initial MFA

In order to identify the key processes, stocks, and flows in a system, an initial MFA has to be carried out. If forthcoming materials accounting is the objective of the MFA, it is important to consider that some procedures will be repeated in the future. Transparency and efficient reproducibility are especially important and should be carefully considered when planning and implementing the initial MFA. Whenever possible, labor-intensive tasks

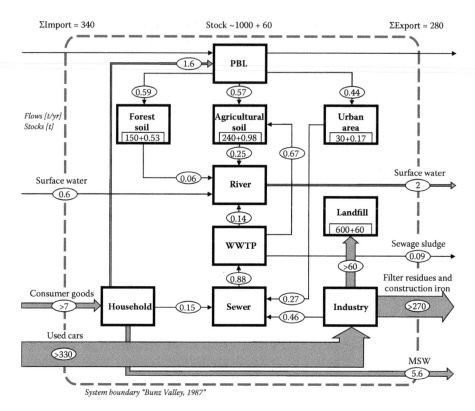

FIGURE 2.20

Exemplary illustration of the result of a regional MFA. Lead flows (t/year) and stocks (t) in the Bunz Valley (see case study 1 in Chapter 3, Section 3.1.1). The system boundaries are given in space and time. The sums of imports and exports are displayed as well as the regional stock and its changes. The size of the arrows is proportional to the mass flows. (From Brunner, P. H. et al., Industrial metabolism at the regional and local level: a case study on a Swiss region, In *Industrial Metabolism—Restructuring for Sustainable Development*, Ayres, R. U. and Simonis, U. E., Eds., United Nations University Press, Tokyo, 1994. With permission.)

such as data acquisition should be automated, and routines should be established. This includes data mining and linkage with existing databases for financial accounting, environmental protection, waste management, water resource management, and the like.

2.2.8.2 Determination of Key Processes, Flows, and Stocks

Based on the results of the initial MFA, key processes, flows, and stocks are identified. First, the goals for future materials management must be defined, such as reduction or increase of single flows, the improvement of pollution control (transfer coefficients) by new or optimized technology, a change in the management of stocks, or combinations of these. The next step is to select those processes and flows that are best suited to determine whether the goals have been achieved. *Best suited* means those processes or flows that deliver the highest accuracy at the lowest cost. *To determine* can mean to measure or to assess using any other method such as indirect calculations based on some intrinsic system property.

In general, measurements are a costly means of monitoring a system. Therefore, it is better to find key processes and flows that are measured for reasons other than MFA, such as gross domestic product (GDP) calculations, national environmental statistics, corporate financial accounting, regional waste management, etc. An important criterion for the selection of flows to be monitored is their homogeneity. The accurate determination of flows with large variations and heterogeneous composition requires intense, expensive monitoring programs. Key flows should be as stable as possible in order to be monitored accurately and cost-effectively. For the same reason, processes with constant transfer coefficients are superior candidates for key processes. If the transfer coefficients are constant for varying throughputs, then knowledge of a single output flow allows calculation of the total input (see Chapter 3, Section 3.3.1.2).

2.2.8.3 Routine Assessment

Once the key processes, stocks, and flows for materials accounting are identified, the system is routinely analyzed and monitored. The generated data sets establish a time series $(t_0, t_1, t_2,..., t_n)$. The results are discussed, measures are evaluated with respect to the given goal (eventually given as a function), and if necessary, adaptations must be made (see Figure 2.21). After several routines, it might be advantageous to carry out another full MFA to check

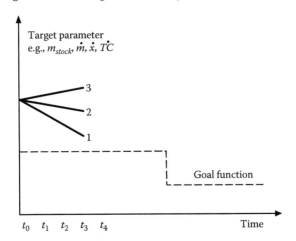

FIGURE 2.21
Time series $(t_0, t_1,..., t_n)$ established through materials accounting. Initial MFA carried out at t_0. The broken line gives the goal function prescribed by the decision makers. Trend 1 is on track. Trend 2 needs adaptation. Trend 3 calls for reconsideration of taken measures.

assumptions and provisions made (Section 2.2.8.2) and to readjust the materials accounting procedures.

PROBLEMS—SECTION 2.2

Problem 2.9:

Design a concept for the management of sewage sludge. First, select the substances you want to include in the study. The average sewage sludge composition is known from a previous study and is given in Table 2.8. (a) Which substances do you include in the study? Give reasons for your choice. (b) Present the concept.

Problem 2.10:

Which time span do you select as a temporal system boundary when your objective is to determine your average daily turnover of materials? Which spatial system boundary do you choose?

Problem 2.11:

Bottom ash is the main solid residue of MSW incineration. In some countries, mechanically processed bottom ash (e.g., crushing, sieving, separation of bulky materials and iron scrap) is used as a substitute for gravel. (a) Assess the contribution of this practice to resource conservation (i.e., conservation of gravel) by an MFA system qualitatively. Assume that all MSW is incinerated. (b) Evaluate the contribution of bottom ash to the conservation of gravel quantitatively. First, determine the data you need. Second, find these data by an Internet search within 15 to 20 min. Third, compare your results with the data given in the solution to this problem.

Problem 2.12:

Concentrations of selected elements in MSW incineration bottom ash and in genuine gravel are given in Table 2.9. Discuss gravel substitution by bottom ash from a substance-based point of view (see Problem 2.11).

Problem 2.13:

Quantify the stock of wastes (mass and volume) that is built up during an average lifetime. Consider all wastes such as sewage sludge,

TABLE 2.8

Mean Substance Concentrations in Sewage Sludge, mg/kg Dry Matter

N	Cl	F	S	PCB	Cd	Hg	As	Co
28,000	360	100	14,000	0.2	2	3	20	15
Ni	Sb	Pb	Cr	Cu	Mn	V	Sn	Zn
800	10	150	100	300	500	30	30	1500

TABLE 2.9

Mean Concentrations of Selected Elements in MSW Incineration Bottom Ash (BA) and Gravel (G)

	Si, %	Al, %	Fe, %	Mg, %	Ca, %	K, %	C_{org}, %	Cu, g/kg	Zn, g/kg	Pb, g/kg	Cd, mg/kg	Hg, mg/kg
BA	20	4	3	1	10	1	2	2	10	6	10	0.5
G	20	4	2	0.16	20	0.17	<<0.1	0.01	0.06	0.015	0.3	0.05

construction and demolition waste, end-of-life vehicles, and the like. (Use data from waste management literature or from the World Wide Web.) Do not consider loss of mass by incineration, biological digestion, and so on. Make three assumptions: (a) waste generation remains constant, (b) waste generation increases linearly, and (c) waste generation increases by 2% per year.

Problem 2.14:

Let us assume that legislation on the application of sewage sludge on land limits the increase of heavy metals to less than 5% (relative) within 100 years. Further, assume for simplicity that erosion of soil, eluviation, leaching, uptake by plants, etc., can be neglected. How much sewage sludge can be applied to land? Discuss the example of zinc: Zn concentration in sewage sludge is 1500 mg.kg dry matter; Zn concentration in soil is 30 mg/kg. Is there enough agricultural soil to spread all sewage sludge?

Problem 2.15:

Consider the system given in Figure 2.22. It was developed to investigate the resource efficiency of metals management. Discuss the system boundary in space: is it chosen appropriately?

Problem 2.16:

Take an environmental report of a regional company of your choice (e.g., pulp and paper mill, electricity supplier, cement manufacturer, car manufacturer, food producer; many reports can be downloaded from the World Wide Web) and analyze it from the point of view of material balances. What are the most important goods and substances (quantitatively and qualitatively)? Is it possible to establish material balances based on the information provided? If not, which data are missing? Find issues for improvement by MFA.

The solutions to the problems are given on the website http://www.MFA -handbook.info.

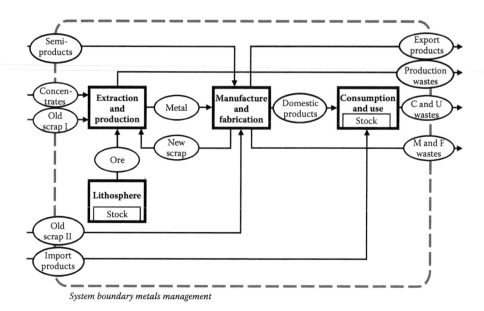

System boundary metals management

FIGURE 2.22
Flowchart of the stocks and flows of a metal.

2.3 Treatment of Data Uncertainties in MFA

Oliver Cencic

A key step of any MFA is to find appropriate data. If direct measurements are not available, data are often taken from alternative sources of varying quality (Laner, Rechberger, and Astrup, 2014). All measured, estimated, or literature values represent variables with at least random deviations. Assuming values to be "precise" leads to conflicts with model constraints, e.g., the law of mass conservation. Only if the uncertainty of the data is considered, can these conflicts be resolved. Thus, complete, reliable, and useful information about data has to include its uncertainty.

Data uncertainties can be grouped into two categories: aleatory variability and epistemic uncertainty (Abrahamson, 2007). Aleatory variability (from the Latin *alea*, meaning a die, or game of dice) arises due to inherent variability, natural stochasticity, environmental or structural variation across space or through time, manufacturing or genetic heterogeneity among components or individuals, and a variety of others sources of randomness. Aleatory variability cannot be reduced but only better understood. On the other hand, epistemic uncertainty (from the Greek επιστήμη, for knowledge or science) arises due to insufficient knowledge about the world, which includes small sample sizes, detection limits, imperfections in scientific understanding, etc.

(Dutta and Ali, 2012; Beven and Young, 2013). Epistemic uncertainty can be reduced by further investigation.

While aleatory variability can be handled by probability theory, epistemic uncertainty is better covered by possibility or fuzzy set theory (Zadeh, 1965; Ferson and Ginzburg, 1996; Laner et al., 2014). Independent of this fact, epistemic errors are often treated in terms of probabilities as if they were aleatory. Consequently, any representation requires subjective assumptions (Beven and Young, 2013). In the following, the probabilistic approach is used for both kinds of uncertainties.

Even though, in uncertainty analysis, epistemic uncertainties (e.g., measurement errors) are frequently assumed to be normally distributed, in scientific models in general and in MFA models in particular, this is often not the case. If, for instance, a process model is correct, mass flows and concentrations cannot take negative values, and transfer coefficients are restricted to the unit interval. However, if uncertainties are assumed to be normally distributed but are not, such negative values might occur. Another example is provided by expert opinions that frequently have to be relied on in MFA due to scarce or missing data. They are often modeled by a uniform, triangular, or trapezoidal distribution depending on the expert's degree of knowledge (Cencic and Frühwirth, 2015).

Despite this fact, in the following sections, the assumption of normality will be used because it simplifies many computations. Information on how to deal with nonnormally distributed data in uncertainty analysis can be found in Sections 2.3.2 and 2.3.3.4. Additionally, uncertain data are assumed to be free of gross and systematic errors, meaning the data contain random errors only.

2.3.1 Data Collection

To collect data means to measure or estimate properties of an entity (e.g., a car, or MSW generated annually in a country) or a collection of entities (e.g., cars used in a region, or inhabitants of a city). There are several methods available that can be grouped into direct and indirect methods.

2.3.1.1 Direct Methods

To apply direct methods, an entity of interest must be accessible to be measured. Two types of direct measurements can be distinguished:

If the investigated entity can be analyzed completely in one step, a single numerical value is obtained. This value has to be combined with the measurement uncertainty (e.g., as stated in the specifications of the measuring device) to complete the data set. Examples for this type of direct measurements are

- Determining the mass of a brick by weighing it on a scale
- Measuring the length of a brick with a tape measure

How to combine repeated measurements or estimates of the same entity is shown in Section 2.3.1.3.

If an entity or a collection of entities is too large to be analyzed completely, it has to be sampled in order to estimate some of its theoretically precise, intensive properties (e.g., the cadmium concentration in MSW, or the average height of persons living in a city). In statistics, a collection of entities is denoted as a population; its individual elements are denoted as population units. Note that also, a single entity can be seen as a population even if it lacks an inherent population unit (countable things like pieces, persons, etc.). To sample a population means to randomly select a manageable number of its units only. The chosen sampling unit is the unit into which the population is divided for the purpose of sampling. In case of a selection of entities, the sampling unit is typically equal to the population unit (e.g., a person) but could also be a multiple of it (e.g., a family). In case of a single entity, the sampling unit can be chosen arbitrarily (e.g., a bucket full of waste). The elements of a sample are denoted as sample units or observations. The number of observations, denoted by n, is called the sample size. Note that the sample size is an integer (it is countable) and has nothing to do with the length, mass, or volume of the individual elements of a sample. The total number of sampling units inside the population is denoted by N.

Each element of the sample is analyzed with respect to the property under investigation to get n individual measurements x_i. From these measurements, the statistics of the sample (sample mean $\bar{x}$ and sample variance s^2) are computed from Equations 2.10 and 2.11.

$$\bar{x} = \frac{1}{n} \cdot \sum_{i=1}^{n} x_i \qquad (2.10)$$

$$s^2 = \frac{1}{n-1} \cdot \sum_{i=1}^{n} (x_i - \bar{x})^2 \qquad (2.11)$$

Using the statistics of the sample, inferences from the sample to the population are made to estimate the population mean μ and population variance σ^2 (=variation among the sampling units). Note that the population variance σ^2 depends on the chosen sampling unit. The bigger the sampling unit chosen, the smaller the population variance σ^2 will be.

Assuming the investigated property to be normally distributed among the sampling units with $\mathcal{N}(\mu, \sigma^2)$ and the measurements to be precise, i.e., there are no measurement errors involved, the error model is

$$x_i \sim \mathcal{N}(\mu, \sigma^2) \qquad (2.12)$$

The population parameters μ and σ^2 can then be estimated from Equations 2.13 and 2.14.

$$\hat{\mu} = \bar{x} \tag{2.13}$$

$$\hat{\sigma}^2 = s^2 \tag{2.14}$$

The "hat" sign $\wedge$ denotes an estimator of a true parameter.

The uncertainty of the estimated parameter $\hat{\mu}$ itself can be calculated from Equations 2.15 and 2.16.

$$var(\hat{\mu}) = \frac{s^2}{n} \tag{2.15}$$

$$std(\hat{\mu}) = ste(\hat{\mu}) = \sqrt{var(\hat{\mu})} \tag{2.16}$$

The standard deviation of the estimated mean $std(\hat{\mu})$ is also known as the standard error of the mean $ste(\hat{\mu})$. Generally, a standard error describes how precisely a parameter of a population (e.g., μ or σ^2) has been estimated from a sample of size n. It is of epistemic nature because it can be reduced if the sample size is increased. If the sample covers the entire population ($n = N$), assuming measurement errors to be zero, the true values of the population mean μ and the population variance σ^2 can be determined exactly from Equations 2.17 and 2.18.

$$\mu = \frac{1}{N} \cdot \sum_{i=1}^{N} x_i \tag{2.17}$$

$$\sigma^2 = \frac{1}{N} \cdot \sum_{i=1}^{N} (x_i - \mu)^2 \tag{2.18}$$

Note that in this case, even though the individual sample values x_i vary around the population mean μ with variance σ^2, the standard errors of μ and σ^2 are zero. This can be seen best for the mean value (Equation 2.19) if the population is assumed to be infinite ($n = N \to \infty$).

$$var(\mu) = \lim_{n \to \infty} \frac{\sigma^2}{n} = 0 \tag{2.19}$$

Whenever the population is finite ($N < \infty$), the finite population correction (FPC) term $(N - n)/n$ has to be applied, ensuring the variance of the population mean μ always to be zero if $n = N$ (Equation 2.20).

$$var(\mu) = \frac{\sigma^2}{n} \cdot \frac{N-n}{N} = \frac{\sigma^2}{N} \cdot \frac{N-N}{N} = 0 \tag{2.20}$$

If n is less than 5% of N ($n \leq 0.05 \cdot N$), FPC is negligible because it is approximately 1.

Example 2.1

The copper content of an ore deposit shall be estimated. Because the deposit is much too large to be analyzed completely (it can be considered as an infinite population), a sample of 10 units (i.e., $n = 10$) with approximately equal mass (= sampling unit) is taken. The sample is analyzed in a laboratory to determine the copper content of each sample unit (Table 2.10).

Assuming the measurement error to be zero, the sample statistics (mean copper content of the sample and the deviation of the unit values around this mean) are computed.

$$\bar{c}_{Cu,sample} = \frac{1}{n} \cdot \sum_{i=1}^{n} c_{Cu,unit\ i} = 27.5\,g/kg$$

$$var\left(c_{Cu,sample}\right) = \frac{1}{n-1} \cdot \sum_{i=1}^{n} (c_{Cu,unit\ i} - \bar{c}_{Cu,sample})^2 = 41.7\,(g/kg)^2$$

$$std(c_{Cu,sample}) = \sqrt{var(c_{Cu,sample})} = 6.5\,g/kg$$

The mean value of the sample is taken as the best estimate of the copper content of the ore deposit:

$$\hat{c}_{Cu,ore} = \bar{c}_{Cu,sample} = 27.5\ g/kg$$

TABLE 2.10

Copper Content in g/kg of 10 Sample Units Randomly Taken from an Ore Deposit

Sample unit	1	2	3	4	5	6	7	8	9	10
Copper content (g/kg)	30.0	15.9	30.4	37.2	28.7	32.8	30.5	22.7	27.2	19.1

To get a measure of the uncertainty of the estimated copper content of the ore deposit, its standard error can be calculated:

$$ste(\hat{c}_{Cu,ore}) = \frac{std(c_{Cu,sample})}{\sqrt{n}} = 2.0 \text{ g/kg}$$

If the standard error of the mean of the sample is of interest instead, then it is obvious that it is zero because the whole population ($n = N = 10$) is taken into account. In this case, with a finite population, the FPC must be applied in order to get the same result in a formal way.

$$ste(\overline{c}_{Cu,sample}) = \frac{std(c_{Cu,sample})}{\sqrt{n}} \cdot \sqrt{\frac{N-n}{N}} = \frac{6.5}{\sqrt{10}} \cdot \sqrt{\frac{10-10}{10}} = 0$$

∎

If the individual measurements are not precise but contain the same assumed or known absolute measurement error τ^2, the error model is modified (Equation 2.21).

$$x_i \sim \mathcal{N}(\mu, \sigma^2 + \tau^2) \tag{2.21}$$

The estimated population mean $\hat{\mu}$ is again calculated from the sample mean $\overline{x}$ (Equation 2.13). To estimate the variance of the sampling units $\hat{\sigma}^2$ and the variance of the estimated population mean $\hat{\mu}$, two cases have to be distinguished.

Case 1. The sample variance s^2 is larger than the measurement errors τ^2 ($s^2 > \tau^2$), meaning a part of the sample variance s^2 can be explained by the natural variability σ^2 of the investigated property in the sampling units.

In case 1, the estimated population variance $\hat{\sigma}^2$ can be calculated from Equation 2.22 and the variance of the estimated population mean $var(\hat{\mu})$ from Equation 2.23.

$$\hat{\sigma}^2 = s^2 - \tau^2 \tag{2.22}$$

$$var(\hat{\mu}) = \frac{1}{n} \cdot (\hat{\sigma}^2 + \tau^2) = \frac{1}{n} \cdot s^2 \tag{2.23}$$

Case 2. The sample variance s^2 is smaller than or equal to the measurement error τ^2 ($s^2 \leq \tau^2$), meaning the observed sample variance s^2 can be explained by the measurement errors τ^2 alone.

In case 2, the population variance $\hat{\sigma}^2$ is assumed to be zero (Equation 2.24), and the variance of the estimated population mean $var(\hat{\mu})$ is calculated from Equation 2.25.

$$\hat{\sigma}^2 = 0 \tag{2.24}$$

$$var(\hat{\mu}) = \frac{1}{n} \cdot (\hat{\sigma}^2 + \tau^2) = \frac{1}{n} \cdot \tau^2 \tag{2.25}$$

Note that in case 2, the estimated variance of the population $\hat{\sigma}^2$ is 0 but not the variance of the estimated population mean $var(\hat{\mu})$.

If each measurement or observation has different absolute measurement errors τ_i^2, the error model has to be modified (Equation 2.26).

$$x_i \sim \mathcal{N}(\mu, \sigma^2 + \tau_i^2) \tag{2.26}$$

Instead of considering each of the individual measurement errors separately, for simplification reasons, the mean of the variance of the measurement errors is calculated from Equation 2.27.

$$\bar{\tau}^2 = \frac{1}{n} \cdot \sum_{i=1}^{n} \tau_i^2 \tag{2.27}$$

The estimated population mean $\hat{\mu}$ is again calculated from the sample mean $\bar{x}$ (Equation 2.13). To compute the estimated population variance $\hat{\sigma}^2$ and the variance of the estimated population mean $var(\hat{\mu})$, τ^2 is replaced by $\bar{\tau}^2$ in Equations 2.22 to 2.25, leading to Equations 2.28 and 2.29.

$$s^2 > \bar{\tau}^2 : \hat{\sigma}^2 = s^2 - \bar{\tau}^2,\ var(\hat{\mu}) = \frac{1}{n} \cdot s^2 \tag{2.28}$$

$$s^2 \leq \bar{\tau}^2 : \hat{\sigma}^2 = 0,\ var(\hat{\mu}) = \frac{1}{n} \cdot \bar{\tau}^2 \tag{2.29}$$

Alternatively, instead of the sample mean $\bar{x}$ (Equation 2.10), the weighted mean of the observations $\bar{x}_w$ can be used. It is computed from Equation 2.30

by using the inverse of the absolute variances τ_i^2 as weighting factors (Equation 2.31).

$$\bar{x}_w = \sum_{i=1}^{n}\left(x_i \cdot \frac{w_i}{\sum_{i=1}^{n} w_i}\right) \tag{2.30}$$

$$w_i = \frac{1}{\tau_i^2} \tag{2.31}$$

The sample variance s_w^2 is then computed with respect to weighted sample mean $\bar{x}_w$ (Equation 2.32).

$$s_w^2 = \frac{1}{n-1} \cdot \sum_{i=1}^{n} (x_i - \bar{x}_w)^2 \tag{2.32}$$

In this case, the population mean μ is estimated from the weighted sample mean $\bar{x}_w$ (Equation 2.33).

$$\hat{\mu} = \bar{x}_w \tag{2.33}$$

To compute the estimated population variance $\hat{\sigma}^2$ and the variance of the estimated population mean $var(\hat{\mu})$, in Equations 2.22 to 2.25, τ^2 is replaced by $\bar{\tau}^2$, and s^2 by s_w^2. This leads to Equations 2.34 and 2.35.

$$s_w^2 > \bar{\tau}^2 : \hat{\sigma}^2 = s_w^2 - \bar{\tau}^2, \ var(\hat{\mu}) = \frac{1}{n} \cdot s_w^2 \tag{2.34}$$

$$s_w^2 \leq \bar{\tau}^2 : \hat{\sigma}^2 = 0, \ var(\hat{\mu}) = \frac{1}{n} \cdot \bar{\tau}^2 \tag{2.35}$$

Note that the population variance $\hat{\sigma}^2$ is still estimated by taking into account the mean $\bar{\tau}^2$ of all τ_i^2 (Equation 2.27). Considering the individual τ_i^2 for the estimation of $\hat{\sigma}^2$ is also feasible but not investigated in the context of this chapter.

2.3.1.2 Indirect Methods

If direct measurements are not available, information from other sources is often taken into account to fill data gaps in MFA (see Section 2.2.4).

One way to get an estimate of a property of interest is to ask experts for their opinions. Their educated guesses often cover intervals given by a minimum and a maximum numerical value. The information "not smaller than" and "not larger than" can be modeled best with a uniform distribution $Unif(a, b)$, where each value in the interval $[a, b]$ is assumed to be equally probable. Because many statistical methods (e.g., data reconciliation) need the assumption of normality, the chosen distribution is often approximated by a normal distribution. This can be achieved, for example, by computing the mean μ and the variance σ^2 of the uniform distribution, and regarding them as the parameters of a normal distribution (Figure 2.23 and Equation 2.36).

$$X \sim Unif(a,b) \rightarrow \mu = \frac{a+b}{2}, \ \sigma^2 = \frac{(b-a)^2}{12} \rightarrow X \sim \mathcal{N}(\mu, \sigma^2) \qquad (2.36)$$

If a value c or subinterval $[c, d]$ within interval $[a, b]$ is more likely to be or to contain the true value, a triangular distribution $tria\ (a, c, b)$ or trapezoidal distribution $trap\ (a, c, d, b)$ could be selected instead of a uniform distribution. The rest of the procedure is the same as demonstrated for the uniform distribution.

This concept can also be applied if multiple estimates of a property without given uncertainty can be obtained from various sources, e.g., take the minimum value as lower boundary a, the maximum value as upper boundary b, and select a proper distribution function depending on the distribution of the other values within the interval $[a, b]$.

Typically, data from different sources are of varying quality, which has to be distinguished from uncertainties of measurement or estimation. While the latter refer to the precision of a measurement or estimation, data quality refers to the degree of usability of the collected data for the performed study.

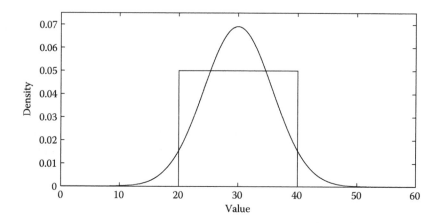

FIGURE 2.23
Approximation of a uniform distribution $Unif(20, 40)$ by a normal distribution with the same parameters mean value μ and variance σ^2.

Data quality is a function of how well the collected data fit in the context of the study, how reliable the data producer is, and so forth. For more detailed information, see Schwab, Laner, and Rechberger (in press). Both measurement uncertainty and data quality represent epistemic uncertainties.

Example 2.2

The concentration of copper in MSW of region A shall be estimated. Because no direct measurement is available for region A, data from similar measurements in region B are taken. Even if the concentration of copper in MSW of region B has been determined precisely, the quality of the data taken as an estimate for region A might be poor.   ■

How to best include the quality of data in an MFA study is still under research. A proper way could be to expand the standard error of measurements and estimates by taking into account a quality measure q that ranges between 0 (not useful at all) and 1 (fits perfectly) and an arbitrarily chosen factor f (Equation 2.37).

$$\sigma_{i,exp} = \sigma_i - f \cdot \log(q) \tag{2.37}$$

Thus, in the worst case ($q = 0$), the expanded standard deviation is ∞; in the best case ($q = 1$), it is σ_i.

If $q = 1$, the estimate fits perfectly in the context of the research question. Thus, its mean value $\hat{\mu}_i$ and its standard error σ_i can be used for further calculations. If $0 < q < 1$, the expanded standard error $\sigma_{i,exp}$, computed with Equation 2.37, can be used instead of σ_i. If $q = 0$, the estimate is virtually useless.

A possible quality measure q (Equation 2.38) can be derived, for instance, from the information defect (*ID*) introduced by Schwab et al. (in press).

$$q = 1 - ID \tag{2.38}$$

Information defects quantify the quality of MFA data on a scale from 0 (no defect—perfectly usable data) to 1 (no data available).

2.3.1.3 Combinations of Multiple Estimates

If multiple estimates $\hat{\mu}_i$ of the investigated property μ of the same population are available, they can be combined by calculating a weighted mean (Equation 2.39).

$$\hat{\mu} = \frac{\sum_{i=1}^{n} (\hat{\mu}_i \cdot w_i)}{\sum_{i=1}^{n} w_i} \tag{2.39}$$

The best linear unbiased estimator (BLUE) of $\hat{\mu}$ is reached if the variance of each estimate is used as the weighting factor (Equation 2.40). The smaller the variance of an estimation, the higher is its weight:

$$w_i = \frac{1}{var(\hat{\mu}_i)} \tag{2.40}$$

The variance of $\hat{\mu}$ can be computed by applying the law of error propagation (Equation 2.46) to Equation 2.39, leading to Equation 2.41.

$$var(\hat{\mu}) = \frac{1}{\sum_{i=1}^{n} w_i} = \frac{1}{\sum_{i=1}^{n} \frac{1}{var(\hat{\mu}_i)}} \tag{2.41}$$

Note that the same results can be obtained by data reconciliation (see Section 2.3.3).

If the variance of all estimates is the same, Equations 2.39 and 2.41 simplify to Equations 2.42 and 2.43.

$$\hat{\mu} = \frac{\sum_{i=1}^{n} \hat{\mu}_i}{n} \tag{2.42}$$

$$var(\hat{\mu}) = \frac{var(\hat{\mu}_i)}{n} \tag{2.43}$$

Example 2.3

The mass of an apple is measured with three different scales (a beam balance, a letter balance, and an electronic balance). The measurement uncertainty of each scale is assumed to be known, given as standard error (Table 2.11).

TABLE 2.11

Mass of an Apple (Mean Value and Standard Error) Measured with Different Scales

Scale	Mass of Apple as Measured by Scale Given as Mean Value, g	Measurement Uncertainty of Scale Given as Standard Error, g
Beam balance	183.0	0.5
Letter balance	185	5
Electronic balance	182.5360	0.0005

The true mass of the apple is estimated with

$$\hat{m} = \frac{\dfrac{183.0}{0.5^2} + \dfrac{185}{5^2} + \dfrac{182.5360}{0.0005^2}}{\dfrac{1}{0.5^2} + \dfrac{1}{5^2} + \dfrac{1}{0.0005^2}} = 182.5360 \text{ g}$$

$$ste(\hat{m}) = \sqrt{var(\hat{m})} = \sqrt{\frac{1}{\dfrac{1}{0.5^2} + \dfrac{1}{5^2} + \dfrac{1}{0.0005^2}}} = 0.0005 \text{ g}$$

It can be seen that the measurements of scales 1 and 2 have almost no influence on the result, which (rounded to four decimals) is identical to the measurement of scale 3. The reason for this behavior is obvious: scale 3 is much more precise than scales 1 and 2.

If only the measurements of scales 1 and 2 were used, the result would be

$$\hat{m} = \frac{\dfrac{183.0}{0.5^2} + \dfrac{185}{5^2}}{\dfrac{1}{0.5^2} + \dfrac{1}{5^2}} = 183.0 \text{ g}$$

$$ste(\hat{m}) = \sqrt{var(\hat{m})} = \sqrt{\frac{1}{\dfrac{1}{0.5^2} + \dfrac{1}{5^2}}} = 0.5 \text{ g}$$

The result (rounded to two decimals) is now almost identical to the measurement of scale 1. This is due to the fact that, in case of two measurements, the distance between the two measurements is split proportional to their variances (Figure 2.24).

$$\hat{x}_m = \hat{x}_1 + (\hat{x}_2 - \hat{x}_1) \cdot \frac{var(\hat{x}_1)}{var(\hat{x}_1) + var(\hat{x}_2)} = \hat{x}_1 + (\hat{x}_2 - \hat{x}_1) \cdot \frac{1}{1 + \dfrac{var(\hat{x}_2)}{var(\hat{x}_1)}}$$

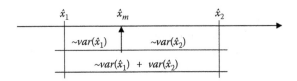

FIGURE 2.24
Splitting of the distance between $\hat{x}_1$ and $\hat{x}_2$ proportional to the variances.

In this example, the ratio of the variances $var(\hat{x}_2)/var(\hat{x}_1) = 100$, leading to

$$\hat{m} = 183.0 + (185 - 183.0) \cdot \frac{1}{1 + 100} = 183.0 \text{ g}$$

2.3.2 Propagation of Uncertainties

If random variables X_i with expectation $E(X_i) = \mu_i$ and variance $var(X_i) = \sigma_i^2$ are combined in a function $Y = f(X_1, X_2,...,X_n)$, the result Y itself is a random variable. If μ_i or σ_i^2 is not known, it can be estimated from a sample. In general, the probability density function (PDF) of the result Y cannot be calculated analytically. In the following sections, two straightforward methods to propagate uncertainties are presented that are frequently used in MFA.

2.3.2.1 Gauss's Law

The propagation of uncertainties can be evaluated by applying Gauss's law of error propagation to a function of interest (Equation 2.44).

$$Y = f(X_1, X_2,...,X_n) \tag{2.44}$$

The expectation of the result Y can be estimated from Equation 2.45 and the variance of the result from Equation 2.46. The latter is called Gauss's law of error propagation.

$$E(Y) \approx f(E(X_1), E(X_2),..., E(X_n)) = f(\mu_1, \mu_2,...,\mu_n) \tag{2.45}$$

$$
\begin{aligned}
var(Y) \approx &\sum_{i=1}^{n} \left(var(X_i) \cdot \left[\frac{\partial Y}{\partial X_i} \right]_{X=\mu}^2 \right) \\
&+ 2 \cdot \sum_{i=1}^{n-1} \sum_{j=i+1}^{n} \left(cov(X_i, X_j) \cdot \left[\frac{\partial Y}{\partial X_i} \right]_{X=\mu} \cdot \left[\frac{\partial Y}{\partial X_j} \right]_{X=\mu} \right)
\end{aligned}
\tag{2.46}
$$

with
$\mu_i = E(X_i) = $ mean of X_i
$\sigma_i = std(X_i) = $ standard deviation of X_i
$\sigma_i^2 = var(X_i) = $ variance of X_i
$\sigma_{ij} = cov(X_i,X_j) = $ covariance of X_i and X_j

If the variables X_i are independent or uncorrelated, Equation 2.46 can be reduced to Equation 2.47.

$$var(Y) \approx \sum_{i=1}^{n} \left(var(X_i) \cdot \left[\frac{\partial Y}{\partial X_i} \right]_{X=\mu}^{2} \right) \tag{2.47}$$

Notes:

- The basis of Gauss's law is a linear approximation of a function with a first-order Taylor series expansion. Thus, for nonlinear functions, reasonable results can be expected only if the nonlinearity is negligible compared to the magnitude of the uncertainties.
- Knowledge or assumptions about the shape of the PDFs of the random variables are not needed.
- If X_i represents a variable with natural variation (e.g., the height of persons), $var(X_i)$ describes its aleatory variability. If X_i represents an estimated parameter (e.g., the average height of people), $var(X_i)$ describes the epistemic uncertainty of the mean value $E(X_i)$ estimated from a sample of size n. For a detailed example, see Example 2.22.

Examples 2.4 and 2.5 demonstrate the application of Gauss's law on a linear and a nonlinear function, respectively.

Example 2.4

The MFA system considered in this example consists of one process, two import flows and one export flow (Figure 2.25).

Due to the law of mass conservation, the output flow Y of the process is the sum of the two input flows X_1 and X_2:

$$Y = X_1 + X_2$$

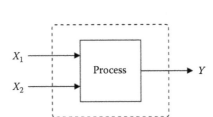

FIGURE 2.25
MFA system with one process, two import, and one export flows.

where

$$E(X_1) = 100, \sigma_{X_1} = 10 \,(= 10\% \text{ of } E(X_1))$$

$$var(X_1) = \sigma_{X_1}^2 = 100$$

and

$$E(X_2) = 200, \sigma_{X_2} = 20 \,(= 10\% \text{ of } E(X_2))$$

$$var(X_2) = \sigma_{X_2}^2 = 400$$

The application of Gauss's law leads to

$$E(Y) = E(X_1 + X_2) = E(X_1) + E(X_2)$$

$$var(Y) = var(X_1 + X_2) = var(X_1) + var(X_2) + 2 \cdot cov(X_1, X_2)$$

If the variables X_1 and X_2 are independent, their covariance is 0. This leads to

$$var(X_1 + X_2) = var(X_1) + var(X_2)$$

Thus, the result is

$$E(Y) = 100 + 200 = 300$$
$$var(Y) = 100 + 400 = 500, \sigma_Y = \sqrt{500} = 22.4 = (7.5 \% \text{ of } E(Y))$$

Example 2.5

The MFA system considered in this example consists of one process, one import flow, and two export flows (Figure 2.26).

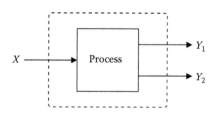

FIGURE 2.26
MFA system with one process, one import, and two export flows.

The output flow Y_1 of the process is a certain fraction (given by the transfer coefficient TC_1) of the input flow X:

$$Y_1 = TC_1 \cdot X$$

where

$$E(X) = 100, \sigma_X = 5 \; (= 5\% \text{ of } E(X))$$

$$var(X) = \sigma_X^2 = 25$$

and

$$E(TC_1) = 0.40, \sigma_{TC_1} = 0.04 \; (= 10\% \text{ of } E(TC_1))$$

$$var(TC_1) = \sigma_{TC_1^2} = 0.0016$$

The application of Gauss's law leads to

$$E(Y_1) = E(TC_1 \cdot X) = E(TC_1) \cdot E(X)$$

$$var(Y_1) \approx var(TC_1 \cdot X) = E(X^2) \cdot var(TC_1) + E(TC_1)^2 \cdot var(X)$$
$$+ 2 \cdot E(TC_1) \cdot E(X) \cdot cov(TC_1, X)$$

If the variables TC_1 and X are independent, their covariance is 0. This leads to

$$var(Y_1) \approx var(TC_1 \cdot X) = E(X)^2 \cdot var(TC_1) + E(TC_1)^2 \cdot var(X)$$

Thus, the result is

$$E(Y_1) = 0.40 \cdot 100 = 40$$

$$var(Y_1) \approx 100^2 \cdot 0.0016 + 0.40^2 \cdot 25 = 20,$$

$$\sigma_{Y_1} \approx \sqrt{20} = 4.5 = (11.4\% \text{ of } E(Y_1))$$

Note that in case of a multiplication, there also exists a precise solution for the variance of multiplications of uncorrelated random variables:

$$var(Y_1) = var(TC_1 \cdot X) = E(X)^2 \cdot var(TC_1) + E(TC_1)^2 \cdot var(X)$$
$$+ var(TC_1) \cdot var(X)$$

In this example, the difference in the resulting variances (= 0.04) is negligible. ∎

For a system of linear equations (Equation 2.48), the expectation of the results (Equation 2.49) and Gauss's law (Equation 2.50) can be written in matrix notation.

$$y = f(x, z) = Ax + Bz = Ax + b \tag{2.48}$$

$$E(y) = f(E(x), z) = AE(x) + Bz = AE(x) + b \tag{2.49}$$

$$Q_y = AQ_xA^T \tag{2.50}$$

with
x = column vector ($i \times 1$) of measured or estimated random variables
y = column vector ($j \times 1$) of unknown random variables
z = column vector ($k \times 1$) of constant variables
A = coefficient matrix ($j \times 1$) of unknown random variables
B = coefficient matrix ($j \times k$) of constant variables
Q_x = covariance matrix ($i \times i$) of random variables
Q_y = covariance matrix ($j \times j$) of random variables

The constant part $b = B \cdot z$ in the equations has no influence on Q_y because the variance is invariant against translations; i.e., if only a constant value is added to a random variable, the shape of the PDF and, with it, the variance will remain the same.

A linear approximation of nonlinear functions (Equation 2.51) can be obtained from a first-order Taylor series expansion at expansion point $E(x)$. The expectations $E(y)$ of the linearized functions can be computed from Equation 2.52 and the variance–covariance matrix Q_y from Equation 2.53.

$$y = f(x, z) \approx J_x(x - E(x)) + f(E(x), z) \tag{2.51}$$

$$E(y) \approx f(E(x), z) \tag{2.52}$$

$$Q_y \approx J_x Q_x J_x^T \tag{2.53}$$

with
$f(x, z)$ = system of nonlinear equations
J_x = Jacobi matrix $\partial f(x, z)/\partial x$ evaluated at expansion point $E(x)$, z

Note that even if the random input variables x are initially not correlated, there are always correlations between x and y.

Examples 2.6, 2.7, and 2.8 demonstrate the necessity to consider existing correlations between input parameters of a model.

Example 2.6

In this example, C is a function of two random parameters A and B:

$$C = f(A, B) = A + B$$

with

$$E(A) = 100, var(A) = 100$$

$$E(B) = 50, var(B) = 25$$

A and B are assumed to be uncorrelated, so

$$cov(A, B) = 0$$

Application of error propagation leads to

$$E(C) = E(A) + E(B) = 150$$

$$var(C) = var(A) + var(B) = 125$$

and

$$cov(A, C) = cov(A, A + B) = cov(A, A) + cov(A, B)$$
$$= var(A) + cov(A, B) = 100 + 0 = 100$$

$$cov(B, C) = cov(B, A + B) = cov(B, A) + cov(B, B)$$
$$= cov(B, A) + var(B) = 0 + 25 = 25$$

Note that due to the functional relation $C = A + B$, C is correlated to A and B, respectively.

Now A is recalculated by using the previous result for C and the initial given B:

$$A = C - B$$

with

$$E(B) = 50, var(B) = 25$$

$$E(C) = 150, var(C) = 125$$

Error propagation without consideration of correlations between B and C leads to

$$E(A) = E(C) - E(B) = 100$$

$$var(A) = var(C) + var(B) = 150$$

Only if the covariance of B and C is considered, the calculation reaches the initial variance of A:

$$var(A) = var(C) + var(B) - 2 \cdot cov(B, C) = 125 + 25 - 2 \cdot 25 = 100$$

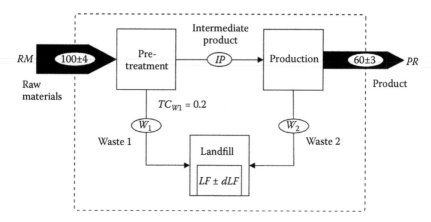

FIGURE 2.27
An MFA system consisting of three processes (one of them containing a stock) and five flows.

Example 2.7

In this example, the change in stock *dLF* in the process Landfill of the MFA system in Figure 2.27 is calculated.

The given input parameters of the model are the mass flows of raw materials *RM* and products *PR*, and the transfer coefficient TC_{W_1} with

$$E(RM) = 100, var(RM) = 4^2$$

$$E(PR) = 60, var(PR) = 3^2$$

$$TC_{W_1} = 0.2$$

If all flows into the landfill are computed sequentially from the given input parameter, ignoring covariances, the resulting variance of *dLF* is wrong:

$$W_1 = TC_{W_1} \cdot RM$$

$$E(W_1) = TC_{W_1} \cdot E(RM) = 0.2 \cdot 100 = 20$$

$$var(W_1) = 0.2^2 \cdot var(RM) = 0.2^2 \cdot 4^2 = 0.64$$

$$IP = (1 - TC_{W_1}) \cdot RM$$

$$E(IP) = (1 - TC_{W_1}) \cdot E(RM) = (1 - 0.2) \cdot 100 = 80$$

$$var(IP) = (1 - 0.2)^2 \cdot var(RM) = 0.8^2 \cdot 4^2 = 10.24$$

$$W_2 = IP - PR$$

$$E(W_2) = E(IP) - E(PR) = 80 - 60 = 20$$

$$var(W_2) = var(IP) + var(PR) = 10.24 + 3^2 = 19.24$$

$$dLF = W_1 + W_2$$

$$E(dLF) = E(W_1) + E(W_2) = 20 + 20 = 40$$

$$var(dLF) = var(W_1) + var(W_2) = 0.64 + 19.24 = 19.88$$

To get the correct variance of *dLF*, the correlation between W_1 and W_2 has to be considered:

$$var(dLF) = var(W_1) + var(W_2) + 2 \cdot cov(W_1, W_2)$$

With the following rules

$$cov(a \cdot X + b \cdot Y + c, d \cdot W + e \cdot V + f)$$
$$= a \cdot d \cdot cov(X,W) + a \cdot e \cdot cov(X,V) + b \cdot d \cdot cov(Y,W) + b \cdot e \cdot cov(Y,V)$$
$$cov(X,X) = var(X)$$

the covariance of W_1 and W_2 can be computed

$$cov(W_1,W_2) = cov(TC_{W_1} \cdot RM, (1 - TC_{W_1}) \cdot RM - PR)$$
$$cov(W_1,W_2) = TC_{W_1} \cdot (1 - TC_{W_1}) \cdot var(RM) + TC_{W_1} \cdot cov(RM,PR)$$
$$cov(W_1,W_2) = 0.2 \cdot 0.8 \cdot 4^2 + 0.2 \cdot 0$$
$$cov(W_1,W_2) = 2.56$$

This leads to

$$var(dLF) = 0.64 + 19.24 + 2 \cdot 2.56 = 25$$

If the variable of interest is expressed directly as a function of the given parameter, which are assumed to be uncorrelated, there is no need to consider covariances at all. In this example, *dLF* can be computed directly from the balance equation of the complete model using the independent variables *RM* and *PR* only:

$$E(dLF) = E(RM) - E(PR) = 100 - 60 = 40$$

$$var(LF) = var(RM) - var(PR) = 4^2 + 3^2 = 5^2 = 25$$

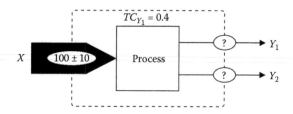

FIGURE 2.28
An MFA system consisting of one process, one import, and two export flows. Input parameters of the model are the input flow X of the process and the transfer coefficient TC_{Y_1}.

Example 2.8

In this example, the two output flows Y_1 and Y_2 of the process in Figure 2.28 are to be calculated from the input flow X and a given transfer coefficient from X to Y_1 (TC_{Y_1}) with

$$E(X) = 100, \, var(X) = 10^2$$

$$TC_{Y_1} = 0.4$$

If the two unknown variables Y_1 and Y_2 are expressed as functions of the two given parameters X and TC_{Y_1}, no covariances have to be considered.

$$Y_1 = TC_{Y_1} \cdot X$$

$$E(Y_1) = TC_{Y_1} \cdot E(X) = 0.4 \cdot 100 = 40$$

$$var(Y_1) = TC_{Y_1^2} \cdot var(X) = 0.4^2 \cdot 10^2 = 16$$

$$Y_2 = \left(1 - TC_{Y_1}\right) \cdot X$$

$$E(Y_2) = \left(1 - TC_{Y_1}\right) \cdot E(X) = 0.6 \cdot 100 = 60$$

$$var(Y_2) = \left(1 - TC_{Y_1}\right)^2 \cdot var(X) = 0.6^2 \cdot 10^2 = 36$$

Alternatively, if Y_2 is computed from X and the result of Y_1, it is necessary to consider the covariance between X and Y_1 due to their functional relation.

$$Y_2 = X - Y_1$$

$$E(Y_2) = E(X) - E(Y_1) = 100 - 40 = 60$$

$$cov(X, Y_1) = cov\left(X, TC_{Y_1} \cdot X\right) = TC_{Y_1} \cdot var(X) = 0.4 \cdot 10^2 = 40$$

$$var(Y_2) = var(X) + var(Y_1) - 2 \cdot cov(X, Y_1) = 10^2 + 16 - 2 \cdot 40 = 36$$

∎

If the covariance is unknown, it can be estimated from pairwise statistics (i.e., statistics that are calculated between pairs of observations in a given data set) by using Equation 2.54.

$$cov(X, Y) = \frac{1}{n-1} \sum_{i=1}^{n} (x_i - \bar{x})(y_i - \bar{y}) \tag{2.54}$$

If the correlation coefficient $\rho(X, Y)$ is given instead of the covariance $cov(X, Y)$, the latter can be calculated from Equation 2.55.

$$cov(X, Y) = \rho(X, Y) \cdot \sqrt{var(X) \cdot var(Y)} \tag{2.55}$$

2.3.2.2 Monte Carlo Simulation

If the actual shape of the resulting PDF is of interest (instead of mean value and variance only), or the uncertainties of input parameters in the case of nonlinear functions are too large to use linear approximations, performing Monte Carlo simulations (MCSs) is the preferred approach to propagate uncertainties. In MCS, computer algorithms are used to create n sets of random numbers for each of the m input parameters of a function (according to their distribution). These n sets of possible values of the m input parameters lead to n possible results of the function, which are statistically evaluated (e.g., mean value, standard deviation, shape of density function). The higher the number of repetitions n, the more precise the resulting distribution function and its parameters will be.

When propagating uncertainties by applying Gauss's law, it is not necessary to make any assumption on the form of the involved PDFs. It is sufficient to know the mean value and the standard deviation of the input parameters. In MCS, however, subjective assumptions about the form of the PDF of the input parameters are required to be able to draw random numbers from it.

Example 2.9 demonstrates the application of MCS on a linear equation.

Example 2.9

In this example, Y is a linear function of two random parameters X_1 and X_2:

$$Y = X_1 - X_2$$

The input parameters are assumed to be uniformly distributed with

$$X_1 \sim Unif(20, 30)$$

$$X_2 \sim Unif(5, 10)$$

Unif(a, b) denotes a uniform distribution in the interval [a, b].

Using, e.g., the MATLAB® function *unifrnd(a,b)*, or the Microsoft Excel® function "*RAND()*(b − a) + a*", a random number is obtained that follows a uniform distribution defined by its interval [*a*, *b*]. In this way, random values for X_1 and X_2 are created. The value of Y is calculated by adding the random values of X_1 and X_2. This procedure is repeated *n* times. In the end, the mean value and the standard deviation of X_1, X_2, and Y can be computed from the values of each variable. Figure 2.29 shows the results of an MCS with 10,000 repetitions.

The higher the number of repetitions, the better the results fit the given parameters of X_1 and X_2, and in the linear case, the calculated results of Y obtained by Gauss's law. Figure 2.30 shows the results of an MCS with 1,000,000 repetitions.

All calculations were performed with MATLAB®. Excel-based alternatives are, for example, the commercial product "@risk", or MonteCarlito (Auer, 2012), a free Excel-based tool developed at TU Wien.

The expectation $E(Y)$ and its variance *var* (Y) obtained with MCS (Table 2.12) are close to the precise values, which can be computed in two ways.

Solving the problem analytically leads to $Y \sim Trap(10, 15, 20, 25)$, where *Trap(a, c, d, b)* denotes a trapezoidal distribution in the interval [*a*, *b*],

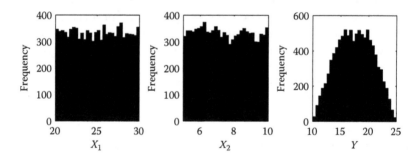

FIGURE 2.29
Results of an MCS with 10,000 repetitions.

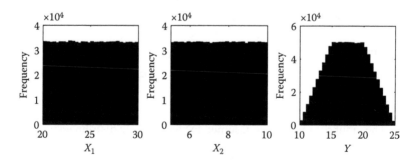

FIGURE 2.30
Results of an MSC with 1,000,000 repetitions.

TABLE 2.12

Results of MCS with 1,000,000 Repetitions

Variables	X_1	X_2	$Y = X_1 - X_2$
Mean	25.00	7.50	17.50
Standard error	0.00	0.00	0.00
Standard deviation	2.88	1.44	3.22
Variance	8.32	2.08	10.40

which is uniform in $[c, d]$. Its expectation and variance can be computed as follows (Raghu and James, 2007):

$$E(Y) = \frac{(b^2 - a^2) + (d^2 - c^2) - ac + bd}{3((b-a) + (d-c))} = 17.5$$

$$var(Y) = \frac{3(r + 2s + t)^4 + 6(r^2 + t^2)(r + 2s + t)^2 - (r^2 - t^2)^2}{(12(r + 2s + t))^2} = 10.41\dot{6}$$

with

$$r = c - a, \quad s = d - c, \quad t = b - d$$

Because Gauss's law can be applied independently of the input parameters' distribution form, it delivers the same results:

$$Y = X_1 - X_2$$

The expectation and the variance of a uniform distribution are

$$E(X_i) = \frac{1}{2}(b_i + a_i)$$

$$var(X_i) = \frac{1}{12}(b_i - a_i)^2$$

With

$$a_1 = 20, b_1 = 30$$

$$a_2 = 5, b_2 = 10$$

this leads to

$$E(X_1) = 1/2 \cdot (30 + 20) = 25$$

$$var(X_1) = 1/12 \cdot (30 - 20)^2 = 8.\dot{3}$$

$$E(X_2) = 1/2 \cdot (10 + 5) = 7.5$$

$$var(X_2) = 1/12 \cdot (10 - 5)^2 = 2.08\dot{3}$$

Because X_1 and X_2 are assumed to be independent, their covariance $cov(X_1, X_2) = 0$. Thus, the application of Gauss's law leads to

$$E(Y) = E(X_1 - X_2) = E(X_1) - E(X_2) = 25 - 7.5 = 17.5$$

$$var(Y) = var(X_1 - X_2) = var(X_1) + var(X_2) = 8.\dot{3} + 2.08\dot{3} = 10.41\dot{6}$$

2.3.3 Data Reconciliation

2.3.3.1 Introduction

Because measurements and estimates are subject to measurement or estimation errors, they virtually always violate model constraints (e.g., the law of mass conservation). The idea of data reconciliation is to statistically adjust the measurements and estimates in order to resolve contradictions and find the values that fit the model the best.

In the following, to simplify the notation, both measured and estimated data will be denoted as measured data as is common in data reconciliation literature.

Data reconciliation is feasible only if the given system of equations (i.e., constraints) can be transformed in such a way that at least one equation without unknown variables can be found. All measured and estimated variables appearing in such equations can be adjusted. Equations containing only constant variables cannot be used for data reconciliation (there are no measured or estimated variables to be reconciled) but can be exploited to perform plausibility checks on constant input data (Example 2.10).

Example 2.10

The MFA system in Figure 2.31 can be described mathematically with one balance equation (i.e., model constraint). Because the process does not contain a stock, according to the law of mass conservation, the import flow must equal the export flow. If both flows are known (so there is no unknown variable involved), and at least one of these flows is measured or estimated (with an uncertainty), data reconciliation can be performed. If both flows have constant, nonmatching values, the law of mass conservation is violated, and the data have to be checked for errors. ∎

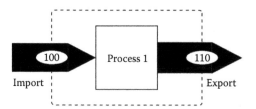

FIGURE 2.31
Example of a system with one process, one import, and one export flow for demonstrating the idea of data reconciliation.

It is often stated that in order to be able to perform data reconciliation, an overdetermined system of equations (i.e., more independent equations than unknown variables) is necessary (Narasimhan and Jordache, 2000). However, this statement is correct only if the system of equations is solvable, meaning all unknown variables can be calculated (see Example 2.11).

Example 2.11

The mathematical model of the MFA system in Figure 2.32 consists of three equations (balance of each process), three measured flows (marked with a tick), and three unknown flows (marked with a question mark).

Although the system of equations is not overdetermined (three equations, three unknown variables), data reconciliation is feasible because one of the involved equations (balance of process 1) contains measured variables only. Thus, these measured variables can be adjusted by data reconciliation. However, the system of equations is not solvable, because for the three unknown variables (flows 3, 4, and 6), only two additional equations (balances of processes 2 and 3) are left. Hence, the unknown variables cannot be calculated.

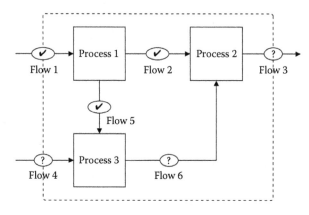

FIGURE 2.32
Example to demonstrate the requirements for data reconciliation.

If additional data for one of these flows can be found, the system of equations becomes overdetermined (three equations, two unknown variables). Data reconciliation is still possible, but this time, the unknown variables also can be calculated.   ■

By data reconciliation, the mean values of measured and estimated data are adjusted in such a way that contradictions in model constraints disappear. The best solution is found when the sum of squares of the necessary adjustments (measurement or estimate $\tilde{x}_i$ – reconciled value x_i) weighted by the inverse of the measurement or estimation variances $var(\tilde{x}_i)$ reaches a minimum:

$$F = \sum_{i=1}^{n} \frac{(\tilde{x}_i - x_i)^2}{var(\tilde{x}_i)} \tag{2.56}$$

This procedure is known as the method of weighted least squares, with F being the objective function (Equation 2.56) to be minimized.

For simplification reasons, all measured or estimated data are assumed to be normally distributed. The reason for this assumption is that in case of linear constraints, the output of data reconciliation is itself normally distributed.

In the following, some basic examples of data reconciliation are presented without considering the data reconciliation algorithm. The mathematical background will be covered in Section 2.3.3.2.

Example 2.12

The following variations of the MFA system in Example 2.10 (Figure 2.31) contain at least one measured or estimated flow, and thus, data reconciliation can be applied. Note that in this simple example, the result of data reconciliation is the weighted mean of the import and the export flow (using the inverse of their variances as weighting factors) because their values can be seen as independent measurements or estimates of the same quantity (compare to Section 2.3.1.3).

In the MFA system in Figure 2.33, the import flow is assumed to be precise, while the export flow has an absolute uncertainty of ±10. Because only uncertain flows can be reconciled, the export flow is the only one allowed to be adjusted. After data reconciliation, both flows must show the same value. Thus, the only possible solution of this problem is 100. Note that the uncertainty of the export flow has been reduced to zero because it has to match the one of the import flow.

In the MFA system in Figure 2.34, the import and the export flow have the same absolute uncertainty. This means both of them can be adjusted to the same extent to resolve the contradiction. Thus, the solution will lie in between 100 and 110. The uncertainty of the results is lower than the originally given ones.

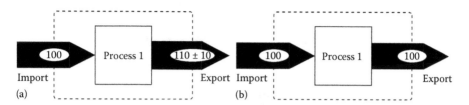

FIGURE 2.33
Data reconciliation in an MFA system with one constant (import) flow and one uncertain
(export) flow. (a) Input data. (b) Reconciled data.

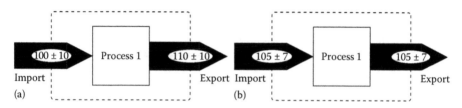

FIGURE 2.34
Data reconciliation in an MFA system with two uncertain flows, both having the same absolute
uncertainty. (a) Input data. (b) Reconciled data.

Note that the stated value ± uncertainty does not define the lower and
upper bound of an interval [value – uncertainty, value + uncertainty] but,
rather, the mean value and the standard deviation of a normal distribution. Because a normal distribution has infinite support (i.e., it stretches to
infinity in both directions), there is always an overlap of the uncertainty
regions of the given data. Thus, data reconciliation is always feasible but,
in case of gross errors, does not always make sense (see Section 2.3.3.3).

If the given data of this example are considered to be intervals instead
of normal distributions, the resulting interval is the region where the
intervals of import flow and export flow overlap. Thus, the lower bound
of the result is the maximum of the lower bounds of the input data, and
the upper bound of the result is the minimum of the upper bounds of the
input data. If the uncertainty intervals do not overlap, data reconciliation
is not feasible.

In the MFA system in Figure 2.35, the absolute uncertainty of the export
flow is twice the uncertainty of the import flow. Because the adjustments

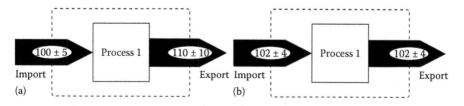

FIGURE 2.35
Data reconciliation in an MFA system with two uncertain flows, both having different absolute
uncertainties. (a) Input data. (b) Reconciled data.

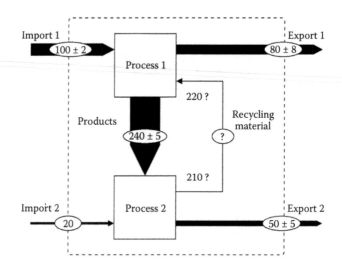

FIGURE 2.36
The MFA system to demonstrate the idea of data reconciliation in the presence of unknown flows.

are proportional to the square of these uncertainties (i.e., variance), the ratio of the adjustments is 1:4.

Example 2.13

Figure 2.36 shows an MFA system containing one unknown flow. The system can be described mathematically by two balance equations both containing the unknown flow recycling material.

The problem in this model is that for a closed balance of process 1, the flow recycling material should be 220 units; for a closed balance of process 2, the same flow should be 210 units. This contradiction can be solved with data reconciliation. Through elementary transformations of the given set of equations (adding the balance equations of processes 1 and 2), one equation can be found that describes the balance equation of the subsystem displayed as a gray rectangle in Figure 2.37.

The balance equation of this subsystem contains no unknown variables. As some of the involved variables are uncertain, they can be reconciled. Thus, the contradiction is resolved, and the mean value of recycling material can be calculated. Its uncertainty is computed by error propagation.

Note that the flow product cannot be reconciled because it is not part of the equation used for data reconciliation.

Example 2.14

In Figure 2.38, a linear data reconciliation problem with two unknown variables x and y is displayed. Each linear equation $y = k_i \cdot x + d_i$ containing these two variables can be represented by a line in a two-dimensional coordinate system.

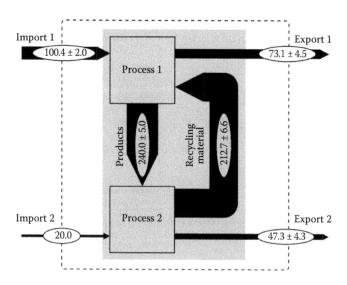

FIGURE 2.37
Data reconciliation applied to the system presented in Figure 2.36.

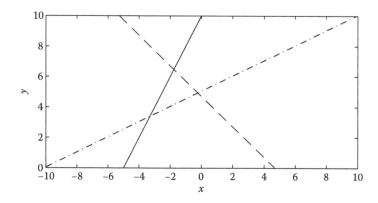

FIGURE 2.38
The three lines do not intersect in a single point. Hence, there is no solution for this system of equations with respect to x and y.

The equations of the three lines displayed in Figure 2.38 are

$$y = 2 \cdot x + 10 \text{ (solid line)}$$

$$y = -1 \cdot x + 5 \text{ (dashed line)}$$

$$y = 0.5 \cdot x + 5 \text{ (dash-dotted line)}$$

with

$$k_1 = 2, d_1 = 10; \quad k_2 = -1, d_2 = 5; \quad k_3 = 0.5, d_3 = 5$$

If the k_i are seen as constant factors and the d_i as uncertain values, the system of equations can be written in matrix form.

$$\begin{pmatrix} 1 & -2 & -1 & 0 & 0 \\ 1 & 1 & 0 & -1 & 0 \\ 1 & -0.5 & 0 & 0 & -1 \end{pmatrix} \begin{pmatrix} y \\ x \\ d_1 \\ d_2 \\ d_3 \end{pmatrix} = 0$$

After applying the Gauss–Jordan elimination to the coefficient matrix to get its reduced row echelon form, a single equation (represented by row 1 of the matrix) can be found containing d_i values only.

$$\begin{pmatrix} 1 & 0 & 0 & -0.\dot{3} & -0.\dot{6} \\ 0 & 1 & 0 & -0.\dot{6} & 0.\dot{6} \\ 0 & 0 & 1 & 1 & -2 \end{pmatrix} \begin{pmatrix} y \\ x \\ d_1 \\ d_2 \\ d_3 \end{pmatrix} = 0$$

If this equation is evaluated with the originally given mean values, a contradiction is encountered.

$$d_1 + d_2 - 2 \cdot d_3 = 0$$

$$10 + 5 - 2 \cdot 5 = 5 \neq 0$$

If the d_i values are not precise but uncertain, the originally given lines can be shifted parallel within the uncertainty ranges to find a point where all lines intersect (Figure 2.39).

The solution with the smallest sum of the squared d_i adjustments is considered the best (Figure 2.40). If all d_i are assumed to have the same absolute uncertainty, no weighting of the d_i adjustments is necessary.

If, also, the k_i are seen as uncertain variables, the lines can be tilted additionally to the parallel movement to find the intersection point that fits best (Figure 2.41). This problem can be approached with nonlinear data reconciliation.

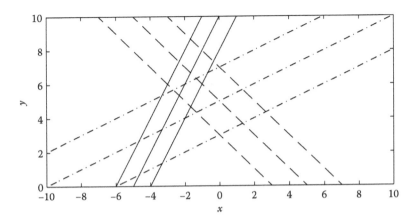

FIGURE 2.39
If d_i are uncertain, the lines can be adjusted parallel to their original position within the uncertainty range of d_i. In this example, the standard deviation of all d_i is assumed to be 1. The ranges shown in this graph are the original d_i +/- two times their standard deviation. If these uncertainty ranges overlap, the best solution can be found in this area. If there is no overlap, the original data are likely to contain large errors.

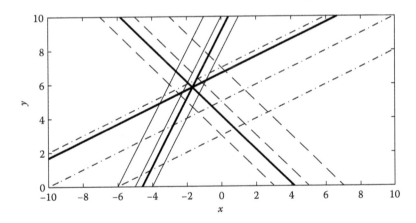

FIGURE 2.40
Solution of data reconciliation (bold lines) if all d_i have the same absolute uncertainty (linear constraints). The standard deviation of d_i is set to 1.

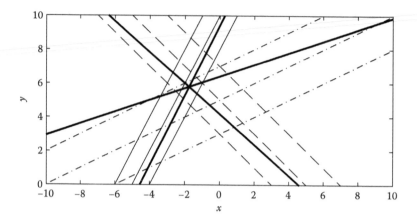

FIGURE 2.41
Solution of data reconciliation (bold lines) if d_i and k_i are uncertain (nonlinear constraints). The standard deviation of d_i is set to 1, the standard deviation of k_i to 0.25.

2.3.3.2 Mathematical Background

In the following, three types of variables are distinguished (Figure 2.42):

- Known quantities with uncertain values (direct measurements, estimates, expert guesses, literature values, etc.), denoted as measured variables
- Known quantities with fixed values (constants), denoted as constant variables
- Unknown quantities (no value available), denoted as unmeasured variables

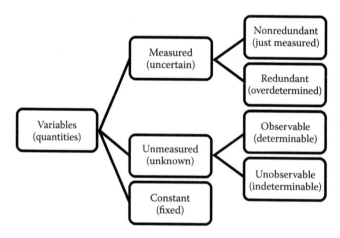

FIGURE 2.42
Classification of variables based on Madron (1992) and Romagnoli and Sánchez (2000).

Additionally, the following definitions are used (Narasimhan and Jordache, 2000):

- An unmeasured variable is said to be observable if it can be estimated by using the measurements and the model constraints. Otherwise, it is said to be unobservable.

- A measured variable is said to be redundant if it is observable when its measurement is removed. Otherwise, it is said to be non-redundant.

REMARKS: The following section contains the mathematical background of data reconciliation with linear constraints. If the reader is not so familiar with matrix notation and linear algebra, it is recommended to proceed directly to Example 2.15, where the data reconciliation procedure is demonstrated step by step. For more detailed information about the nonlinear data reconciliation algorithm implemented in the software STAN (Section 2.4), see Cencic (2016 in press).

The linear data reconciliation challenge can be formulated as a weighted least-square optimization problem by minimizing Equation 2.57 subject to Equation 2.58.

$$F(x) = (\tilde{x} - x)^T Q^{-1} (\tilde{x} - x) \qquad (2.57)$$

$$f(y, x, z) = By + Ax + Cz = (B\ A\ C) \begin{pmatrix} y \\ x \\ z \end{pmatrix} = 0 \qquad (2.58)$$

In the objective function (Equation 2.57), $\tilde{x}$ is the vector of measurements, x is the vector of reconciled measured values, and Q is the variance–covariance matrix of the measurement errors that are assumed to be normally distributed (Gaussian), with zero mean (Johnston and Kramer, 1995). The equality constraint functions (Equation 2.58) are linear combinations of unmeasured variables y (i.e., unknown), reconciled measured variables x (i.e., known with an uncertainty), and fixed or constant variables z (i.e., known without uncertainty). A, B, and C denote coefficient matrices with, in the linear case, constant entries.

If by elementary transformation of the linear set of equations (i.e., equality constraints), at least one equation can be found that contains no unknown and at least one measured variable, data reconciliation can be performed to improve the accuracy of the measurements.

Madron (1992) proposed to apply the Gauss–Jordan elimination to the linear constraint coefficient matrix $(B\ A\ C)$ in order to decouple the unmeasured variables from the data reconciliation process (Equation 2.59).

$$M = rref(B\ A\ C) = (M_y\ M_x\ M_z) = \begin{pmatrix} M_{cy} & M_{cx} & M_{cz} \\ O & M_{rx} & M_{rz} \\ O & O & M_{tz} \\ O & O & O \end{pmatrix} = \begin{pmatrix} M_c \\ M_r \\ M_t \\ O \end{pmatrix} \quad (2.59)$$

The structure of the resulting matrix M, known as the reduced row echelon form (rref) or canonical form, can also be used to classify the involved variables, detect contradictions in constant input data, and eliminate dependent equations from the constraints.

Each zero row at the bottom of M originates from a dependent equation in the originally given system of equations that is eliminated automatically during the Gauss–Jordan elimination process. Dependent equations contain no additional information about the model because they can be derived from the others. Thus, the zero rows of M do not have to be considered anymore. If there are no zero rows in M, all given equations are independent.

If $M_t \neq O$ (more specifically, $M_{tz} \neq O$) exists, meaning some of the transformed equations contain constant variables only, M_{tz} can be used to check the constant variables for contradictions because $M_{tz}z \overset{\text{def}}{=} 0$. If there are contradictions, they have to be resolved before being able to proceed. When the contradictions are resolved, M_t does not have to be considered anymore.

If M_z does not exist, the problem does not contain any constant variables. Note that in this case, there is also no M_t.

If $M_r \neq O$ (more specifically, $M_{rx} \neq O$) exists, meaning some of the transformed equations contain at least one measured but no unmeasured variable, the submatrix $(M_{rx}\ M_{rz})$ can be used to adjust the measured variables. The constraints for data reconciliation have then been reduced to a set of equations that no longer contains any unmeasured variables (Equation 2.60).

$$M_{rx}x + M_{rz}z = 0 \quad (2.60)$$

If M_r does not exist, but there is an $M_x \neq O$ (more specifically, $M_{cx} \neq O$), given measurements cannot be reconciled.

If M_x does not exist, the problem does not contain any measured variables. Note that in this case, there is also no M_r.

The solution of minimizing the objective function (Equation 2.57) subject to the now reduced set of constraints (Equation 2.60) can be derived by using the classical method of Lagrange multipliers, leading to Equation 2.61.

$$x = \tilde{x} - QM_{rx}^T \left(M_{rx} Q M_{rx}^T \right)^{-1} \left(M_{rx} \tilde{x} + M_{rz} z \right) \qquad (2.61)$$

The covariance matrix Q_x of the reconciled measured variables x can be computed by applying Gauss's law of error propagation to Equation 2.61, leading to Equation 2.62.

$$Q_x = \left(I - QM_{rx}^T \left(M_{rx} Q M_{rx}^T \right)^{-1} M_{rx} \right) Q \qquad (2.62)$$

If M_c does not exist, there are no unmeasured variables involved in the problem.

If $M_{cy} = I$ exists, all unknown variables are observable and therefore can be calculated $\left(M_{cy}^* = M_{cy} = I, \ M_{cx}^* = M_{cx}, \ M_{cz}^* = M_{cz}, \ y^* = y \right)$.

If $M_{cy} \neq I$ exists, matrix M must be altered in order to be able to calculate the observable unknown variables. Therefore, all rows in M_c that contain more than one nonzero entry in M_{cy} and all columns in M_y that have nonzero entries in these rows have to be deleted $\left(M_{cy} \rightarrow M_{cy}^* = I, \ M_{cx} \rightarrow M_{cx}^*, \ M_{cz} \rightarrow M_{cz}^* \right)$. The deleted columns of M_{cy} refer to unobservable unknown variables that also have to be removed from y ($y \rightarrow y^*$).

If M_y does not exist, the problem does not contain any unmeasured variables. Note that in this case, there is also no M_c.

After the elimination of unobservable unmeasured variables, the observable variables can be calculated from Equation 2.63.

$$Iy^* + M_{cx}^* x + M_{cz}^* z = 0 \qquad (2.63)$$

This leads to Equation 2.64.

$$y^* = -M_{cx}^* x - M_{cz}^* z \qquad (2.64)$$

The covariance matrix Q_{y^*} of the observable unmeasured variables y^* can be computed by applying Gauss's law of error propagation to Equation 2.64, leading to Equation 2.65.

$$Q_{y^*} = A_{cx}^* Q_x A_{cx}^{*T}$$

(2.65)

A review of other methods to classify variables can be found in Narasimhan and Jordache (2000) and Romagnoli and Sánchez (2000).

Example 2.15

In this example, the data reconciliation procedure is demonstrated step by step with the help of an MFA system (Figure 2.43) that consists of 4 processes and 10 mass flows. To simplify the notation, the mass flows ($\dot{m}$) are denoted by a capital letter followed by a number. The letter Y represents an unmeasured flow, X a measured flow, and Z a constant flow assumed to be known precisely.

The constraints f_i of this model without stock are the mass balances of the four processes and additionally, for demonstration reasons, the mass balance of the total system (sum of imports = sum of exports).

f_1:	$X1 + Y1 + X3 - X4 = 0$	=>	balance equation of process $P1$
f_2:	$X2 - Y1 - X3 - Z1 = 0$	=>	balance equation of process $P2$
f_3:	$X4 + Z2 - Y2 - Y3 = 0$	=>	balance equation of process $P3$
f_4:	$Z1 - Z2 - Z3 = 0$	=>	balance equation of process $P4$
f_5:	$X1 + X2 - Y2 - Y3 - Z3 = 0$	=>	balance equation of the total system

In this example, inputs into a process have a positive sign, while outputs have a negative sign.

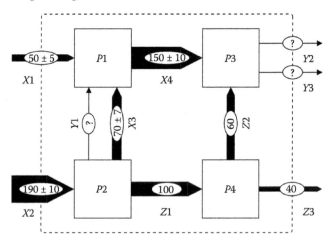

FIGURE 2.43
MFA system used to demonstrate variable classification and data reconciliation.

Written in matrix notation, the system of equations is

$$
\begin{pmatrix}
1 & 0 & 0 & 1 & 0 & 1 & -1 & 0 & 0 & 0 \\
-1 & 0 & 0 & 0 & 1 & -1 & 0 & -1 & 0 & 0 \\
0 & -1 & -1 & 0 & 0 & 0 & 1 & 0 & 1 & 0 \\
0 & 0 & 0 & 0 & 0 & 0 & 0 & 1 & -1 & -1 \\
0 & -1 & -1 & 1 & 1 & 0 & 0 & 0 & 0 & -1
\end{pmatrix}
\begin{pmatrix}
Y1 \\ Y2 \\ Y3 \\ X1 \\ X2 \\ X3 \\ X4 \\ Z1 \\ Z2 \\ Z3
\end{pmatrix}
=
\begin{pmatrix}
0 \\ 0 \\ 0 \\ 0 \\ 0 \\ 0
\end{pmatrix}
$$

Note that the entries of the variable vector are sorted in a way such that first, the unmeasured, then the measured, and finally, the constant variables are located.

Generally, this can be written as

$$
(B \ A \ C)\begin{pmatrix} y \\ x \\ z \end{pmatrix} = By + Ax + Cz = 0
$$

with

$$
B = \begin{pmatrix}
1 & 0 & 0 \\
-1 & 0 & 0 \\
0 & -1 & -1 \\
0 & 0 & 0 \\
0 & -1 & -1
\end{pmatrix}
\qquad
y = \begin{pmatrix} Y1 \\ Y2 \\ Y3 \end{pmatrix}
$$

$$
A = \begin{pmatrix}
1 & 0 & 1 & -1 \\
0 & 1 & -1 & 0 \\
0 & 0 & 0 & 1 \\
0 & 0 & 0 & 0 \\
1 & 1 & 0 & 0
\end{pmatrix}
\qquad
x = \begin{pmatrix} X1 \\ X2 \\ X3 \\ X4 \end{pmatrix}
$$

$$
C = \begin{pmatrix}
0 & 0 & 0 \\
-1 & 0 & 0 \\
0 & 1 & 0 \\
1 & -1 & -1 \\
0 & 0 & -1
\end{pmatrix}
\qquad
z = \begin{pmatrix} Z1 \\ Z2 \\ Z3 \end{pmatrix}
$$

Applying the Gauss–Jordan elimination to coefficient matrix $(B\ A\ C)$, leads to

$$M = rref(B\ A\ C) = \begin{pmatrix} 1 & 0 & 0 & 0 & -1 & 1 & 0 & 0 & 1 & 1 \\ 0 & 1 & 1 & 0 & 0 & 0 & -1 & 0 & -1 & 0 \\ 0 & 0 & 0 & 1 & 1 & 0 & -1 & 0 & -1 & -1 \\ 0 & 0 & 0 & 0 & 0 & 0 & 0 & 1 & -1 & -1 \\ 0 & 0 & 0 & 0 & 0 & 0 & 0 & 0 & 0 & 0 \end{pmatrix}$$

The corresponding system of equations is

$$\begin{pmatrix} 1 & 0 & 0 & 0 & -1 & 1 & 0 & 0 & 1 & 1 \\ 0 & 1 & 1 & 0 & 0 & 0 & -1 & 0 & -1 & 0 \\ 0 & 0 & 0 & 1 & 1 & 0 & -1 & 0 & -1 & -1 \\ 0 & 0 & 0 & 0 & 0 & 0 & 0 & 1 & -1 & -1 \\ 0 & 0 & 0 & 0 & 0 & 0 & 0 & 0 & 0 & 0 \end{pmatrix} \begin{pmatrix} Y1 \\ Y2 \\ Y3 \\ X1 \\ X2 \\ X3 \\ X4 \\ Z1 \\ Z2 \\ Z3 \end{pmatrix} = \begin{pmatrix} 0 \\ 0 \\ 0 \\ 0 \\ 0 \end{pmatrix}$$

The transformed equations f_i^* written in regular notation are as follows:

f_1^*:	$Y1 - X2 + X3 + Z2 + Z3 = 0$	=>	balance equation of process P2 + P4
f_2^*:	$Y2 + Y3 - X4 - Z2 = 0$	=>	balance equation of process P3
f_3^*:	$X1 + X2 - X4 - Z2 - Z3 = 0$	=>	balance equation of process P1 + P2 + P4
f_4^*:	$Z1 - Z2 - Z3 = 0$	=>	balance equation of process P4
f_5^*:	$0 = 0$	=>	one redundant equation was eliminated!

REMARKS: due to the performed transformations, in equations f_1^* and f_2^*, inputs of a process or a collection of processes have a negative sign, while outputs have a positive sign.

The equations f_1^* to f_5^* have the following features:

Equation f_5^* has been eliminated during the Gauss–Jordan elimination because the originally given system of equations contains only four independent equationsl e.g., the balance equation of the total system can be derived by adding the balance equations of the single processes. Hence, the total balance of the system represents redundant information and could be removed from the initially given system of equations without loss of information if the balances of all processes are included. Equation f_5^* does not have to be considered any more.

Equation f_4^* contains only fixed variables. It is used to verify the values of the fixed variables. If there is a conflict, the values of the fixed variables have to be checked for potential errors. In this example, there is a contradiction because $Z1 - Z2 - Z3 = 10 - 60 - 40 = -90 \neq 0$. It is assumed that upon checking the data, the value of $Z1$ is detected to be incorrectly given as 10 instead of 100. When the value is corrected, the contradiction is resolved, and equation f_4^* does not have to be considered anymore.

Equation f_3^* contains measured and fixed variables only. It can be used to adjust the involved measured variables that usually, due to measurement errors, do not fulfill the constraints. Because the measured variable $X3$ is not included in this equation, it will not be adjusted and is said to be nonredundant.

Only equations f_1^* and f_2^* contain unmeasured variables. Equation f_1^* can be used to calculate $Y1$ because $Y1$ is the only unknown variable in this equation. Equation f_2^* contains two unmeasured variables ($Y2$ and $Y3$) and thus cannot be used to calculate either of them (there is only one equation for two unknown variables). Therefore, equation f_2^* and the variables $Y2$ and $Y3$ are eliminated from the equation system. The variables $Y2$ and $Y3$ are classified as unobservable.

After eliminating the unobservable variables $Y2$ and $Y3$ as well as the equations f_2^*, f_4^*, and f_5^*, the remaining system of equations is

$$\left(\begin{array}{cc|cccc|cc} 1 & 0 & -1 & 1 & 0 & 0 & 1 & 1 \\ 0 & 1 & 1 & 0 & -1 & 0 & -1 & -1 \end{array}\right) \begin{pmatrix} Y1 \\ \hline X1 \\ X2 \\ X3 \\ X4 \\ \hline Z1 \\ Z2 \\ Z3 \end{pmatrix} = \begin{pmatrix} 0 \\ 0 \end{pmatrix}$$

which can be written generally as

$$\begin{pmatrix} I & D_1 & E_1 \\ O & D_2 & E_2 \end{pmatrix} \begin{pmatrix} y^* \\ x \\ z \end{pmatrix} = \begin{pmatrix} 0 \\ 0 \end{pmatrix}$$

with

    $I = (1)$ ... identity matrix,

    $O = (0)$ ... zero matrix,

    $0 = (0)$ ... zero vector,

$$D_1 = (0 \;\; -1 \;\; 1 \;\;\;\; 0), \quad E_1 = (0 \;\; 1 \;\; 1)$$
$$D_2 = (1 \;\;\;\; 1 \;\; 0 \;\; -1), \quad E_2 = (0 \;\; 1 \;\; 1)$$
$$y^* = (Y1)$$

The reconciled values of the measured variables x and their variance–covariance matrix Q_x are calculated according to Equations 2.61 and 2.62 from

$$x = \tilde{x} - QD_2^T \left(D_2 Q D_2^T\right)^{-1} (D_2\tilde{x} + E_2 z),$$

$$Q_x = \left(I - QD_2^T \left(D_2 Q D_2^T\right)^{-1} D_2\right)Q$$

With

$$\tilde{x} = \begin{pmatrix} \tilde{X}1 \\ \tilde{X}2 \\ \tilde{X}3 \\ \tilde{X}4 \end{pmatrix} = \begin{pmatrix} 50 \\ 190 \\ 70 \\ 150 \end{pmatrix} \quad Q = \begin{pmatrix} 5^2 & 0 & 0 & 0 \\ 0 & 10^2 & 0 & 0 \\ 0 & 0 & 7^2 & 0 \\ 0 & 0 & 0 & 10^2 \end{pmatrix}$$

$$z = \begin{pmatrix} Z1 \\ Z2 \\ Z3 \end{pmatrix} = \begin{pmatrix} 100 \\ 60 \\ 40 \end{pmatrix}$$

follows

$$x = \begin{pmatrix} X1 \\ X2 \\ X3 \\ X4 \end{pmatrix} = \begin{pmatrix} 51.1 \\ 194.4 \\ 70.0 \\ 145.6 \end{pmatrix} \quad Q_x = \begin{pmatrix} 22.2 & -11.1 & 0 & 11.1 \\ -11.1 & 55.6 & 0 & 44.4 \\ 0 & 0 & 49.0 & 0 \\ 11.1 & 44.4 & 0 & 55.6 \end{pmatrix}$$

The observable unknown variables y^* and their variance–covariance matrix Q_{y^*} are calculated according to Equations 2.64 and 2.65 from

$$y^* = -D_1 x - E_1 z$$

$$Q_{y^*} = D_1 Q_x D_1^T$$

leading to

$$y^* = (Y1) = (24.4) \quad Q_y = (104.6).$$

Taking only the variances from the variance–covariance matrices into account, the results are (Figure 2.44)

$$X1 = 51.1 \pm \sqrt{22.2} = 51.1 \pm 4.7$$

$$X2 = 194.4 \pm \sqrt{55.6} = 194.4 \pm 7.5$$

$$X3 = 70.0 \pm \sqrt{49.0} = 70.0 \pm 7.0$$

$$X4 = 145.6 \pm \sqrt{55.6} = 145.6 \pm 7.5$$

$$Y1 = 24.4 \pm \sqrt{104.6} = 24.4 \pm 10.2$$

REMARKS: While the uncertainties of the measured variables were initially assumed to be uncorrelated (Q contains only entries in the diagonal), the uncertainties of the reconciled variables are correlated due to the constraints (Q_x contains also nonzero entries off the diagonal).

The measurement of flow X3 is not reconciled, because it is included in the balance equation that is used for data reconciliation. Thus, X3 remains unchanged and is still not correlated with other measured variables (see 0 entries in Q_x). ■

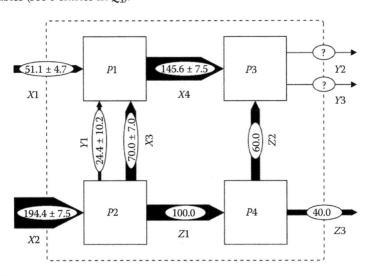

FIGURE 2.44
Results computed from available input data. To establish the complete MFA, information about one of the unknown flows is required.

To see how measured values are adjusted during data reconciliation, the procedure is exemplified for a single linear equation without matrix notation (Equation 2.66).

$$\sum_{i=1}^{n}(a_i \cdot x_i) + c = 0 \qquad (2.66)$$

a_i is a constant coefficient, and c is a constant term (e.g., the sum of constant parameters).

Because the individual measurements $\tilde{x}_i$ are subject to measurement errors, the constraint is usually not fulfilled, i.e., there is a residual r (Equation 2.67).

$$\sum_{i=1}^{n}(a_i \cdot \tilde{x}_i) + c = r \qquad (2.67)$$

The necessary measurement adjustment Δ_{x_i}, derived from Equation 2.61, can be computed from Equations 2.68 and 2.69.

$$\Delta_{x_i} = \frac{a_i \cdot var(\tilde{x}_i) \cdot r}{\sum_{i=1}^{n} a_i^2 \cdot var(\tilde{x}_i)} \qquad (2.68)$$

$$\Delta_{x_i} \propto a_i \cdot var(\tilde{x}_i) \qquad (2.69)$$

The reconciled measurements x_i (Equation 2.70) finally fulfill Equation 2.66, meaning there are no residuals left.

$$x_i = \tilde{x}_i - \Delta_{x_i} \qquad (2.70)$$

The variance of the reconciled measurements x_i can be computed with Equation 2.71 or Equation 2.72.

$$var(x_i) = var(\tilde{x}_i) - \frac{a_i^2 \cdot var(\tilde{x}_i)^2}{\sum_{j=1}^{n} a_j^2 \cdot var(\tilde{x}_j)} \qquad (2.71)$$

$$var(x_i) = var(\tilde{x}_i) \cdot \frac{\sum_{j=1,i \neq j}^{n} a_j^2 \cdot var(\tilde{x}_j)}{\sum_{j=1}^{n} a_j^2 \cdot var(\tilde{x}_j)} \qquad (2.72)$$

If all $var(\tilde{x}_i)$ have the same absolute value, Equations 2.68 and 2.72 simplify to Equations 2.73 and 2.74.

$$\Delta_{x_i} = \frac{a_i \cdot r}{\displaystyle\sum_{i=1}^{n} a_i^2} \alpha\; a_i \tag{2.73}$$

$$var(x_i) = var(\tilde{x}_i) \cdot \frac{\displaystyle\sum_{j=1,i\neq j}^{n} a_j^2}{\displaystyle\sum_{j=1}^{n} a_j^2} \tag{2.74}$$

Example 2.16

In the following equation (constraint)

$$A + B - 2 \cdot C = 0$$

the coefficients a_i of the variables are

$$a_A = 1,\; a_B = 1,\; a_C = -2$$

All variables are assumed to be measured with

$$E(\tilde{A}) = 100,\; E(\tilde{B}) = 200,\; E(\tilde{C}) = 135,\; var(\tilde{A}) = var(\tilde{B}) = var(\tilde{C}) = 10^2$$

The mean values of the measurements do not fulfill the constraint. Thus, there is a residual.

$$r = 1 \cdot 100 + 1 \cdot 200 - 2 \cdot 135 = 30$$

Because all measurement uncertainties have the same absolute value, the necessary adjustments are given with

$$\Delta_{x_i} = a_i \cdot \frac{30}{1^2 + 1^2 + (-2)^2} = a_i \cdot 5$$

$$\Delta_A = 1 \cdot 5 = 5 \rightarrow E(A) = E(\tilde{A}) + \Delta_A = 100 - 5 = 95$$

$$\Delta_B = 1 \cdot 5 = 5 \rightarrow E(B) = E(\tilde{B}) + \Delta_B = 200 - 5 = 195$$

$$\Delta_C = -2 = -10 \rightarrow E(C) = E(\tilde{C}) + \Delta_C = 135 - 10 = 145$$

Note that even if the original variance is the same, the mean values will be adjusted proportional to their coefficients.

After data reconciliation, the residuals computed from the reconciled values are 0.

$$r = 1 \cdot 95 + 1 \cdot 195 - 2 \cdot 145 = 0$$

The variances of the reconciled values are

$$var(A) = var(\tilde{A}) \cdot \frac{\left(a_B^2 + a_c^2\right)}{\left(a_A^2 + a_B^2 + a_c^2\right)} = 10^2 \cdot \frac{5}{6} = 9.1^2$$

$$var(B) = var(\tilde{B}) \cdot \frac{\left(a_A^2 + a_c^2\right)}{\left(a_A^2 + a_B^2 + a_c^2\right)} = 10^2 \cdot \frac{5}{6} = 9.1^2$$

$$var(C) = var(\tilde{C}) \cdot \frac{\left(a_A^2 + a_B^2\right)}{\left(a_A^2 + a_B^2 + a_c^2\right)} = 10^2 \cdot \frac{2}{6} = 5.8^2$$

2.3.3.3 Gross Error Detection

If the input data contain gross errors, e.g., errors that cause a measurement to be far from the expected result, the results of data reconciliation are biased (see Example 2.17).

Example 2.17

In Figure 2.45, the import and the export flow of the MFA system have the same absolute uncertainty, but the measurement of the import flow contains a gross error (1000 instead of 100). Applying data reconciliation to the problem, the solution is found to be the average of the import flow

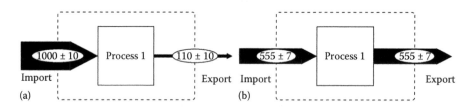

FIGURE 2.45
Data reconciliation in an MFA system with two uncertain flows, both having the same absolute uncertainty, the measurement of the import flow containing a gross error. (a) Input data. (b) Reconciled data.

and the export flow (as in Figure 2.34). However, compared to the originally given uncertainties, the adjustments are much too large. ∎

To detect gross errors as the one in Figure 2.45, a basic statistical test can be performed. The idea is to check if the reconciled values are in the interval that covers 95% of the assumed normal distribution of the originally given data ($= \tilde{x}_i \pm 1.96 \cdot ste(\tilde{x}_i)$). If this check fails, the input data are likely to contain gross errors.

If the model is free of gross errors, an overall data reconciliation quality *DRQ* ranging between 0 (bad) and 1 (very good) can be computed using the same idea (Equations 2.75 through 2.77).

$$D_i = \left| x_i - \tilde{x}_i \right| \tag{2.75}$$

$$D_{i,\max} = 1.96 \cdot S_{\tilde{x}_i} \tag{2.76}$$

$$DRQ = 1 - \frac{1}{n} \cdot \sum_{i=1}^{n} \frac{D_i}{D_{i,\max}} = 1 - \frac{1}{n \cdot 1.96} \cdot \sum_{i=1}^{n} \frac{\left| x_i - \tilde{x}_i \right|}{S_{\tilde{x}_i}} \tag{2.77}$$

D_i is the absolute value of a measurement adjustment. $D_{i,\max}$ is the maximum allowed adjustment, which is assumed to be ±1.96 times the standard deviation of the measurement. *DRQ* is the difference of mean value of all $D_i/D_{i,\max}$ to 1.

Other more sophisticated gross error detection tests can be found in Narasimhan and Jordache (2000).

2.3.3.4 Nonnormally Distributed Data

The measured variables of a model can be seen as a multivariate random variable. Its PDF, denoted as the joint prior probability distribution, covers the complete variable space, i.e., for all combinations of variable values, a density is given. The main idea of data reconciliation is to restrict the joint prior probability distribution with model constraints to get the joint posterior probability distribution, i.e., the resulting PDF covers only those combinations of variable values that are feasible with respect to the model constraints. The joint posterior probability distribution is then used to compute the reconciled PDFs of the individual measured variables (Cencic and Frühwirth, 2015). Note that this procedure can be visualized for two- and three-dimensional problems only. In Example 2.18, a two-dimensional problem is presented.

Example 2.18

Two independent discrete variables A and B are given by their probability mass functions (PMFs). In case of independence, their joint prior probability distribution (Figure 2.46) can be computed from

$$P(A, B) = P_A(A) \cdot P_B(B)$$

If it is known that A equals B (i.e., the model constraint is $A = B$), only those combinations of A and B are feasible that are situated on the diagonal of plane A-B. Normalizing the probabilities along $A = B$, the joint posterior probability distribution is determined. Computing the marginal distributions of A and B, the reconciled PMFs of A and B (Figure 2.47) are found. ∎

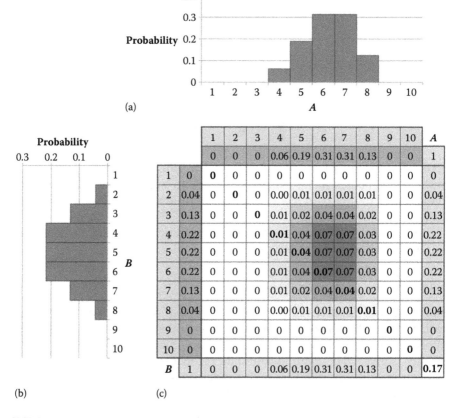

FIGURE 2.46

(a) Prior probability distribution of A. (b) Prior probability distribution of B. (c) Joint probability distribution of A and B (= prior) under the assumption of independence. If the joint probability distribution is cut along the diagonal ($A = B$), the squares in the diagonal (with bold font) represent the nonnormalized joint posterior probability distribution. It is nonnormalized because the sum of the squares in the diagonal is not 1.

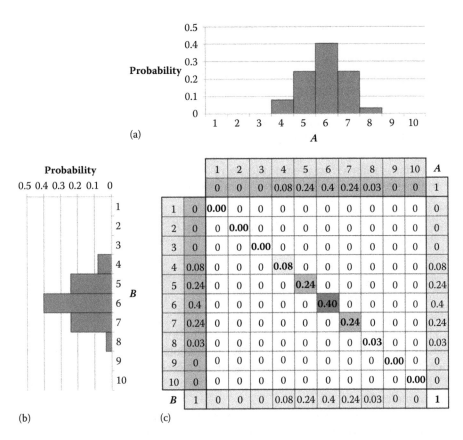

FIGURE 2.47
(a) Posterior probability distribution of *A*. (b) Posterior probability distribution of *B*. (c) Normalized joint posterior probability distribution. The sum of the squares in the diagonal is now 1.

In the case of normally distributed prior PDFs and linear (or linearized) constraints, the posterior PDFs themselves are also normally distributed. That is the reason why the assumption of normality is used in the classical data reconciliation approach as presented in Section 2.3.3.2.

Detailed instructions on how to perform data reconciliation with non-normally distributed data in the case of linear constraints can be found in Cencic and Frühwirth (2015). The case of nonlinear constraints will be covered in future work.

2.3.4 Sensitivity Analysis

Sensitivity analysis explores the effect of input variations on the output of a mathematical model. It can be distinguished between local and global approaches of sensitivity analysis. So far, mainly local sensitivity analyses have been used in MFA, where the importance of factors is investigated by the derivative of the output with respect to that factor (Saltelli and Annoni, 2010). In this type of analysis, all derivatives are taken at a single point (development point), which is why only a small part of the possible parameter space is explored. On the contrary, global sensitivity analysis is concerned with exploring the effect of input variation on the model results for plausible ranges of the parameter space. It aims at apportioning the uncertainty of the output to the different sources of uncertainty in the inputs (Saltelli et al., 2008). In this chapter, a simple analytical approach for local sensitivity analysis is presented. For more information on elaborate forms of sensitivity analysis, the reader is referred to specific literature (Saltelli et al., 2008).

In this section, the following two local sensitivity measures are covered:

$$\text{Absolute sensitivity: } S_{X_j}^{Y_i} = \frac{\Delta Y_i(X_i)}{\Delta X_j} \tag{2.78}$$

$$\text{Relative sensitivity: } \overline{S}_{X_j}^{Y_i} = \frac{\Delta Y_i(X_i)}{\Delta X_j} \cdot \frac{X_j}{Y_i} = S_{X_j}^{Y_i} \cdot \frac{X_j}{Y_i} \tag{2.79}$$

The absolute sensitivity coefficient $S_{X_j}^{Y_i}$ (Equation 2.78) shows the absolute change ΔY_i of variable Y_i in response to an absolute change $\Delta X_j = 1$ of parameter X_j. The relative sensitivity $\overline{S}_{X_j}^{Y_i}$ (Equation 2.79) shows the relative change $\Delta Y_i/Y_i$ of the variable Y_i in response to a relative change $\Delta X_j/X_j = 1$ (= 100%) of parameter X_j.

In many nonlinear cases, it is sufficient to calculate the partial derivative $\partial Y/\partial X$ instead of the exact value of $\Delta Y/\Delta X$ (Baccini and Bader, 1996).

Within a sensitivity analysis, all input parameters are altered systematically. The procedure can be visualized as a machine with wheels to adjust the parameters. By turning one wheel, while all others remain fixed, the sensitivity of the result concerning this parameter can be identified.

Example 2.19

The output flow Y_1 of the process in Figure 2.48 is a certain percentage of the input flow X_1 (given by the transfer coefficient TC_1) and a certain

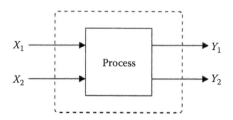

FIGURE 2.48
MFA system with two import flows and two export flows.

percentage of the input flow X_2 (given by the transfer coefficient TC_2). The transfer coefficients are considered to be constant (Figure 2.48).

Question: What is the absolute sensitivity of Y_1 with respect to X_1 and X_2?

Because TC_1 and TC_2 are assumed to be constant, Y_1 is a linear function of X_1 and X_2.

$$Y_1 = f(X_1, X_2) = TC_1 \cdot X_1 + TC_2 \cdot X_2$$

with

$$TC_1 = 0.5, \quad TC_2 = 0.3, \quad X_1 = 50, \quad X_2 = 200$$

$$S_{X_1}^{Y_1} = \frac{\Delta Y_1}{\Delta X_1} = \frac{\partial Y_1}{\partial X_1} = TC_1 = 0.5$$

$$S_{X_2}^{Y_1} = \frac{\Delta Y_1}{\Delta X_2} = \frac{\partial Y_1}{\partial X_2} = TC_2 = 0.3$$

The partial derivative (here the transfer coefficient) with the largest gradient shows the parameter to which the result is most sensitive. Thus, in the previous example, a change in X_1 has a greater effect on the result of Y_1 than a change in X_2.

Example 2.20

For the MFA system in Figure 2.49, the following question shall be answered: what impact on the consumption of raw materials RA has a

   a. 10% reduction of product PR?
   b. 10% increase of recycling rate RR?

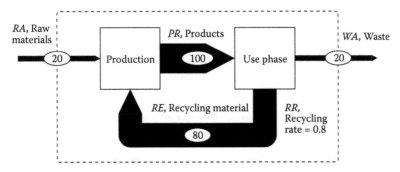

FIGURE 2.49
MFA system with two processes to demonstrate sensitivity analysis.

The MFA system can be described completely with the following three equations:

f_1:	$RA + RE = PR$	=>	balance of process production
f_2:	$PR = WA + RE$	=>	balance of process consumption
f_3:	$RE = RR \cdot PR$	=>	relation of recycling material to product flow

The system of equations has to be transformed to be able to compute the unknown variables as functions $f(PR, RR)$ of the chosen parameter PR and RR. From equation f_1 and f_3 follows

$$RA = PR \cdot (1 - RR)$$

Because the parameter PR and RR are assumed to be variable, $RA = PR \cdot (1 - RR)$ is a nonlinear equation. Thus, the exact value of $\Delta Y/\Delta X$ is approximated by the partial derivative $\partial Y/\partial X$ at the expansion point RR, PR.
Answer (a):
The relative sensitivity coefficient of RA with respect to parameter PR is

$$\bar{S}_{PR}^{RO} \approx \left.\frac{\partial RA}{\partial PR}\right|_{RR,PR} \cdot \frac{PR}{RA} = (1-RR)\cdot\frac{PR}{RA} = (1-0.8)\cdot\frac{100}{20} = 1$$

Thus, the relative change of RA in response to a 10% reduction of PR is

$$\frac{dRA}{RA} \approx \bar{S}_{PR}^{RO}\cdot\frac{dPR}{PR} = 1\cdot(-0.1) = -0.1 \rightarrow -10\%$$

That is, a 10% reduction of product PR causes a 10% reduction of the required raw materials RA.
Answer (b):
The relative sensitivity coefficient of RA with respect to parameter RR is

$$\bar{S}_{RR}^{RA} \approx \left.\frac{\partial RA}{\partial RR}\right|_{RR,PR} \cdot \frac{RR}{RA} = -PR\cdot\frac{RR}{RA} = -100\cdot\frac{0.8}{20} = -4$$

Thus, the relative change of RA in response to a 10% increase of RR is

$$\frac{dRA}{RA} \approx \bar{S}_{RR}^{RA}\cdot\frac{dRR}{RA} = -4\cdot0.1 = -0.4 \rightarrow -40\%$$

That is, a 10% increase of the recycling rate RR causes a 40% reduction of the required raw materials RA.

The exact solutions with $RA = f_{RA}(PR, RR) = PR \cdot (1 - RR)$ are

$$\frac{dRA}{RA} = \frac{f_{RA}((1-0.1)\cdot PR, RR) - f_{RA}(PR, RR)}{f_{RA}(PR, RR)} = \frac{18-20}{20} = -0.1$$

$$\frac{dRA}{RA} = \frac{f_{RA}(PR,(1+0.1)\cdot RR) - f_{RA}(PR, RR)}{f_{RA}(PR, RR)} = \frac{12-20}{20} = -0.4$$

In this case (multiplication of variables), the approximate solution is identical to the exact solution: the partial derivation with respect to one of the variables is a function of the other variable only, which is assumed to be fixed. Thus, the sensitivity coefficients are constant.

$$S_{PR}^{RA} = \frac{\partial RA}{\partial PR} = \frac{\partial f_{RA}(PR, RR)}{\partial PR} = g(RR) = (1 - RR)$$

$$S_{RR}^{RA} = \frac{\partial RA}{\partial RR} = \frac{\partial f_{RA}(PR, RR)}{\partial RR} = h(PR) = -PR$$

2.3.5 Examples

Example 2.21

The total mass of end-of-life vehicles (ELVs) shall be estimated from the reported number of ELVs (assumed to be constant) and the estimated mass of an ELV.

1. Number of ELVs: $N = 10,000$ ELV/year
2. Mass of one ELV: $m = 1,500 \pm 250$ kg [$= \bar{m} \pm s_m$ = normally distributed with $\mathcal{N}(\bar{m}, s_m^2)$]

The mean value of the total mass can be calculated from

$$\bar{m}_{tot} = \bar{m}_1 + \bar{m}_2 + \ldots + \bar{m}_N = N \cdot \bar{m} = 10,000 \cdot 1,500 = 15,000,000 \text{ kg} = 15,000 \text{ t}$$

The standard deviation of total mass can be estimated by applying Gauss's law to the previous equation

$$s_{m_{tot}} = \sqrt{s_{m_1}^2 + s_{m_2}^2 + \ldots + s_{m_N}^2} = \sqrt{N} \cdot s_m = \sqrt{10,000} \cdot 250 = 25,000 \text{ kg} = 25 \text{ t}$$

The results can be verified by a MCS (Table 2.13).

TABLE 2.13

Results of MCS (Correct Approach)

MCS	m_1	m_2	...	$m_{10,000}$	m_{tot}
1	1,020 kg	1,467 kg	...	1,307 kg	14,992 t
2	1,431 kg	1,706 kg	...	2,371 kg	15,021 t
...	...	...	...	...	...
∞	1,304 kg	1,621 kg	...	1,245 kg	14,970 t
$\bar{m}$	1,500 kg	1,500 kg	...	1,500 kg	15,000 t
s_m	250 kg	250 kg	...	250 kg	25 t

TABLE 2.14

Results of MCS (Wrong Approach)

MCS	m_1	m_2	...	$m_{10,000}$	m_{tot}
1	1,025 kg	1,025 kg	...	1,025 kg	10,250 t
2	1,683 kg	1,683 kg	...	1,683 kg	16,830 t
...	...	...	...	...	...
∞	1,554 kg	1,554 kg	...	1,554 kg	15,540 t
$\bar{m}$	1,500 kg	1,500 kg	...	1,500 kg	15,000 t
s_m	250 kg	250 kg	...	250 kg	2,500 t

Note that a wrong solution is obtained if the standard deviation of total mass is calculated by applying Gauss's law directly to $\bar{m}_{tot} = N \cdot \bar{m}$:

$$s_{m_{tot}} = \sqrt{N^2 \cdot s_m^2} = N \cdot s_m = 10,000 \cdot 250 = 2,500,000 \text{ kg} = 2,500 \text{ t}$$

In this approach, all ELVs are assumed to have the same unknown mass, which is wrong (see results of MCS in Table 2.14).

When multiplying a number with an uncertain variable, the correct uncertainty of the result is obtained only if the standard error of the mean of this variable is used.

In the presented example, the correct result can be obtained if the 10,000 ELVs are considered to be a sample of a larger population of ELVs. Then the standard error of the mean mass of an ELV in this population can be calculated from

$$s_{\bar{m}} = \frac{s_m}{\sqrt{N}} = \frac{250}{\sqrt{10,000}} = 2.5 \text{ kg}$$

leading to

$$s_{m_{tot}} = \sqrt{N^2 \cdot s_{\bar{m}}^2} = N \cdot s_{\bar{m}} = 10,000 \cdot 2.5 = 25,000 \text{ kg} = 25 \text{ t}$$

Example 2.22

The following research questions are to be investigated:

1. How much Cu is in a car of arbitrary mass?
2. How much Cu is in a car of average mass?
3. How much Cu is there on average in a car of arbitrary mass?
4. How much Cu is there on average in a car of average mass?

While all these questions sound similar, they represent different perspectives on uncertainty.

The random variable G (short for *good*) describes the mass of a car. Because the parameters μ_G and σ_G of its theoretically precise PDF are not known, they are estimated from a sample of size n_G:

$$\hat{\mu}_G = \bar{g} = \frac{1}{n_G} \sum_{i=1}^{n} g_i$$

$$\hat{\sigma}_G^2 = s_G^2 = \frac{1}{n_G - 1} \sum_{i=1}^{n_G} (g_i - \bar{g})^2$$

The random variable $\bar{G}$ describes the average mass of a car. Its parameters can be estimated from

$$\hat{\mu}_{\bar{G}} = \hat{\mu}_G = \bar{g}$$

$$\hat{\sigma}_{\bar{G}}^2 = var(\hat{\mu}_G) = \frac{\hat{\sigma}_G^2}{n_G} = \frac{s_G^2}{n_G}$$

G is subject to aleatory variability because the population contains elements (cars) with different mass. $\bar{G}$ is subject to epistemic uncertainty only because there is just one true value of the average mass of a car in a given population, which cannot be determined precisely.

The random variable C (short for *concentration*) describes the concentration of Cu in a car of arbitrary mass. Because the parameters μ_C and σ_C of its theoretically precise PDF are not known, they are estimated from a sample of size n_C:

$$\hat{\mu}_C = \bar{c} = \frac{1}{n_C} \sum_{i=1}^{n} c_i$$

$$\hat{\sigma}_C^2 = s_C^2 = \frac{1}{n_C - 1} \sum_{i=1}^{n_C} (c_i - \bar{c})^2$$

The random variable $\bar{C}$ describes the average concentration of Cu in a car of arbitrary mass. Its parameters can be estimated from

$$\hat{\mu}_{\bar{C}} = \hat{\mu}_C = \bar{c}$$

$$\hat{\sigma}_{\bar{C}}^2 = var(\hat{\mu}_C) = \frac{\hat{\sigma}_C^2}{n_C} = \frac{s_C^2}{n_C}$$

C is subject to aleatory variability because the population contains elements (cars) with different copper concentration. $\bar{C}$ is subject to epistemic uncertainty only because there is just one true value of the average copper concentration of a car in a given population, which cannot be determined precisely.

It is further assumed that there is no correlation between the mass of a car and its copper concentration.

With the random variable S_i (short for *substance*), the aforementioned questions can be reformulated:

1. How much Cu is in a car of arbitrary mass? $\rightarrow S_1 = G \cdot C$
2. How much Cu is in a car of average mass? $\rightarrow S_2 = \bar{G} \cdot C$
3. How much Cu is there on average in a car of arbitrary mass? $\rightarrow S_3 = G \cdot \bar{C}$
4. How much Cu is there on average in a car of average mass? $\rightarrow S_4 = \bar{G} \cdot \bar{C}$

In all cases, the location parameter $\hat{\mu}_{S_i}$ can be estimated by

$$\hat{\mu}_{S_i} = \bar{g} \cdot \bar{c}$$

However, the size of the scale parameter $\hat{\sigma}_{S_i}^2$ (computed with error propagation) is different:

1. $\hat{\sigma}_{S_1}^2 \approx \bar{g}^2 \cdot \hat{\sigma}_C^2 + \bar{c}^2 \cdot \hat{\sigma}_G^2 = \bar{g}^2 \cdot s_C^2 + \bar{c}^2 \cdot s_G^2$
2. $\hat{\sigma}_{S_2}^2 \approx \bar{g}^2 \cdot \hat{\sigma}_C^2 + \bar{c}^2 \cdot \hat{\sigma}_{\bar{G}}^2 = \bar{g}^2 \cdot s_C^2 + \bar{c}^2 \cdot s_G^2 / n_G$
3. $\hat{\sigma}_{S_3}^2 \approx \bar{g}^2 \cdot \hat{\sigma}_{\bar{C}}^2 + \bar{c}^2 \cdot \hat{\sigma}_G^2 = \bar{g}^2 \cdot s_C^2 / n_C + \bar{c}^2 \cdot s_G^2$
4. $\hat{\sigma}_{S_3}^2 \approx \bar{g}^2 \cdot \hat{\sigma}_{\bar{C}}^2 + \bar{c}^2 \cdot \hat{\sigma}_{\bar{G}}^2 = \bar{g}^2 \cdot \dfrac{s_C^2}{n_C} + \bar{c}^2 \cdot \dfrac{s_G^2}{n_G}$

2.4 MFA with Software STAN

Oliver Cencic

The use of appropriate software to support MFA has many advantages: graphical modelling with automatic translation into a mathematical model; database of collected data including source documentation; consideration of data uncertainties; reconciliation of contradicting data; computation of unknown quantities of mass flows and stocks; computerized display of resulting flows as Sankey-style diagrams, and others. Thus, several software packages have been developed to facilitate MFA, particularly on the level of goods (Eyerer, 1996; Schmidt and Häuslein, 1997). For a short comparison of Umberto 4.0, GaBi 4, and Microsoft Excel, see Cencic (2004). On the level of substances, only a few products (SIMBOX, STAN) exist that are able to combine flows and stocks of goods with those of substances (Baccini and Bader, 1996; Cencic and Rechberger, 2008). In the following, the application of the software STAN, a development of Technische Universität Wien and Inka software (Cencic, Rechberger and Kovacs, 2006) is described in detail. The reason for selecting STAN is that—with more than 13,000 users registering between October 2006 and March 2016—this freeware is the most widely used software tool for performing MFA on the level of both goods and substances.

STAN, short for subSTance flow ANalysis, is a freely available software (http://www.stan2web.net) that helps to perform MFA according to the Austrian standard ÖNORM S 2096, Material flow analysis—Application in waste management (Austrian Standard, 2005). It offers the following features that are tailor-made for MFA users:

- Predefined components (process, flow, system boundary, text field) can be used to build a graphical model that is allowed to contain subsystems.
- Different data types can be handled: flows, stocks, concentrations, as well as transfer coefficients of masses for multiple layers (goods, substances, and energy) and periods. Additionally, on the layer of goods, volume flows and stocks, and densities can be considered.
- Data can be entered manually or imported semiautomatically from Microsoft Excel.
- Predefined or user defined units (for mass, volume, energy, and time) can be assigned to data.
- Data uncertainties can be considered.
- It is possible (and recommended) to document every component and value used in the STAN model (e.g., the source of data).
- The calculation algorithm uses mathematical statistical tools such as nonlinear data reconciliation (to adjust conflicting measurements or estimates) and error propagation (to compute the uncertainty of unknown quantities that can be calculated), and performs basic statistical tests to detect gross errors. A description of the nonlinear data reconciliation algorithm implemented in STAN can be found in the work of Cencic (2016 in press).
- The results of the MFA can be normalized and scaled with respect to the sum of import flows, the sum of export flows, or an arbitrary chosen factor.

- The results of the MFA are displayed as Sankey-style diagram. The graph can be printed and exported in various graphical formats.
- Data can be exported using MS Excel as an interface.
- The database of STAN files on http://www.stan2web.net can be accessed directly from the user interface of STAN.

In the following, three examples are presented that are subsequently modeled and analyzed with the help of STAN. After each example description, first, the required features of STAN are explained, and second, detailed instructions on how to implement the examples are provided, indicated by the ✿ sign.

Example 2.23 demonstrates how to build a graphical model, enter data, and perform a calculation.

Example 2.23

The model in Figure 2.50 shows the simplified material flows and stocks in a region XY related to a product under investigation: Imported raw materials (*RA*) and intraregional recycling material (*RE*) are used to

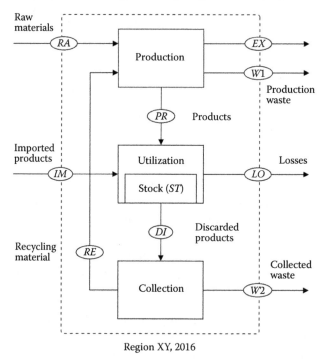

Region XY, 2016

FIGURE 2.50
Material flows, stocks, and processes of region XY, 2016 (main system).

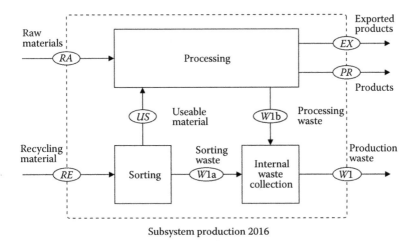

Subsystem production 2016

FIGURE 2.51
Material flows and processes of subsystem production in region XY, 2016.

create products that are either used within the region (*PR*) or exported (*EX*). The waste (*W1*) that is generated during production is treated by a company outside of the region. To satisfy the demand of the region, additional products have to be imported (*IM*). During utilization, a fraction of the material in stock (*ST*) is lost as wear, emissions, and the like. After utilization, the discarded products (*DI*) are collected. A part of these collected products is reused as recycling material (*RE*) in the production, while the other part is exported as waste (*W2*) to a waste treatment plant outside of the region.

What is happening inside of process production is to be modeled as a subsystem (Figure 2.51): Only a fraction of the recycling material (*RE*) is usable material (*US*) to be processed together with the imported raw material (*RA*). The rest is sorting waste (*W1a*). This sorting waste has to be combined with the waste that is generated during processing (*W1b*) to be exported from the subsystem as production waste (*W1*).

Some of the flows are measured (*RA, PR, IM, EX*) or estimated (*RE*). Additionally, data about the recycling rate (TC_{RE}), the fraction of recycling material that can be used in processing (TC_{US}), the fraction of waste generated during processing (TC_{W1}), and the fraction of material in stock lost during utilization (TC_{LO}^*) are available, while the material in stock of process utilization (*ST*) is estimated. Note that TC_{LO}^* is not a regular transfer coefficient, because it is related to the stock and not to the input flows of process Utilization. Hence, it is marked with an asterisk.

2.4.1 Prerequisites for Using STAN

In this chapter, the necessary preparations to be able to work with STAN are listed (registration, download, installation, start).

2.4.1.1 Registration

1. Open the website http://www.stan2web.net.
2. Click Register in the Login window.
3. Fill out the registration form, which includes choosing a user name and a password.
4. Retype the displayed security code and click the Register button.

   An e-mail with further instructions on how to complete your registration is sent to the e-mail address you provided.
5. Check your e-mail inbox and click the link in the received e-mail to complete the registration process.

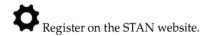 Register on the STAN website.

2.4.1.2 Download the Freeware STAN

1. Open the website http://www.stan2web.net.
2. In the Login window, enter your chosen user name and password and click the Login button.
3. In the menu, select Downloads, then STAN.
4. Click STAN-*-EN.zip to download the packed setup program of STAN in English language.

 Download STAN.

2.4.1.3 Installation of STAN

1. Unzip the downloaded zip file into a directory of your choice.
2. Double-click STAN-*-EN.exe in this folder.
3. Follow the installation instructions.

To install STAN in C:\Programs, you need administrator rights. If you have only restricted rights, you have to choose a different installation path that you are allowed to write to.

 Install STAN.

2.4.1.4 Start STAN

- On the desktop, double-click the STAN icon.

 Start STAN.

2.4.2 Graphical User Interface

The graphical user interface (GUI) of STAN consists of several elements (menu, toolbars, windows) that can be arranged arbitrarily (Figure 2.52):

1. The menu offers access to the dialogs for editing and controlling the functions of STAN.
2. The various toolbars contain shortcuts to the most frequently used functions of STAN.
3. The Model-Explorer shows the hierarchical structure of your MFA system (processes, flows, subsystems).
4. The elements (process, system boundary, legend, text field, import flow, export flow, internal flow) in the Shapes window can be used for building and editing a model on the Drawing Area. These elements can be seen as the building blocks of your model.
5. The Drawing Area is the place where you construct your model with the predefined shapes from the Shapes window.
6. In the Properties window, you can change the properties of a shape selected on the Drawing Area.
7. In the Trace Output, you find messages and warnings that appear during calculation.

i The following description of the features of STAN is only an excerpt of existing possibilities. For a complete list of step-by-step instructions, consult the user manual of STAN, which can be opened by pressing F1 in the GUI.

2.4.3 Create a Graphical Model

In a first step, a graphical model of the system under study is created. For this purpose, the predefined objects from the Shapes window (processes, flows, etc.) are used as building blocks.

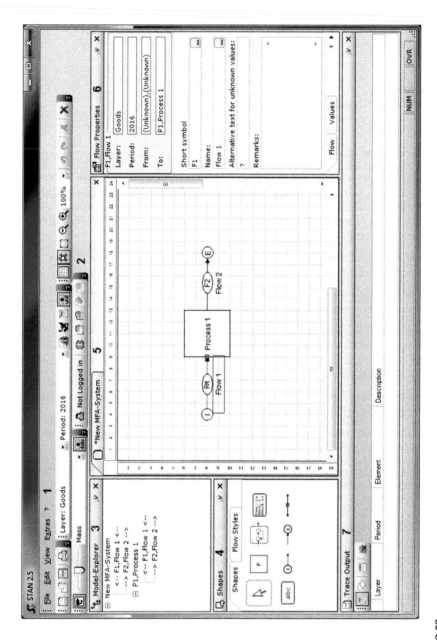

FIGURE 2.52
The graphical user interface of STAN.

2.4.3.1 Graphical Aids

- To create a new model, click 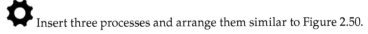 in the toolbar.
- To switch on the grid, activate ▦ in the toolbar.
- To align inserted objects to the grid (Snap to Grid), activate 🛱. If this option is disabled, it is hard to draw parallel or perpendicular flow arrows, or align processes.
- To select an object on the Drawing Area, click it. Multiple selections can be made by additionally pressing the CTRL button on the keyboard, or by erecting a frame around the objects to be selected.
- To delete an object, select it on the Drawing Area. Then press Delete on the keyboard or click ✗ in the toolbar.

⚙ Open a new empty document (not necessary if STAN has just been started). In the toolbar, make sure that the grid is visible and that the option Snap to Grid is activated.

2.4.3.2 Insert a Process

1. In the Shapes window, click the Process icon ⌐P⌐.
2. To insert a process with a predefined size, click the desired position of its left upper corner on the Drawing Area.
3. Drag the process to the desired position.
4. To change the size of a process, select it and drag one of the appearing green handles on the border of the process to the desired position.

ⓘ On the border of each process, anchor points are situated, which can be used to connect flows to processes. If a process is aligned to the grid, these anchor points are identical with the intersection points of the grid that coincide with the border of the process. The only exceptions are the corners of a process where it is not allowed to attach flows. When selecting anchor points, they do not have to be hit precisely.

⚙ Insert three processes and arrange them similar to Figure 2.50.

2.4.3.3 Rename a Process/Flow

1. Select a process/flow.
2. In the Properties window, select the Process/Flow tab.
3. Edit the short symbol and name of the process/flow.

✱ Denote the upper of the inserted processes as Production, the middle process as Utilization, and the lower process as Collection. Choose proper short symbols for the processes.

2.4.3.4 Change Process Type

1. Select a process.
2. In the Properties window, select the Process tab.
3. Set or delete the tick in the desired checkbox (subsystem, stock, TCs).

- If TCs is selected, all possible sets of transfer coefficients (with respect to sum of inputs and individual inputs) of this process will be displayed in the Data Explorer (see Section 2.4.8). Otherwise, only those sets with assigned values are displayed.
- Processes containing a subsystem will be displayed with a blue frame.

✱ Define process Production as a subsystem, process Utilization as a stock, and leave process Collection as it is.

2.4.3.5 Insert Internal Flow

An internal flow connects two processes inside of the system boundary.

1. In the Shapes window, click the Flow icon ●━━━━▣━━━→ .
2. Click a free anchor point on the source process (you do not have to hit it exactly) to define the starting point of the internal flow. Keep the mouse button pressed and drop the endpoint of the appearing arrow (i.e., mouse pointer) onto a free anchor point on the target process.

𝒊 An internal flow that is not properly connected with two processes will be displayed with a dashed line.

✱ Insert and denote the following three internal flows (source process → target process/name/short symbol):

- Production → Utilization/Products/PR
- Utilization → Collection/Discarded products/DI
- Collection → Production/Recycling material/RE

2.4.3.6 Change Flow Course

1. Select a flow.
2. Drag one of the appearing green handles to change its position. Use the handle in the middle of a straight line to insert additional turning points, or the handles on both ends of the arrow to connect the flow to a different anchor point.

ⓘ As general advice, try to connect your processes by flows that require as few turning points as possible.

⚙ Add additional turning points to the flow Recycling material in order to avoid the arrow crossing process Utilization.

2.4.3.7 Insert Import or Export Flow

Import and export flows represent flows that connect processes inside of the system boundary with the outside world (displayed as a circle labeled with I or E).

1. In the Shapes window, select the Shapes tab and click the Import flow icon ⓘ⟶ to insert an import flow, or the Export flow icon ⟶Ⓔ to insert an export flow.
2. To insert an import flow, click a point on the Drawing Area where you want the flow to start. Keep the mouse button pressed and drop the endpoint of the appearing arrow (i.e., mouse pointer) onto a free anchor point of the target process.
3. To insert an export flow, click a free anchor point on the source process. Keep the mouse button pressed and drop the endpoint of the appearing arrow (i.e., mouse pointer) on the point on the Drawing Area where you want the flow to end.

- An import or export flow that is not connected properly to a process will be displayed with a dashed line.
- A good practice is to draw import flows to enter a system from the left side, and export flows to leave a system to the right side.
- To hide the circles with the I and E (recommended for presentations and reports), click Extras > Options in the menu. Select tab Processes and Flows and untick Show shapes (I, E) for import and export flows.

Insert and denote the following import flows (target process/ name/short symbol):

- Production/Raw materials/RA
- Utilization/Imported products/IM

Insert and denote the following export flows (source process/name/ short symbol):

- Production/Exported products/EX
- Production/Production waste/W1
- Utilization/Losses/LO
- Collection/Collected waste/W2

2.4.3.8 Edit Subsystem

In order to be able to edit a process as a subsystem, the process has to be defined first as subsystem (see Section 2.4.3.4).

1. Double-click a subsystem. A new Drawing Area opens, in which all the flows connected to the subsystem already exist as import and export flows.
2. Insert additional processes and internal flows to model the subsystem.
3. Connect the displayed import and export flows of the subsystem to the inserted processes.

You cannot add or delete import and export flows on the subsystem layer. This has always to be done on the main layer.

Model subsystem Production:

1. Open (double-click) subsystem Production.
2. Insert three additional processes and denote them as Sorting, Processing, and Internal waste collection. Arange them similar to Figure 2.51.
3. Insert and denote the following three internal flows (source process → target process/name/short symbol):
   - Sorting → Processing/Useable material/US
   - Sorting → Internal waste collection/Sorting waste/W1a
   - Processing → Internal waste collection/Processing waste/ W1b

4. Connect the displayed import flows of the subsystem to the processes (flow name/target process):
   - Raw materials/Processing
   - Recycling material/Sorting
5. Connect the displayed export flows of the subsystem to the processes (flow name/source process):
   - Exported products/Processing
   - Products/Processing
   - Production waste/Internal waste collection

2.4.3.9 Insert/Refresh System Boundary

1. On the Shapes window, click the system boundary icon .
2. Click on the Drawing Area.

The system boundary (a rectangle with rounded corners displayed in a dashed line) will be inserted/refreshed automatically going through the oval shapes of the import and export flows. Additionally to the system boundary, a couple of text fields are inserted (sum of imports, sum of exports, sum of change in stocks, name of the system, and displayed period of time).

- If the system boundary is selected, objects inside cannot be selected. To deselect the system boundary, click a point outside of the system boundary on the Drawing Area.
- A system boundary can also be inserted for subsystems.

2.4.3.10 Insert Text Field or Legend

1. On the Shapes window, click the Text field icon abc or the Legend icon .
2. On the Drawing Area, click the position where you want to insert a text field with predefined size.
3. Resize the text field until the complete content is visible.

- A legend is a text field that contains the chosen display units of flows and stocks.
- In the Properties window, you can edit the content (free text and system parameters) of a selected text field (or legend).

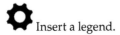 Insert a legend.

2.4.3.11 Rename System

1. In the Model-Explorer window, click the name of the system (first entry).
2. In the Properties window, enter the name of the system.

 Rename your model (e.g., MFA Region XY).

2.4.3.12 Save the Model

- On the Standard toolbar, click 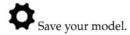.
- Press CTRL+S.
- On the File menu, click Save.

Save your model.

2.4.4 Enter Data

The layer of goods, which is already predefined, is the layer to enter data about flows and stocks of goods. For how to define additional layers for substances or energy, see Section 2.4.6.

The default setting of the period length (i.e., temporal system boundary) to be considered is 1 year, starting with January 1 of the current year. For how to change the length, number, and starting date of periods, see Section 2.4.7.

2.4.4.1 Enter Flow Values and/or Concentrations

1. Select a flow.
2. In the Properties window, select the Values tab.

3. Enter the values and, if known, their uncertainties (in text field right of "±" sign).

4. If you want use a different unit than the displayed one, type it directly after the number (e.g., 100 kg/d) or click the unit button to choose from predefined units.

- On the layer of goods, you can optionally enter mass flows, volume flows, and/or densities.

- On the layer of a substance, you can optionally enter mass flows and/or concentrations (mass substance per mass or volume good, respectively).

- On the layer of energy, you can optionally enter energy flows and/ or concentrations (energy per mass or volume good, respectively).

- If you want to consider data uncertainties, you can enter the range of uncertainty in the form of the standard error in the text field right of the "±" sign. The data set will be interpreted as mean value and standard error of an assumed normal distribution.

- Uncertainties can also be entered in percent if a mean value is already given (e.g., 10%). The entry will be automatically transformed into an absolute value.

- By default, data in the flow diagram will be displayed rounded to two digits after the decimal separator. This setting can (and should) be changed under Extras > Options > tab Number formats. When determining the total number of digits to be presented, the uncertainty of the results has to be taken into account. In general, more than two to three significant numbers are rarely justified for real-world MFA.

1. In STAN, the default abbreviation for 1 year is "a" instead of "yr". To define the unit "yr," press CTRL+u to open the Edit Units dialog window. On tab Units, select Time as the physical dimension and enter "1yr = 1a" as the relation. Click Insert.

2. On tab Display units, change the time-related units from "a" to "yr" to get "yr" as the new default unit for displaying and entering data. The current display units are shown in the caption.

3. Enter the mass flow values (mean and standard error) from Table 2.15 by using the flow properties window.

TABLE 2.15

Measurements of Mass Flows of Goods

Flow Name	Short Symbol	Mean, t/Year	Standard Error, t/Year
Raw materials	RA	60	6
Imported products	IM	30	3
Exported products	EX	40	4
Products	PR	100	10
Recycling material	RE	90	18

2.4.4.2 Enter Stock, Change in Stock and/or Concentration

In order to be able to enter stock data, the process has to be defined first to contain a stock (see Section 2.4.3.4).

1. Double-click a process with stock.
2. Select tab Stock or dStock (= change in stock).

or

1. Select a process with stock.
2. In the Properties window, select tab Stock/dStock.

3. Enter a value in one of the text fields (mean value and optional uncertainty).
4. If you want use a different unit than the displayed one, enter it directly after the number (e.g., 100 kg/year) or click the unit button to choose from predefined units.

- Stock refers to the material/energy in stock at the beginning of the considered period of time.
- dStock refers to the change of material/energy in stock during the considered period of time.
- For more information about the data entry procedure see Section 2.4.4.1.

 Define the stock of process Utilization to be 100 t ± 10 t (= 10%).

2.4.4.3 Enter Transfer Coefficient

Transfer coefficients can be entered only if the process contains no subsystem or stock. The reason for the latter is that in the case of a stock, it is not always clear whether the change in stock is to be considered as an additional virtual input (depletion of stock) or an additional virtual output (accumulation of stock) for the definition/calculation of transfer ceofficients.

1. Double-click a normal process (without stock and subsystem). In the Process dialog window on the tab TC, you see a given matrix where you can enter transfer coefficients with respect to the sum of inputs or single inputs.
2. Enter transfer coefficients directly into the fitting cells. You can choose between two different ways: a number between 0 and 1 (e.g., 0.6) or a percentage number (e.g., 60%).
3. To enter data uncertainties of transfer coefficients (in the form of the standard error of an assumed normal distribution), write "+-" directly after the mean value and enter the standard error (e.g., 0.60+-0.06 or 60%+-6%). Alternatively, click the [...] button to open an Edit dialog. Only there is it possible to enter literature references and remarks.
4. Click OK or Apply.

- If you want STAN to calculate transfer coefficients with respect to the sum of inputs, enter a question mark in one of the cells of the first matrix row.
- If you hover over a cell of the matrix, a tooltip with information about the input and output flows will be displayed.

Enter the transfer coefficients (mean and standard error) from Table 2.16.

2.4.4.4 Enter Additional Linear Relation

Additional linear relation between similar data types (e.g., mass/mass flows, volume/volume flows, concentrations in mass per mass, concentrations in mass per volume, transfer coefficients) can be added.

1. On the Edit menu, select Edit Relations.
2. Click the empty field under the column heading Expression and then the [...] button.

TABLE 2.16

Estimates of Transfer Coefficients

ID	Process	Input	Output	Mean	Standard Error
TC_{RE}	Collection	Σ Inputs	Recycling material	0.8	0.08
TC_{US}	Sorting	Σ Inputs	Useable material	0.95	0.005
TC_{W1}	Processing	Σ Inputs	Processing waste	0.05	0.005

3. Choose the Value Type (e.g., masses), Layer, Period, Subtype (e.g., flow-mass [mass/time], stock-mass [mass]), and Source of the variables A and B.

4. Define the proportion factor.

5. Confirm with OK.

⚙ Enter the following information as a linear relation: 1% (= TC_{LO}^{*}) of the material in stock of process Utilization is lost during the period under consideration (1 year):

$$\dot{m}_{Losses} = TC_{LO}^{*} \cdot m_{stock,P2}$$

Remember that TC_{LO}^{*} is not a regular transfer coefficient, because it is related to the stock of process Utilization and not to its input flows (see Section 2.4, Example 2.23).

2.4.5 Perform Calculation

Whenever data have been entered, a calculation can be started to try to reconcile conflicting data and compute unknown values, subsequently.

1. For a quick start of the calculation, click 🖩 in the Data Input toolbar or press F5 on the keyboard.

Or

1. To set additional calculation options (calculation method, trace level), choose Calculation Dialog from Edit menu or press ALT + F5.

2. Choose the calculation method (Cencic 2012 or IAL-IMPL 2013). If necessary, click Settings to change the parameter of the selected calculation method.

3. Choose the trace level (none < errors < warning < information < details) of logging. The default setting is information.

4. Click Start.

5. After calculation, click Close to display the results in the system diagram.

- After the calculation, in the Trace Output window, a list of information, warnings, and errors is displayed. Click any entry in this list to highlight the corresponding object in the system diagram.

- In the diagram, you can toggle between displaying input data and calculation results by clicking **data** in the toolbar.

Start the calculation. Figure 2.53 shows the resulting mass flows of the computation for the mainsystem, and Figure 2.54, the results for the subsystem Production.

The measurements of the flows Raw materials, Recycling materials, Products, and Exported products, and the given estimates of the transfer coefficients in the processes Sorting and Processing have been reconciled because they contain redundant information. The measurement of flow Imported products and the given estimate of the transfer coefficient in process Collection remain unchanged (they are nonredundant) because there is not enough information given to improve their values. After data reconciliation, the values of the unknown flows are computed, and their uncertainties are estimated by error propagation.

Example 2.24 demonstrates how to perform a substance flow analysis by using an additional substance layer.

Example 2.24

The model in Figure 2.55 shows a transformation process with one heterogeneous input and two homogeneous output flows. Because the process is part of a larger MFA system, no system boundary is shown.

The concentration of a hazardous substance A contained in Input 1 is to be determined. Because the input flow I1 is too heterogeneous to be sampled and analyzed with respect to substance A, the two resulting output flows Output 1 and Output 2 of a chemical transformation, which are much more homogeneous than the input flow, are sampled and analyzed instead. The concentration of substance A in Input 1 can be computed from the mass balance of substance A and the mass balance on the layer of goods. All mass flows on the layer of goods and the concentrations of Output 1 and 2 on the layer of substance A have been measured.

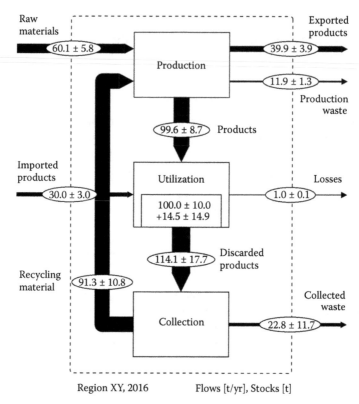

FIGURE 2.53
Numerical results of MFA in region XY, 2016 (main system).

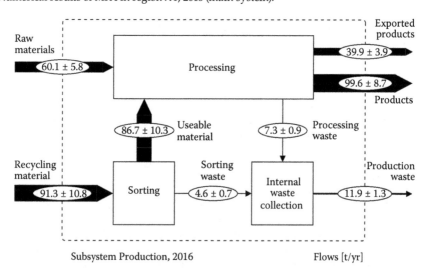

FIGURE 2.54
Numerical results of MFA of subsystem Production in region XY, 2016.

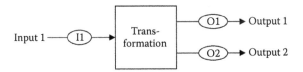

FIGURE 2.55
Model of a process that transforms a heterogeneous input flow into two homogeneous output flows.

TABLE 2.17

Measurements of Mass Flows and Concentrations

| Flow | Mass Flow (Goods) | | Concentration (Sub. A in Goods) | |
Name	Mean, kg/d	Standard Error, kg/d	Mean, mg/kg	Standard Error, mg/kg
Input 1	200	5	?	?
Output 1	185	5	1.0	0.1
Output 2	20	5	40.0	0.1

1. Reproduce the model from Figure 2.55.
2. On the layer of goods, enter the given mass flows from Table 2.17 in the flow property window.

2.4.6 Set Substance Layers

To be able to enter data about substance flows and stocks, additional substance layers have to be created.

2.4.6.1 Define Substances

A selection of predefined substances (from the table of elements) and energy is available. You can customize this selection with your own entries.

1. On the Edit menu, click Define Material.
2. Enter Short Symbol and Name of the new material.
3. Click OK or Apply.

Define a new substance with short symbol A and name Substance A.

2.4.6.2 Create New Layer

Choose which substances you want to consider in your substance flow analysis.

1. On the Edit menu, click Layers and Periods.
2. On the Layers tab, click the empty field under the column heading Short symbol.
3. Choose a material from the drop-down list.
4. Click OK or Apply.

Create a new layer for Substance A.

2.4.6.3 Select Layer

Select the layer for which you want to enter data or show the results.

- On the Input-Data toolbar, select a layer from the Layer drop-down list.

Enter data with respect to Substance A and start a computation:

1. Select layer Substance A to be able to enter data for this layer.
2. On the layer of Substance A, enter the given concentrations from Table 2.17 as mass fraction in the flow properties window.
3. Start the calculation.
4. Insert a caption.
5. Instead of the default display unit, which is t/a, set proper display units for each layer by pressing CTRL+u:
   a. Mass flows of goods: kg/d
   b. Mass flows of Substance A: mg/d
   c. Concentration of Substance A: mg/kg
6. Figure 2.56 shows the result of the MFA for the layer of goods. Because the original measurements of the mass flows did not match, they have been reconciled.
7. Figure 2.57 shows the result of the MFA for the layer of Substance A. The substance concentrations of the flows Output 1 and Output 2 have been combined with the respective reconciled mass flows on the layer of goods to compute the mass flows on the layer of substance A. The mass flow of Substance A in Input 1 can be computed from the mass balance of process Transformation.
8. The resulting concentration of Substance A in Input 1 can be found in the flow properties window on tab Flow: 4.5 ± 0.9 mg/kg. Alternatively, select mass fraction instead of mass from the drop-down menu on the Sankey toolbar to display mass fractions in the system diagram (Figure 2.58).

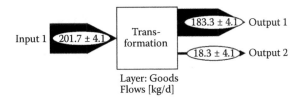

FIGURE 2.56
Resulting mass flows for the transformation process on the layer of goods.

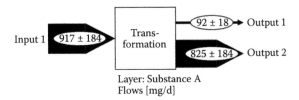

FIGURE 2.57
Resulting mass flows for the transformation process on the layer of Substance A.

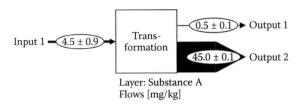

FIGURE 2.58
Resulting mass fractions for transformation process on the layer of Substance A.

> The results of the substance flow analysis show that, while the mass of Output 2 is only approximately 9% of the input, it contains nearly 90% of Substance A included in the input. Thus, if a rough estimation of the mass of Substance A incorporated in Input 1 is needed, it is sufficient to measure the mass flow of Output 2 and the concentration of Substance A in it.

Example 2.25 demonstrates how to consider multiple periods of time and how to import data semiautomatically from MS Excel.

Example 2.25

The model in Figure 2.59 shows a splitting process with two input and three output flows. Because the process is part of a larger MFA system, no system boundary is shown.

The mass flows of Input 1 and Input 2 are given based on time series for the years 2011 to 2015. These values are to be imported semiautomatically from an Excel file. The transfer coefficients with respect to individual input flows are assumed to be known and constant over time. The time series of the mass flows of Output 1 to Output 3 are to be computed.

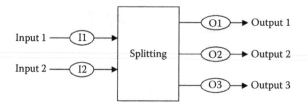

FIGURE 2.59
Model of a process that splits two input flows into three output flows.

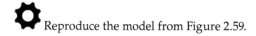 Reproduce the model from Figure 2.59.

2.4.7 Set Periods

If you want to model the behavior of a system for multiple periods of time, it is necessary to define additional periods.

2.4.7.1 Edit Period

1. On the Edit menu, click Layers and Periods.
2. On the Periods tab, edit the Period duration, Number of periods, and/or the Start of first period.
3. Click OK.
4. If only the number of periods has been changed, confirm the appearing warning by clicking OK. If the new number of periods is smaller than the old one, obsolete periods and their data will be deleted. If the new number is larger than the old one, new empty periods will be attached to the end of the existing interval.
5. If the period length and/or the start of the first period has been changed, confirm the appearing warning by clicking Keep or Rename. If Keep is chosen, only those data will be kept that have a matching period in the new interval. If Rename is chosen, the existing periods will be renamed keeping the original data. If additionally, the number of periods has been changed, periods will be deleted or added at the end of the interval.

- For each new MFA document, the default period is the current year.
- If a record of periods shall be considered, those periods must have the same duration (e.g., 1 year), and they must be attached to each other (e.g., 2000–2005).

- Example: The periods 2000 to 2006 have been defined. Data for period 2002 is given. The starting point for the first period is changed to 1998. If the periods are renamed, the periods 1998 to 2004 are available with given data in period 2000. If the periods are kept, the periods 1998 to 2004 are also available, but the given data remain in period 2002.

⚙ Create a record of five periods, with Period length 1 year, ranging from 2011 to 2015.

2.4.7.2 Select Period

- To enter data or show the results for a period, select that period from the Period drop-down list on the Input-Data toolbar.

⚙ Select period 2011.

Enter the following transfer coefficients of process Splitting for 2011 (input flow → export flow: value):

- Input 1 → Output 1: 0.5
- Input 1 → Output 2: 0.3
- Input 1 → Output 3: 0.2
- Input 2 → Output 1: 0.7
- Input 2 → Output 2: 0.3
- Input 2 → Output 3: 0

Figure 2.60 shows the resulting transfer coefficient matrix.

2.4.8 Use of the Data Explorer

The Data Explorer can be used to import or export data semiautomatically (by copy–paste actions) from or to Microsoft Excel. The Data Explorer displays all existing data sets at once. These data sets can be grouped hierarchically using existing common attributes (periods, layers, names).

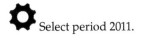

	O1	O2	O3
Σ Inputs ▶			
I1	0.5	0.3	0.2
I2	0.7	0.3	0

FIGURE 2.60
Transfer coefficient matrix of process Splitting.

2.4.8.1 Open Data Explorer

- On the Edit menu, click Data Explorer or press CTRL + D.

2.4.8.2 Select Data Sets

- Click the row marker (i.e., square) in front of a row to select a data set.
- For multiple selections, click additional row markers by simultaneously pressing CTRL.
- To select a data block, click the row marker of the first data set and then, by simultaneously pressing SHIFT, the row marker of the last data set of the block.
- Alternatively, click the row marker of the first data set and sweep the mouse pointer to the last row marker while keeping the mouse button pressed.

2.4.8.3 Copy STAN Data to Clipboard

1. Select the desired data sets.
2. To copy the rows to the clipboard, click or press CTRL+C.

This copy action also includes columns of the data grid that are not displayed (e.g., calculated values).

2.4.8.4 Insert Data from Clipboard into STAN Data Table

1. Select the target area (rules see select data sets).
2. To insert the data sets from the clipboard into the STAN data table, click or press CTRL+V.
3. Click OK or Apply.

- Only those parts of the data sets of the clipboard will be inserted that fit into the target area.
- This action also includes columns of the data grid that are not displayed (e.g., calculated values).

� Copy the transfer coefficients for 2011 of process Splitting to all other periods (2012 to 2015):

1. Select process Splitting.
2. Tick TCs in the process properties to display all possible sets of transfer coefficients in the Data Explorer.
3. Open the Data Explorer and select tab Transfer coefficients.
4. Select and copy the data sets containing the transfer coefficients for period 2011.
5. Select the target area in the periods 2012 to 2015 sequentially and paste the copied data.

2.4.8.5 Group Data Sets

Data sets can be grouped hierarchically by using common attributes (periods, layers, names).

1. Drag a desired grouping attribute (i.e., column heading) into the gray grouping area above the column headings.
2. If a grouping hierarchy already exists, decide on which position to insert the attribute.
3. Repeat the procedure until you have created the desired hierarchy.
4. To change the hierarchy, drag an attribute to another position in the grouping area.
5. To remove a grouping attribute from the hierarchy, drag it from the grouping area.

i The default setting of the grouping hierarchy is periods > layers. By clicking ▣ in the toolbar, you can restore this setting anytime.

2.4.8.6 Sort Data Sets

1. To sort data sets alphabetically by certain fields/attributes, click the respective column headings.
2. To include other fields in the sorting procedure, click additional column headings by simultaneously pressing SHIFT.

i Each click on a column heading changes the sorting order (alphabetically rising or falling).

2.4.8.7 Import Data from Excel

1. In the Excel table, select a block of cells containing data.
2. Press CTRL+C to copy this data block to the clipboard.
3. In the STAN data table, click the upper left cell of the region where you want to insert the data block.
4. Click or press CTRL + V to insert the data block from the clipboard into the STAN data table.
5. Click Apply or OK.

- Pay attention to the correct assignment of data blocks. i.e., data sets in STAN and/or Excel must have the same order.
- The imported values are interpreted as text. Thus, only those digits that are visible in Excel will be imported.
- If the text inside of the cells does not contain a unit, it will be imported with the currently chosen display unit.
- If the imported data block contains a mixture of different data types (mass flows, volume flows, mass fractions, mass concentrations) that can be recognized by their units, they will automatically be assigned to the correct columns. If the block is inserted into a mean value column, the data will be moved to the correct mean value columns. Inserting into an uncertainty column works the same way.
- In the Data Explorer, data can only be imported if input values are displayed (data).
- If the imported data contain errors (e.g., unknown units), the cells containing the errors will be marked with a ◈ after clicking Apply or OK.

Create an Excel file with time series for Input 1 and Input 2 as shown in the first three columns (A to C) of Figure 2.61.

Copy the time series of mass flows for Input 1 and Input 2 (columns B and C) from Microsoft Excel to STAN.

Because the data given in the Excel table contain the time unit "yr," which is not known in STAN per default, it has to be defined as in Example 2.23:

- To define the unit "yr," press CTRL+u to open the Edit Units dialog window. On tab Units, select Time as the physical dimension and enter "1yr = 1a" as the relation. Click Insert and then OK.

	A	B	C	D	E
1	**Flows**	**Input 1**	**Input 2**	**Input 1**	**Input 2**
2	**Period**	**t/yr**	**kg/yr**	**text**	**text**
3	**2011**	100	100000	=B3&B2	100000kg/yr
4	**2012**	110	90000	110t/yr	90000t/yr
5	**2013**	90	110000	90t/yr	110000t/yr
6	**2014**	80	85000	80t/yr	85000t/yr
7	**2015**	120	100000	120t/yr	100000t/yr

FIGURE 2.61
Excel table containing time series of Input 1 and Input 2 for the years 2011 to 2015. The columns D and E contain a combination of values and units that is necessary to import the data into STAN.

Preparation of Excel sheet:

1. Check if the decimal separator (comma or point) is the same in STAN and Excel. If not,
   a. Change (temporarily) the language of STAN or
   b. Change the regional setting of your computer
2. Combine each given value with its units ("= cell_value & cell_unit"). The results are cells that contain text (see columns D and E of Figure 2.61).

Preparation of Data Explorer:

1. Open the Data Explorer.
2. Edit the grouping hierarchy to Layers > Flow names.
3. Sort the displayed data sets with rising periods.

Import data from Excel to STAN:

1. In Excel, select the range of cells (i.e., data block) that contains the edited mass flows of Input 1 for the years 2011 to 2015 (D3:D7).
2. Press CTRL+C to copy the data block to the clipboard.
3. Insert the copied data block into the Data Explorer of STAN.
   a. Select the upper left cell where you want to insert the data block.
   b. Press CTRL+V or ⧉ in the toolbar.
4. Repeat the import procedure for the edited mass flows of Input 2 (E3:E7).
5. Click OK.

Figure 2.62 shows the content of the Data Explorer after you have imported the data from MS Excel.

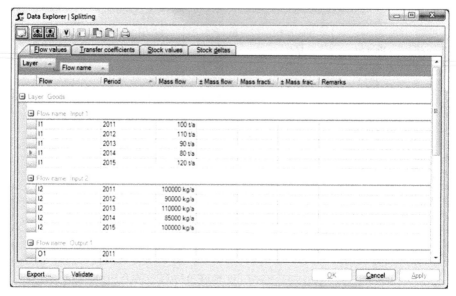

FIGURE 2.62
Data Explorer after import of data block from MS Excel.

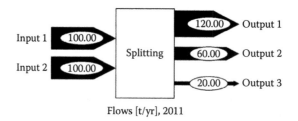

Flows [t/yr], 2011

FIGURE 2.63
Resulting mass flows of splitting process for the year 2011.

Start a calculation. To display the results of the individual periods, select the desired period in the toolbar or open the Data Explorer to see all the results at once. Figure 2.63 shows the result of the MFA for the period 2011.

2.5 Evaluation Methods for MFA Results

2.5.1 Introduction

The results of an MFA are quantities of flows and stocks of materials for the system of study. Aside from analytical and numerical uncertainties, these are *objective* quantities derived from analyses, measurements, and the principle of mass conservation. On the assumption that the study has been carried out in detail, carefully, and comprehensibly, there is usually little or no discussion about the numerical results. On the other hand, and in contrast to the measurement of mass flows and concentrations, the interpretation and evaluation of MFA results is a *subjective* process, too: it is based on social, moral, and political values. For instance, a depletion time for a reservoir of a nonrenewable resource of, say, 50 years may be considered as sufficiently long or alarmingly short, depending on the person one may consult. Another example: the "eco-indicator 95" evaluation method attributes the same weight to the death of one human being per million and to the damage of 5% of an ecosystem (Goedkoop, 1995). It is obvious that such valuations and weightings cannot be based exclusively on scientific/technical principles. Social and ethical aspects play an important role, too. Assessment is a matter of values, and values can change over time and may vary among societies and cultures. Hence, assessment is and will remain a dynamic process that must be considered as a result of the according era.

Another problem when dealing with MFA results arises when alternative scenarios for a single system or from different systems are to be compared. A common situation is the following: consider a system with 5 processes and 20 flows. As the result of an optimization step, 10 flows and 3 processes are changed. Some flows may have become "better"; some have become "worse." How much has the system improved, relative and/or absolute? A measure and a scale are needed.

To bring objectivity and comparability into the evaluation process, so-called indicators need to be applied. The Organization for Economic Cooperation and Development (OECD) provides the following definition for an (environmental) indicator: a parameter or a value derived from parameters that provide information about a phenomenon. The indicator has significance that extends beyond the properties directly associated with the parameter value. Indicators possess a synthetic meaning and are developed for a specific purpose (OECD, 1994). According to the World Resource Institute, indicators have two defining characteristics: (1) They quantify information so that its significance is more readily apparent. (2) They simplify information about complex phenomena to improve communication (Hammond, Adriaanse, Rodenburg, Bryant, and Woodward, 1995). In a wide sense, an indicator can be considered as a metric that provides condensed information about the state of a system. When applied to time series, the information is about the development of a system. Indicators should convey information that is meaningful to decision makers, and it should be in a form that they and the public find readily understandable. This implies policy relevance and, in cases of complex systems, a certain degree of aggregation. The more an indicator is based on appropriate scientific laws and principles, the more it can be considered as objective. However, as will be shown in Section 2.5.2, none of the available evaluation methods fully satisfies these demands at present.

Generally, to be comprehensive, assessment should consider resource as well as human and ecotoxicological aspects. From this requirement, it can already be concluded that "the one and only" indicator may not exist. As a result of continuously increasing knowledge, the rating of nonrenewable resources and the toxicity of substances are constantly being revised. The same is true for the weighting between the importance of resources and toxicity. It is certain that 1 kg of zinc and 1 kg of dioxin will be rated differently in 50 years compared with today. Hence, it is clear that indicators cannot relieve decision making in environmental, resource, and waste management from all types of subjectivity.

Selected evaluation methods are briefly described and discussed in this section. The selection is based on the potential for application to MFA results. These methods are based on different ideas, philosophies, or concepts, and therefore, each has certain advantages and shortcomings. In most cases, none of them can be considered complete and sufficient for a comprehensive assessment. On the other hand, most of the introduced methods are constantly undergoing further development regarding standardization, reliability, and completeness. The choice to apply a certain method is usually determined by the kind of specific problem to be investigated. Another motive may be the preference of the client and the performer of the study. In cases where the results of the evaluation process give reason for doubts, the application of another, complementary method is advisable. Generally, the MFA results themselves are the best starting point to analyze and evaluate a system. This requires some practice and experience and will be demonstrated in the case studies presented in Chapter 3.

2.5.2 Material Intensity per Service Unit

The concept of material intensity per service unit (MIPS) was developed by Schmidt-Bleek and colleagues from the Wuppertal Institute, Germany (Schmidt-Bleek, 1994, 1998; Hinterberger, Luks, and Schmidt-Bleek, 1997; Schmidt-Bleek et al., 1998). MIPS measures the total mass flow of materials caused by production, consumption (e.g., maintenance), and waste deposal/recycling of a defined service unit or product. Examples for a service unit are a haircut, the washing cycle of a dishwasher, a person-kilometer, the fabrication of a kitchen, and a power pole (Schmidt-Bleek, 1993; Schmidt-Bleek et al., 1998; Ritthoff, Rohn, and Liedtke, 2002). The total mass flow for a service unit can consist of overburden, minerals, ores, fossil fuels, water, air, and biomass; i.e., MIPS employs a life-cycle perspective and also considers the "hidden" flows of a service unit. This "ecological rucksack" comprises that part of the material input that is not incorporated within the products or materials directly associated with the service unit. The material intensity of 1 t of copper from primary production is 350 t of abiotic materials, 365 t of water, and 1.6 t of air. Other examples are the 3000 t of soil that have to be moved to produce 1 kg gold in the United States (Schmidt-Bleek, 1994) or the 14 t of processed materials that are necessary to produce a personal computer (Schmidt-Bleek, 1998). Table 2.18 gives examples of the material intensity for various materials and products.

MIPS only considers input flows to avoid double counting, since input equals output. Also, there are fewer inputs than outputs for the industrial economy, which facilitates accounting. In order to get more structure into the approach, Schmidt-Bleek and colleagues suggest grouping inputs into five categories, namely, biotic and abiotic materials, Earth movements, water, and air (Hinterberger et al., 1997; Schmidt-Bleek, 1998). Energy demands for the supply of the service unit are also accounted for on a mass basis. In later works, electricity and fuels are listed as a sixth category to provide further information (see Table 2.18). Another characteristic is that MIPS does not discriminate among different materials. The indicator assigns the same relevance to 1 kg of gravel and 1 kg of plutonium. A rationale for this assumption is given in Schmidt-Bleek (1993) and Hinterberger et al. (1997). It is mainly based on the insight that it is difficult to determine the ecotoxicity of a given substance due to unknown long-term impacts and unknown synergistic and antagonistic effects of substances. Hence, it is impossible to determine the ecotoxicity of 100,000 or more chemicals. A detailed discussion of this rationale is provided by Cleveland and Ruth (1998).

TABLE 2.18

Material Intensity for Materials and Products Compiled Using the MIPS Concept

	Abiotic Materials, t/t	Biotic Materials, t/t	Water, t/t	Air, t/t	Soil, t/t	Electricity, kWh/t
Aluminum	85	0	1380	9.8	0	16,300
Pig iron	5.6	0	22	1	0	190
Steel (mix)	6.4	0	47	1.2	0	480
Copper	500	0	260	2	0	3000
Diamonds[a]	5,300,000	0	0	0	0	n.d.
Brown coal	9.7	0	9.3	0.02	0	39
Hard coal	2.4	0	9.1	0.05	0	80
Concrete	1.3	0	3.4	0.04	0.02	24
Cement (Portland)	3.22	0	17	0.33	0	170
Plateglass	2.9	0	12	0.74	0.13	86
Wood (spruce)	0.68	4.7	9.4	0.16	0	109
Paper clip	0.008	0	0.06	0.002	n.d.	n.d.
Shirt	1.6	0.6	400	0.06	n.d.	n.d.
Jeans	5.1	1.6	1200	0.15	n.d.	n.d.
Toilet paper	0.3	0	3	0.13	n.d.	n.d.
Tooth brush	0.12	0	1.5	0.028	n.d.	n.d.

Source: Schmidt-Bleek, F., *Das MIPS-Konzept*, Droemer Knaur, Munich, 1998.

Note: n.d. = not determined. Updated data may soon be available at http://www.mips-online .info.

[a] Overburden and mining.

MIPS plays an important role in the discussion about dematerialization. Since one-fifth of the world's population consumes some 80% of the resources, the developed economies have to cut down their turnover of materials and energy if equality for all societies and countries is a goal and if less developed countries are to have similar chances to prosper. Discussions about dematerialization are also known as eco-efficiency and the factor-X debate. Results suggest that reduction factors from 4 to 50 are needed (Reijnders, 1998). Applied to the material flows of large economies as well as to single services and products, the MIPS concept is regarded as a useful tool for monitoring progress in dematerialization. MIPS in MFA can be applied to mass balances at the level of goods. The input into a system can be aggregated according the aforementioned rules. In most cases, it will be possible to derive a reasonable service unit from the system investigated.

2.5.3 Sustainable Process Index

The sustainable process index (SPI) was developed by Narodoslawsky and Krotscheck (1995), and Krotscheck and Narodoslawsky (1996) from Graz University of Technology, Austria. The basic concept of the SPI is to calculate the area that is necessary to embed a process or service into the biosphere under the constraint of sustainability. The idea is that all mass and energy flows that the process extracts or emits can be translated into area quantities by a precisely defined procedure and, ultimately, aggregated to a final value (A_{tot}). The lower the A_{tot} for a given process, the lesser is the impact on the environment. The rationale for using area as the normative value is that, in a sustainable economy, the only real input that can be utilized over an indefinite period of time is solar energy. The utilization of solar energy is bound to the surface of the Earth. Furthermore, area can be considered as the limiting resource for supply and disposal in a world of growing population.

The SPI concept considers the consumption of raw materials (A_R); energy (A_E); the requirements for infrastructure (again mass and energy) and the area to set up the infrastructure for the process (A_I); as well as the necessary area to assimilate the products, wastes, and emissions of the process (A_P). In cases where labor-intensive processes are investigated, an area for the staff can also be allocated (A_{ST}).

$$A_{tot} = A_R + A_E + A_I + A_P + (A_{ST})   \qquad (2.80)$$

Three different types of raw materials are distinguished: renewable raw materials, fossil raw materials, and mineral raw materials. A_{RR} is the area for renewable raw materials and is given as

$$A_{RR} = \frac{F_R \cdot (1 + f_R)}{Y_R}   \qquad (2.81)$$

F_R is the consumed flow (mass per unit of time, normally 1 year) of the considered raw material. The factor f_R takes into account how much "gray" area has been exerted downstream to provide F_R. The factor f_R is sometimes designated as cumulative expenditure or "rucksack" (see Section 2.5.1). Y_R is the yield for the renewable material given as mass per area and time.

The area for fossil raw materials (A_{FR}) is derived from a formally identical equation as A_{RR}. F_F (instead of F_R) is the flow of fossil raw materials into the process, and f_F considers the area of the rucksack (e.g., energy expenditure for refining and transporting fossil fuels). Y_F stands for the "yield" of sedimentation of carbon in the oceans (ca. 0.002 kg/m²/year). The rationale for the "sedimentation yield" is that as long as no more carbon is emitted as can be fixed by oceans, the global carbon cycle is not changed relevantly, and sustainability is guaranteed.

A_{MR}, the area for minerals, is defined by the following equation:

$$A_{MR} = \frac{F_M \cdot e_D}{Y_E} \tag{2.82}$$

F_M is the flow of mineral raw material consumed by the process. The energy demand to provide one mass unit of the considered mineral is e_D. Y_E, the yield for industrial energy, is dependent on the mix of energy-transformation technologies in a country (e.g., hydropower or fossil or nuclear sources). For a sustainable energy system, Y_E is approximately 0.16 kWh/(m²·year). A_E, the area for electricity consumption, is given by

$$A_E = \frac{F_E}{Y_E} \tag{2.83}$$

with F_E representing the electricity demand in kWh/year.

The infrastructure area A_I often contributes only a small part to A_{tot}. Hence a rough assessment usually is sufficient. In contrast, the process-dissipation area A_P is usually decisive. The SPI concept assumes a renewal rate for the assimilation capacity of any environmental compartment. Sustainable assimilation occurs when the emissions of the process are outweighed by the renewal rate and the elemental composition of the compartment is not changed. Note the similarity to the A/G approach (Section 2.5.8) in this point. The following equation calculates A_P:

$$A_{Pci} = \frac{F_{Pi}}{R_c \cdot c_{ci}} \tag{2.84}$$

where F_{Pi} denotes the mass per year of substance i in product/emission flow P (e.g., kg Cd/year). R_c stands for the renewal rate of compartment c in mass per area and year (e.g., kg soil/m²/year), and c_{ci} is the natural (geogenic) concentration of substance i in compartment c (e.g., kg Cd/kg soil).

The final step is to relate A_{tot} to the product or service provided by the process. The SPI has been applied to various processes such as transport, to aluminum and steel production (Krozer, 1996), to pulp and paper production, to energy from biomass (Krotscheck, König, and Obernberger, 2000), as well as to entire regional economies (Eder and Narodoslawsky, 1996; Steinmüller and Krotscheck, 1997). Figure 2.64 gives a qualitative example of SPIs for energy-supply systems.

The SPI can be applied to any MFA result. The consumed area ("footprint") of the investigated system can be compiled if data about the various yield factors and other nonspecific MFA data such as energy demand are available. The advantage of the SPI is that resource consumption is considered in a more differentiated way than is done by MIPS. Emissions and wastes are also included in the assessment. Determination of the SPI can be demanding and labor intensive, but the indicator can be regarded as one of the most universal, holistic, and comprehensive metrics.

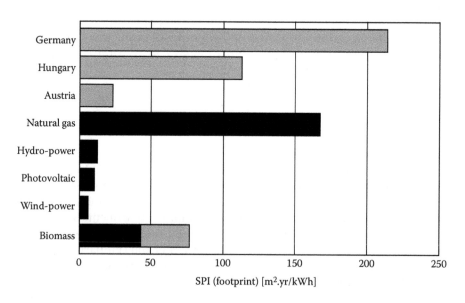

FIGURE 2.64
Sustainable process index (SPI) of the energy supply of selected national economies and various energy systems (Krotscheck, König, and Obernberger, 2000). Differences between Germany, Hungary, and Austria are due to the mix of hydropower, fossil fuel, and nuclear energy that make up the national energy supply. The range for biomass stems from different technologies such as pyrolysis, gasification, and combustion. (From Krotscheck C. et al., *Biomass Bioenergy*, 18, 341, 2000. With permission.)

2.5.4 Life-Cycle Assessment

Life-cycle assessment (LCA) is a tool that was developed during the 1980s and 1990s in Europe, and the US Society of Environmental Toxicology and Chemistry (SETAC) soon served as an umbrella organization with the aim of further developing LCA and standardizing and harmonizing procedures (Consoli et al., 1993; Udo de Haes, 1996). These activities finally led to the development of a series of ISO LCA standards (the 14040 series of the International Organization for Standardization). Accordingly, LCA consists of four steps (Guinée et al., 2001):

1. *Goal and scope definition*, where the goal of the study is formulated. The scope is defined in terms of temporal, spatial, and technological coverage, and the level of sophistication in relation to the goals is fixed. Additionally, the product(s) of study are described, and the functional unit is determined.

2. The *inventory analysis*, which results in a table that lists inputs from and outputs to the environment (*environmental interventions*) associated with the functional unit. This requires the setting of system boundaries, selection of processes, collection of data, and performing allocation steps for multifunctional processes (e.g., a power plant producing energy not only for a single product).

3. The *impact assessment*, during which the inventory table is further processed and interpreted in terms of environmental impacts and societal preferences. This means that impact categories such as depletion of resources, climate change, human toxicity and ecotoxicity, noise, etc. have to be selected. *Classification* designates the step where the entries of the inventory table are qualitatively assigned to the preselected impact categories. In the *characterization* step, the environmental interventions are quantified in terms of a common unit for that category (e.g., kg CO_2 equivalents for climate change), allowing aggregation into a single score for that category: the category indicator result. Additional and optional steps are *normalization* and *weighting* of impact categories that lead to a single final score.

4. The *interpretation* of the results, which comprises an evaluation in terms of soundness, robustness, consistency, completeness, etc., as well as the formulation of conclusions and recommendations.

LCAs have been carried out for a multitude of goods ranging from batteries (Rydh and Karlström, 2002), PET bottles (Song and Hyun, 1999), paper (Finnveden and Ekvall, 1989), tomato ketchup (Andersson, Ohlson, and Olsson, 1998), catalytic converters for passenger cars (Amatayakul and Ramnäs, 2001), fuel products (Furuholt, 1995), different floor coverings (Potting and Blok, 1995), and a rock crusher (Landfield and Karra, 2000) to steel bridges (Widman, 1998). One of the first LCAs was for packaging materials (Bundesamt für Umweltschutz, 1984). Fewer studies have been undertaken on the process (Schleisner, 2000; Burgess and Brennan, 2001; Lenzen and Munksgaard, 2002) and system level (Myer and Chaffee, 1997; Seppälä, Melanen, Jouttijärvi, Kauppi, and Leikola, 1998; Finnveden, 1999; Brentrup, Küsters, Kuhlmann, and Lammel, 2001; Haas, Wetterich, and Köpke, 2001; Seppälä, Koskela, Melanen, and Palperi, 2002). Despite well-defined rules and recipes on how to execute an LCA, studies are sometimes disputed. Most objections concern data consistency and the reliability of the impact assessment (Krozer and Vis, 1998). Ayres (1995) provides a concise discussion about potential problems concerning LCA. He concludes that often, studies are too focused on the impact assessment, and analysis and control of basis data are neglected. Since LCAs can be labor-intensive and therefore costly, Graedel (1998) suggests an approach on how to streamline LCA in order to make it more attractive to companies.

MFA can be regarded as a method to establish the inventory for an LCA. This is especially true when LCA is applied to systems rather than to single goods. Hence, the impact assessment of LCA can be applied to MFA results. A certain discrepancy will be that LCA strives for assessing as many substances and compounds as possible to guarantee completeness, while MFA is directed toward reducing the number of substances of study as much as possible to maintain transparency and manageability.

2.5.5 Swiss Ecopoints

The Swiss ecopoints (SEP) approach belongs to the family of impact-assessment methodologies in LCA such as the SETAC method (Udo de Haes, 1996), the CML method (Heijungs et al., 1992), or the eco-indicators by Goedkoop (Goedkoop, 1995). SEP is based on the idea of critical pollution loads, an idea first published by Müller-Wenk (1978). Later, it was further developed and concretized by members of the service sector, the industry, the administration, and the academia in Switzerland (Ahbe, Braunschweig, and Müller-Wenk, 1990; Braunschweig and Müller-Wenk, 1993; Brand, Braunschweig, Scheidegger, and Schwank, 1998). The SEP score for an environmental stressor (emissions to air, water, and soil) is calculated using the following formula:

$$SEP_i = F_i \cdot \frac{1}{F_{crit}} \cdot \frac{F_{sys}}{F_{crit}} \cdot 10^{12} \qquad (2.85)$$

F_{crit} stands for the critical (in the meaning of maximal acceptable) flow of the stressor in a defined region (e.g., Switzerland). F_{sys} is the actual flow of the stressor within the same region. The first ratio in Equation 2.85 normalizes the stressor flow F_i of the source of study and determines its importance. The second ratio weights the stressor with regard to its importance for the region. The prominent introduction of F_{crit}—it appears two times as a reference factor—also brings in social and political aspects. This is the main difference of the SEP approach compared with other toxicity- and effects-based impact assessment methodologies. F_{crit} can be fixed differently from region to region with regard to time, condition of the environment, technological standards, economic development, etc. (e.g., see the different reduction targets for countries in the Kyoto protocol) and stands for an environmental-quality goal. Scores of different stressors F_i can be added up to a final score. The higher the score, the higher the environmental burden of the investigated product or process for the system.

Besides emissions to the environmental compartments, the SEP approach also considers the quantity of waste produced on a mass basis and the consumption, energetically and as feedstock, of scarce energy resources (mainly fossil fuels, uranium, and potential energy). Other resource consumption is not considered. The rationale is that minerals are not scarce, since matter does not vanish. However, the availability of minerals can diminish (e.g., through declining ore grades), which results in increasing environmental impacts when such resources are mined and processed. Those impacts (emissions) are considered in the impact analysis. Hertwich and colleagues (Hertwich, Pease, and Koshland, 1997) give an illustrative example for a discrepancy between the SEP approach and the established rating method for greenhouse gases: The ecopoints calculated for the United States for 1 kg of CO_2, CH_4, and N_2O are 1.14, 59,700, and 3,890, respectively. The global warming potentials (GWP) for a time span of 20 years are, according to the IPCC, 1, 63, and 270, respectively. The authors impute the difference to the equal and linear valuation of the different stressors. Ahbe et al. (1990) discuss the pros and cons of other nonlinear valuation functions, such as logistic and parabolic functions for ecopoints.

The SEP method has been applied to a multitude of problems ranging from packaging (Habersatter, 1991) to MSW incineration (Hellweg, 2000). Table 2.19 gives SEP scores for selected air pollutants and compares the results with the human toxicity potential (as used in LCA) and with the exergy concept, which is described in Section 2.5.6.

2.5.6 Exergy

Exergy is a measure of the maximum amount of work that can theoretically be obtained by bringing a resource (energy or material) into equilibrium with its surroundings through a reversible process (i.e., a process working without losses such as friction, waste heat, etc.). The surroundings, the reference environment, or simply the reference state must be specified, i.e., temperature and pressure. In cases in which materials are considered, the chemical composition must be known, too. For material flow studies, the environment usually consists of the atmosphere, the ocean, and the Earth's

TABLE 2.19

Impact Assessment of Selected Air Pollutants by SEP, by HTP as an Impact Category in LCA, and by Exergy

	Swiss Ecopoints (SEP)[c] SEP/g	Human Toxicity Potential (HTP)[b] kg 1,4-DCB eq./kg[a]	Exergy,[d,e] kJ/g
Particulates	60.5	0.82	7.9
C_6H_6	32	1900	42.3
NH_3	63	0.1	19.8
HCl	47	0.5	2.3
HF	85	2900	4.0
H_2S	50	0.22	23.8
SO_2	53	0.096	4.9
NO_2	67	1.2	1.2
Pb	2900	470	–
Cd	120,000	150,000	–
Hg	120,000	6000	–
Zn	520	100	–

Note: Different results may be obtained, depending on which assessment method is applied.
[a] 1,4-dichlorobenzene equivalent.
[b] Huijbregts, 2000.
[c] Stahel, Fecker, Förster, Maillefer, and Reusser, 1998.
[d] Costa, Schaeffer, and Worrell, 2001.
[e] Ayres, Martinás, and Ayres, 1996.

crust, as suggested by Szargut, Morris, and Steward (1988). The term *exergy* for "technical working capacity" was coined by Rant in 1956 (Rant, 1956). Other terms have been used synonymously, such as *available work, availability*, and *essergy* (essence of energy). Exergy is an extensive property and has the same unit as energy (e.g., J/g). [An extensive property is dependent on the size (mass, volume) of the system. Intensive properties are, for example, temperature, pressure, and chemical potentials.] Unlike energy, there is no conservation law for exergy. Rather, exergy is consumed or destroyed, due to irreversibility in any real process.

Consider the following example: Approximately 3.1 kWh of electricity (P) is required at minimum to heat water from 10°C to 37°C for a 100-l bathtub. This is calculated using the first law of thermodynamics, i.e., $P = Q_W + L$, with loss (*L*) assumed to be ≈0. The energy content (Q_W) of the water (*m* = 100 kg) in the tub is $Q_W = mc_p(T_W - T_0) = 11,300$ kJ ≈ 3.1 kWh (specific heat capacity of water: c_p = 4.18 kJ/kg/K; T_W = 310.15 K; reference temperature: T_0 = 283.15 K). While the exergy content of electricity is 100% ($P = E_1$), which means that electricity can be transformed into all other kinds of energy (heat, mechanical work, etc.), the exergy content of the water in the tub is $E_2 = Q_W[1 - (T_0/T_W)] = 0.27$ kWh. The difference between E_1 and E_2 (≈91% of E_1) is the exergy loss of the process *water heating*. E_2 is the maximum amount of energy that could be transformed back into work (e.g., again electricity) from the water.

E_1 and E_2 can be regarded as measures that quantify the "usefulness" or "availability" of 3.1 kWh of electricity or warm water. First-law efficiency of the process is $\eta_I = Q_W/P = 100\%$; second-law efficiency is $\eta_{II} = E_2/E_1 = 8.7\%$.

The exergy content of a solid material can be compiled from standard chemical exergy values e_{chj}^0 as introduced by Szargut, Morris, and Steward (1988). The e_{chj}^0 values are substance-specific values (j) that are calculated for the standard state (T_0, p_0) and related to the mean concentration of the reference species of substance j in the environment. The assumption is that there is only one reference species for each element. Consider the example Fe_3O_4. The reference species of Fe is assumed to be Fe_2O_3 in the Earth's crust. The standard chemical exergy of reference species in the Earth's crust is calculated using Equation 2.86.

$$e_{chj}^0 = -R \cdot T_0 \cdot \ln x_j \qquad (2.86)$$

with x_j being the average mole fraction of the reference species in the Earth's crust ($x_{Fe_2O_3} = 1.3 \cdot 10^{-3}$); $e_{ch,Fe_2O_3}^0 = -8.31 \cdot 298.15 \cdot \ln 1.3 \cdot 10^{-3} = 16.5$ kJ/mol = 0.1033 kJ/g. The reference species for O_2 is O_2 in the atmosphere and is calculated using Equation 2.87.

$$e_{chj}^0 = R \cdot T \cdot \ln \frac{p_0}{p_{j,0}} \qquad (2.87)$$

with $p_0 = 101.325$ kPa (mean atmospheric pressure) and $p_{j,0} = 20.4$ kPa (partial pressure of O_2 in the reference state); $e_{ch,O_2}^0 = 3.97$ kJ/mol. The standard chemical exergy of Fe can now be calculated from

$$2Fe + 3/2\, O_2 \rightarrow Fe_2O_3$$

and

$$e_{ch}^0 = \Delta G_f^0 + \sum_{el} n_{el} \cdot e_{ch,el}^0 \qquad (2.88)$$

where e_{ch}^0 is the standard chemical exergy of the target compound (e.g., Fe_2O_3), ΔG_f^0 is Gibbs's energy of formation (e.g., tabulated in Barin (1989), $\Delta G_{f,Fe_2O_3}^0 = -742.3$ kJ/mol), n_{el} is the number of moles of the elements in the target compound, and $e_{ch,el}^0$ are the standard chemical exergy values of the elements. Equation 2.88 yields $16.5 = -742.3 + 2 \cdot e_{ch,Fe}^0 + 3/2 \cdot 3.97$ and $e_{ch,Fe}^0 = 376.4$ kJ/mol.

The standard chemical exergy of Fe_3O_4 is now (again using Equation 2.88, $\Delta G_{f,Fe_3O_4}^0 = -1015.2$ kJ/mol)

$$3Fe + 2O_2 \rightarrow Fe_3O_4$$

$$e^0_{ch,Fe_3O_4} = -1015.227 + (3 \times 376.4) + (2 \times 3.97) = 121.9 \text{ kJ/mol} = 0.5265 \text{ kJ/g}$$

If a hypothetical type of iron ore consists of 60% Fe_2O_3, 30% Fe_3O_4, and 10% other minerals (having a standard chemical exergy of ≈ 2 kJ/g), then the standard chemical exergy of the iron ore is

$$e^0_{ch,\text{iron ore}} = (0.6 \times 0.1033) + (0.3 \times 0.5265) + (0.1 \times 2) = 0.42 \text{ kJ/g}$$

Applying the standard chemical exergy values, which are tabulated for many common substances (Szardgut, Morris, and Steward, 1988; Ayres and Ayres, 1999), and having information about the chemical composition of materials, the exergy of materials can be calculated, and exergy balances for combined material/energy systems can be established.

Initially, the exergy concept was applied to energetic systems such as heat and turbo engines in order to understand which processes cause major losses (e.g., cooling, throttling) and to learn how to improve energy efficiency. Since exergy can be calculated theoretically for all materials and energy flows, it can be applied to any materials balance. Hence, it is a useful tool for resource accounting because it aggregates materials and energy to one final exergy quantity. For instance, the production and fabrication industry can be described as a system that uses exergy in the form of fossil fuels and raw materials to produce consumer goods and wastes of lower exergy. Moreover, the technical efficiency of any system can be expressed as exergy efficiency. Studies for single branches of industry, as well as for entire national economies, have been carried out (Wall, 1977; Wall, Sciubba, and Naso, 1994; Ertesvag and Mielnik, 2000; Michaelis and Jackson, 2000a,b; Costa, Schaeffer, and Worrell, 2001).

In addition, exergy is considered to be a useful indicator for environmental impacts of emissions and wastes (Wall, 1993; Rosen and Dincer, 2001). The rationale for this assertion is as follows: the higher the exergy of a material or energy flow, the more the flow deviates from the thermodynamic and chemical state of the environment, and the higher the potential to cause environmental harm. On the other hand, the correlation between exergy and environmental impact is not very strong. For example, the exergy values of substances emitted to the atmosphere are not proportional to their toxicity (Connelly and Koshland, 2001). The exergy value for PCDD/F (dioxins and furans) is ca. 13.0 kJ/g, and for carbon monoxide, it is 9.8 kJ/g (Costa, Schaeffer, and Worrell, 2001). Hence, exergy of both substances ranges in the same order of magnitude. Yet the emission limits for MSW incinerators within the European Union are 0.1 ng/m³ for PCDD/F and 50 mg/m³ for CO (European Union, 2000), a difference of eight orders of magnitude.

Another characteristic of the exergy concept is that exergy balances are often dominated by energy flows, and materials (e.g., wastes, emissions) seem to play a minor role. Consider the following example. The emission of 1 kg PCDD/F corresponds to an exergy value of 13 MJ. [For comparison, the estimated total dioxin and furan emissions for Germany in 1990 were between 70 and 950 g TEQ/year (Fiedler and Hutzinger, 1992). The abbreviation *TEQ* stands for *toxic equivalents* and is instrumental in assessing toxicities of blends. For example, PCDD/F occurs as a mixture of different individual compounds (congeners) that have different degrees of toxicity. The emission of each congener is multiplied by a weighting factor (referred to as a toxic equivalent factor (TEF)]. The weighted values are then added together to give the TEQ of the mixture.

70 and 950 g TEQ/year is equivalent to the release of some 500 l of warm water ($Q_W = [4.18 \times 500 \times (55 - 10)]/1000 = 94$ MJ; $E_W = Q_W[1 - (T_0/T_W)] = 13$ MJ). Such examples show that the exergy concept must be carefully considered when applied to materials as well as combined materials and energy systems. Detailed information about the theory of exergy and application in resource accounting can be found in the work of Wall (1986), Baehr (1989), Ayres and Ayres (1999), and Szargut, Morris, and Steward (1988).

2.5.7 Cost–Benefit Analysis

The concept of cost–benefit analysis (CBA), sometimes referred to as "benefit–cost analysis" dates back more than 150 years to the work of J. Dupuit, who was concerned with the benefits and costs of constructing a bridge (Johansson, 1993). Since then, the concept of CBA has been constantly refined and focused. In the late 1950s, an extensive literature on the foundations of CBA emerged. Most of the published works focused on how to assess the net economic value of public works projects. Of special interest were water-resource developments that withdrew productive factor inputs such as land, labor, capital, and materials from the economy to produce tangible outputs such as water, hydroelectric power, and transportation (Johansson, 1993; Hanley and Spash, 1995).

CBA has its roots in welfare economics, in the theory of public goods, and in microeconomic investment appraisal (Schönbäck, Kosz, and Madreiter, 1997). Generally, CBA is a tool to determine quantitatively the total advantages (benefits) and disadvantages (costs) of alternative projects or measures. The goal is to determine whether and how much a public project can contribute to national economic welfare, which of several options should be selected for action, and when the investment is to be executed. Benefits and costs are quantified in monetary units (e.g., $, €) and can therefore be balanced against each other. This is the crucial advantage of the method, since decision makers are already familiar with the measure [in contrast to metrics such as exergy efficiency (%), ecopoints (—), eutrophication (kg PO_4 equivalents), etc.]. CBA also has some obvious shortcomings. Many effects, be they costs or benefits, cannot be exactly quantified in monetary terms (e.g., the beauty of a landscape or the life of a human being). On the other hand, several methods such as the contingent valuation method, the hedonic price method, and the travel cost method have been developed to convert problematic effects and environmental impacts into costs (Johansson, 1993; Hanley and Spash, 1995).

Another approach to overcome such deficiencies—developed at *Technische Universität Wien* (Döberl, Huber, Brunner, Eder, Pierrard, Schönbäck, Frühwirt, and Hutterer, 2002)—combines cost-effectiveness analysis and multicriteria analysis in a method known as modified cost-effectiveness analysis (MCEA). MCEA subdivides general goals into concrete subgoals. For example, the general goal "protection of human health and the environment" can be subdivided in a first step into (1.1) protection of air, (1.2) protection of water, and (1.3) protection of soil quality. In a second step, goal 1.1 can be subdivided into the subgoals (1.1.1) reduction of impact by regionally important pollutants, (1.1.2) reduction of the anthropogenic greenhouse effect, and (1.1.3) reduction of damage to the ozone layer. In contrast to the abstract goal "protection of human health and the environment," each of the latter subgoals can be described by single indicators (e.g., global-warming potential for 1.1.2 and ozone-depletion potential for 1.1.3), and targets for reduction can be quantified. This procedure may result in a multitude of subgoals and indicators of different importance and public preference. One way to make them comparable and amenable to aggregation is to assign a specific weight to each indicator. The weights can be obtained from a ranking process carried out by a group of experts or by stakeholders from a variety of interests. Finally, MCEA compares costs with the efficiency of reaching the defined targets.

According to Hanley and Spash (1995), a CBA comprises the following eight steps:

1. Definition of the project, which includes identifying the boundaries of the analysis and determining the population over which costs and benefits are to be aggregated.

2. Identification of all impacts resulting from the implementation of the project [required resources (materials, labor), effects on local unemployment levels, effects on local property prices, emissions to the environment, change to the landscape, etc.].

3. Determination of which impacts are to be counted based on certain rules and conventions.

4. Determination of the physical amounts of cost and benefit flows for a project and identification of when they will occur in time.

5. Valuation of the physical measures of impact flows in monetary units. (This includes predicting prices for value flows extending into the future, correcting market prices where necessary, and calculating prices where none exist.)

6. Conversion of the monetary amounts of all relevant costs and benefits into present money values. (This is achieved by discounting,

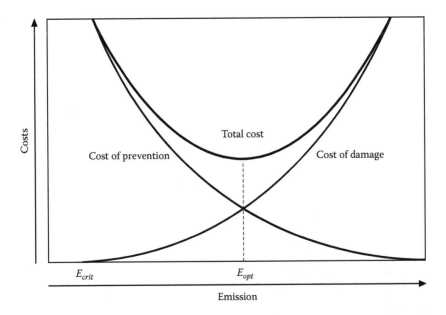

FIGURE 2.65

CBA of an emission problem: E_{crit} stands for the minimum emission load where environmental damage is expected to occur. E_{opt} is the emission where the total of costs of prevention (e.g., for filter technology) and costs of damage (e.g., treatment of respiratory diseases) are minimal. Note that the economically optimized emission (E_{opt}) accepts a certain degree of environmental burden.

a method that makes costs and benefits comparable regardless of when they occur.)

7. Comparison of total costs (C) and total benefits (B). (If $B > C$, the project is qualified for acceptance or at least improves social welfare in the theory of neoclassical welfare economics.)

8. Performance of a sensitivity analysis to assess the relevance of uncertainties.

The latter step is a requirement for all assessment methods. The combination of CBA and MFA has been further developed and applied to challenges mainly in waste management (Schönbäck, Kosz, and Madreiter, 1997; Hutterer, Pilz, Angst, and Musial-Mencik, 2000; Döberl et al., 2002; Brunner et al., 2016). Figure 2.65 gives an example of a typical CBA solution.

2.5.8 Anthropogenic versus Geogenic Flows

The anthropogenic versus geogenic flow (A/G) approach is derived from the *precautionary principle* (P2) and is a possible definition for the concept of sustainability. The Wingspread Conference in 1998 defined the P2 as follows

(Hileman, 1999): "When an activity raises threats of harm to human health or the environment, precautionary measures should be taken, even if some cause and effect relationships are not fully established scientifically."

Aspects of the P2 can be traced back centuries, even millennia. For example, precaution as a management guideline can be found in the historical oral traditions of indigenous people of Eurasia, Africa, the Americas, Oceania, and Australia (Martin, 1997). Haigh (1994) mentions the British Alkali Act of 1874, which required that emissions of noxious gases from certain plants should be prevented without any need to demonstrate that the gases were actually causing harm in any particular case. In the twentieth century, the principle emerged in Scandinavian legislation in the early 1970s (Sand, 2000) and, a few years later, in Germany, when large-scale environmental challenges such as acid rain, pollution of the North Sea (seals dying, carpets of algae, bans on swimming, etc.), and global climate change became a matter of public debate. Since then, the P2 has been used in other international agreements and legislation (Sand, 2000), notably the Rio Declaration on Environment and Development of 1992 (United Nations Conference on Environment and Development, 1992). In that document, principle 15 says that "in order to protect the environment, the precautionary approach shall be widely applied by States according to their capabilities. Where there are threats of serious or irreversible damage, lack of full scientific certainty shall not be used as a reason for postponing cost-effective measures to prevent environmental degradation." The P2 is more widely accepted in Europe, where it has been regularly used in cases with less-than-certain scientific information for decision makers, especially in E.U. legislation (e.g., the Maastricht Treaty). In the United States, the P2 has often been criticized on the basis of implementation costs (deFur and Kaszuba, 2002), and the lack of comprehensive and authoritative definition (Sandin, 1999). Nevertheless, elements of the principle can be found in US environmental laws (Applegate, 2000).

The A/G approach, first mentioned by Güttinger and Stumm (1990), is also part of the definition of ecological sustainability as given in SUSTAIN (1994) and cited in Narodoslawsky and Krotscheck (1995). For more information, see also Daly (1993):

1. Anthropogenic material flows must not exceed the local assimilation capacity and should be smaller than natural fluctuations of geogenic flows.

2. Anthropogenic material flows must not alter the quality and the quantity of global material cycles.

3. The natural variety of species and landscapes must be sustained or improved.

Applying the A/G approach to an MFA system means to determine both the materials balances of the anthropogenic system and the corresponding materials balances of the environment into which the anthropogenic system is embedded. Of particular relevance is the extent to which anthropogenic flows alter geogenic flows and stocks. Examples are given in Figure 2.66. Since many cause/effect relationships between external material flows and environmental compartments are not entirely known and understood, the P2 is applied. Exclusively conservative (i.e., small) alterations of geogenic flows and stocks are considered to be acceptable. A trivialized summary of the A/G approach might go as follows: "As long as a human activity does not affect the environmental compartments significantly, there is no harm to the environment. The activity can be regarded as ecologically sustainable." Since the cause/effect relationship is not known, a "nonsignificant" change cannot be determined on scientific terms. Significance has to be defined, rather, on political and ethical grounds. This may be regarded as the weak point of the A/G approach. On the other hand, this reflects the inherent problem of evaluation as a subjective process.

In order to determine the ratios AF/GF and AF/m_{stock} as illustrated in Figure 2.66, the natural concentrations of geogenic flows and stocks have to be known. Sometimes it is difficult to determine real geogenic concentrations, since virtually all ecosystems show traces of anthropogenic activities. Moreover, geogenic concentrations can vary considerably between regions. On the other hand, it is not essential to know the exact concentration to apply the approach, and abundant data are available on trace element concentrations in unpolluted air, surface water and groundwater, soils, etc. Generally, the A/G approach is a reliable and handy tool for MFA evaluation, since its application requires only a little additional work. Figure 2.67 shows how the A/G approach can be employed to evaluate an MFA system.

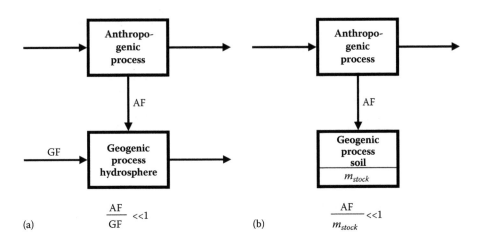

FIGURE 2.66
Anthropogenic versus geogenic flow approach: (a) alteration of a geogenic flow by an anthropogenic process; (b) alteration of a geogenic stock. The smaller the ratios of AF/GF and AF/ m_{stock}, the greater the sustainability of the anthropogenic process.

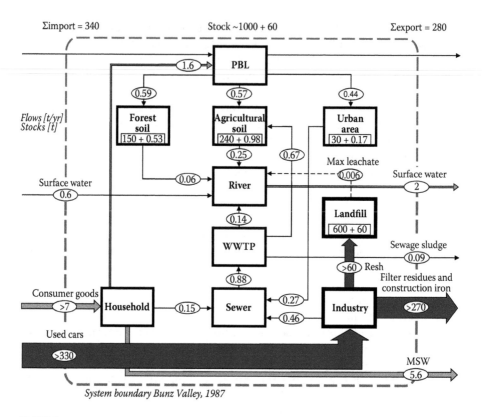

FIGURE 2.67
Lead balance (in t/year) for a 66 km² region in Switzerland with 28,000 inhabitants (Brunner, Daxbeck, and Baccini, 1994). The main flow of lead is induced by the import of used cars that are treated in a large shredder. Some 60 t of lead are contained in the waste stream of the shredder (resh), which is landfilled. The product (scrap) from the shredder is processed in a regional steel mill, where lead is concentrated in the filter dust and exported. The stock of lead in the landfill is assessed at some 600 t and represents the largest and fastest-growing reservoir of lead within the region. Assuming that the river entering the system can be regarded as unpolluted (geogenic concentration), the A/G approach limits the leachate from the landfill to a nonrelevant impact of, say, 1%, which equals 0.006 t/year. Such an emission means that a maximum of 0.001% of the lead in the landfill may be emitted per year. It is evident that the lead flows from wastewater treatment plant (WWTP) and soils exceed this limit (see discussion in Chapter 3, Sections 3.1.1 and 3.4.1).

2.5.9 Statistical Entropy Analysis

2.5.9.1 Background

The idea of applying statistical entropy to MFA results was developed at the Vienna University of Technology (Rechberger, 1999). So far, it is the only evaluation method that has been tailor-made for MFA. While other methods do not consider all aspects of an MFA or are capable of processing more information about the system than is provided by the MFA (e.g., SPI or CBA), statistical entropy analysis (SEA) uses all information and requires little additional computing. There is one exception: the magnitude of a stock is not considered in SEA. This is discussed in Chapter 3, Section 3.2.2.

SEA is a method that quantifies the power of a system to concentrate or dilute substances. As will be shown in Section 3.2.2, this is an essential feature of any material flow system. Today, especially in waste management, the importance of concentrating resources as well as pollutants is not yet fully understood (the term *concentrating* is used instead of the more common term *concentration* to stress that concentrating designates a process or action). For example, one argument against state-of-the-art incineration is that it produces a concentrate of hazardous substances: the fly ash. On the other hand, another concentrating process is commonly and rightly considered as positive, namely, the concentrating of paper, plastics, metals, glass, etc. Indeed, this process is designated as separate *collection* of valuable resources for recycling purposes, though it is a typical *concentrating process*. An incinerator also collects resources in the fly ash and can be designated as a valuable *collection process* (see Chapter 3, Section 3.3.2.3). However, the metric that is derived in this chapter is designated as the "substance concentrating efficiency" (SCE) of a process, albeit "substance collection efficiency" would be possible as well. This metric is the only existing indicator that measures concentrating and diluting effects. For the moment, the method is derived for "one-process" systems (taken from Rechberger and Brunner, 2002b). Chapter 3, Section 3.2.2 gives an example of how to apply SEA to an MFA system consisting of several processes. The discussion there also makes a case for the need to establish a concentrating waste-management system to achieve sustainable management for conservative substances.

As shown earlier in this chapter, a material balance is established by combining mass flows of all goods with the substance concentrations in those goods. In Figure 2.68, an exemplary MFA is displayed for a selected substance. For simplicity, the system is in an ideal steady state. There is neither an exchange with an internal stock ($\dot{m}_{storage} = 0$) nor a stock ($m_{stock} = 0$). The input and output goods of the system are defined by a set of element concentrations ($[c_I], [c_O]$) and a set of mass flows ($[\dot{m}_I], [\dot{m}_O]$). Hence, a system can be regarded as a procedure that transforms an input set of concentrations into an output set of concentrations; the same applies to the mass flows. Each system can be viewed as a unit that concentrates, dilutes, or leaves unchanged its throughput of a substance. In order to measure this transformation, an appropriate function that quantifies the various sets is required. The transformation can be defined as the difference between the quantities for the input (X) and the output (Y). This allows determination of whether a system concentrates ($X - Y > 0$) or dilutes ($X - Y < 0$) substances.

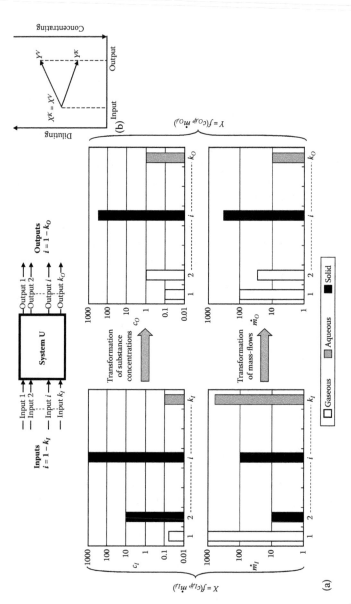

FIGURE 2.68

(a) A balance for substance j of a system U is determined by substance concentrations (c_{ij}) and mass flows ($\dot{m}_i$) of all input and output goods. For all j and for steady-state conditions, the applicable equation is $\sum_{i=1}^{k_I} \dot{m}_{I,i} \cdot c_{I,ij} = \sum_{i=1}^{k_O} \dot{m}_{O,i} \cdot c_{O,ij}$. The system can be regarded as an algorithm that transforms sets of concentrations ($[c_I],[c_O]$) and mass flows ($[\dot{m}_I],[\dot{m}_O]$) from the input (I) to the output (O). X and Y are functions that quantify the sets of concentrations and mass flows. (b) Comparison of a concentrating system K ($X^K - Y^K > 0$) and a diluting system V ($X^V - Y^V < 0$). (From Rechberger, H. and Brunner, P. H., *Environ. Sci. Technol.*, 36, 809, 2002. With permission.)

2.5.9.2 Information Theory

In order to calculate X and Y, a mathematical function from the field of information theory is used (Vogel, 1996). This function originates from Boltzmann's statistical description of entropy. It is formally and mathematically identical with Boltzmann's well-known H-theorem (Boltzmann, 1923). Information theory, developed by Shannon in the 1940s (Shannon, 1948), is used to measure the loss or gain of information within a system. In statistics, the so-called Shannon entropy is used to measure the variance of a probability distribution: the greater the variance, the less the information about the quantity of interest. Note that the thermodynamic entropy denoted "S" (J/mole/K), as introduced by Clausius (1856), is *formally* identical with the statistical entropy "H" (bit) of Shannon; however, although there is no physical relationship between the two entropy terms, there is phenomenological congruence. The following is based on Shannon's statistical entropy and not on thermodynamic entropy.

The statistical entropy H of a finite probability distribution is defined by Equations 2.89 and 2.90:

$$H(P_i) = -\lambda \cdot \sum_{i=1}^{k} P_i \cdot \ln(P_i) \geq 0 \tag{2.89}$$

$$\sum_{i=1}^{k} P_i = 1 \tag{2.90}$$

where P_i is the probability that event i happens.

In statistical mechanics, λ is defined by Boltzmann as the ratio of the gas constant per mole (R) to Avogadro's number (N_0): $k_B = R/N_0$, unit J/K. In information theory, λ is replaced by the term $1/(\ln 2)$. This converts the natural logarithm in Equation 2.89 to the logarithm to the base 2 [indicated as $ld(x)$ in the following equations]. The unit of H then becomes 1 bit (short for "binary digit"). For two events with equal probability ($P_1 = P_2 = 1/2$), H is 1 bit, and the connection to coding and information theory becomes evident. The term $0 \times ld(0)$ is defined to be zero. Figure 2.69 illustrates three different distributions with extreme as well as arbitrary values of H. A case is presented in which one of three events (E_i) can happen. In Figure 2.69a, the probability of event 2 is unity ($P_2 = 1$). The statistical entropy of such a distribution is 0. In Figure 2.69c, the probabilities for all three events are identical. The entropy of such a distribution becomes a maximum. This can be proven using the Lagrange multiplier theorem. Since H is a positive definite function, the distribution in Figure 2.69a must yield the minimum value of H. Hence, all other possible combinations of probabilities (e.g., the distribution in Figure 2.69b) must yield a value of H in the range between 0 and max.

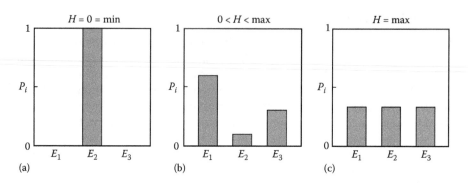

FIGURE 2.69
(a–c) Probability distributions for a case where one of three events E_i can happen: extreme (a, c) and arbitrary (b) values of H. P_i, probability for event i, $\Sigma P_i = 1$ (Rechberger and Brunner, 2002).

2.5.9.3 *Transformation of Statistical Entropy Function*

In order to be applicable to sets of concentrations and mass flows, the statistical entropy function is transformed in three consecutive steps.

First Transformation:

The statistical entropy function is applied to both the input and the output of the investigated system (see Figure 2.68). During this first step, it is assumed that the mass flows of the investigated set of goods are identical and equal to unity ($\dot{m}_i = 1$). This simplification is necessary to understand the analogy between probability and concentration. To quantify the variance of the attribute *concentration* (instead of *probability*), Equation 2.89 with $\lambda = 1/(\ln 2)$ is transformed to Equation 2.91, replacing P_i with c_{ij}/c_j. The relative concentrations c_{ij}/c_j range between 0 and 1, and they can be regarded as a measure of the distribution of substance j among the goods. Equation 2.92 has no physical meaning and only serves to normalize the concentrations c_{ij}. (Since all $\dot{m}_i = 1$, the variable c_{ij} represents normalized substance flows.)

$$H^I(c_{ij}) = \mathrm{ld}(c_j) - \frac{1}{c_j} \cdot \sum_{i=1}^{k} c_{ij} \cdot \mathrm{ld}(c_{ij}) \tag{2.91}$$

$$c_j = \sum_{i=1}^{k} c_{ij} \tag{2.92}$$

where
$i = 1,\ldots, k$
$j = 1,\ldots, n$
c_{ij} = concentration of substance j in good i

Index k gives the number of goods in the set, and n gives the number of investigated substances. As statede, the maximum of H^I is found for $c_{1j} = c_{2j} = \ldots = c_{kj} = c_{j/k}$. Equation 2.91 consequently changes to

$$H^I_{\max} = \mathrm{ld}(k) \qquad (2.93)$$

Thus, H^I ranges from 0 to $\mathrm{ld}(k)$ for all possible sets of concentrations. Since the maximum of H^I is a function of the number of goods (k), the *relative* statistical entropy H^I_{rel} is used to compare sets with different numbers of goods. H^I_{rel} is defined as follows:

$$H(c) = \frac{H^I(c_{ij})}{H^I_{\max}} = \frac{H^I(c_{ij})}{\mathrm{ld}(k)} \qquad (2.94)$$

The value of H^I_{rel} ranges between 0 and 1 for all possible sets of concentrations; it is dimensionless.

Second Transformation:

To further quantify the attribute *mass flow*, Equation 2.91 is modified. The mean concentration in a good (c_{ij}) is weighted with the normalized mass flow ($\hat{m}_i$) of this good; $\hat{m}_i$ can be regarded as the frequency of "occurrence" of the concentration c_{ij} (see Figure 2.70). The entropy H^{II} of a mass-weighted set of concentrations therefore is given by Equations 2.95 and 2.96, which correspond to Equations 2.91 and 2.92, respectively.

$$H^{II}(c_{ij}, \hat{m}_i) = -\sum_{i=1}^{k} \hat{m}_i \cdot c_{ij} \cdot \mathrm{ld}(c_{ij}) \qquad (2.95)$$

$$\dot{X}_j = \sum_{i=1}^{k} \dot{m}_i \cdot c_{ij} \qquad (2.96)$$

$$\hat{m}_i = \frac{\dot{m}_i}{\dot{X}_j} \qquad (2.97)$$

where
  $\dot{m}_i$ = mass flow of good i
  $\dot{X}_j$ = total substance flow induced by the set of goods
  $\hat{m}_i$ = normalized mass flow of good i

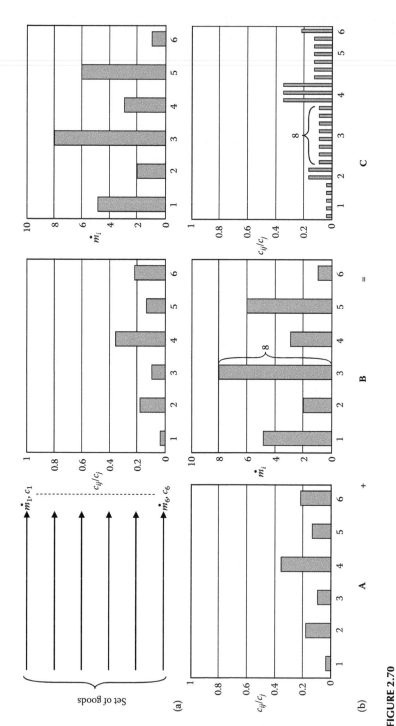

FIGURE 2.70

(a) Exemplary illustration of a set of $k = 6$ goods. The normalized concentrations c_{ij}/c_j and the mass flows $\dot{m}_i$ represent the distribution of substance j among the goods. (b) The mass flow of a good can be interpreted as the frequency of a concentration. The combination of the distribution of concentrations (A) and mass flows (B) can be defined as a weighted distribution of concentrations (C). Distribution C can be quantified by applying Equation 2.27. (From Rechberger, H. and Brunner, P. H., *Environ. Sci. Technol.*, 36, 809, 2002. With permission.)

Third Transformation:

The last modification of the initial function concerns the gaseous and aqueous output goods (emissions). In contrast to the products (e.g., solid residues) of the investigated system, the emissions are diluted in air and receiving waters, which results in an increase in entropy. In Equations 2.99 and 2.100, the index "geog" indicates the "natural" or geogenic concentrations of substances in the atmosphere and the hydrosphere. These concentrations serve as reference values to describe the dilution process. Factor "100" in Equations 2.99 and 2.100 means that the emission (c_{ij}, $\dot{m}_i$; e.g., measured in a stack or wastewater pipe) is mixed with a geogenic flow ($c_{j,geog}$, $\dot{m}_{geog}$), so that the concentration of the resulting flow ($\dot{m}_i + \dot{m}_{geog}$) is 1% above $c_{j,geog}$. It has been demonstrated that this approximation reflects the actual (unlimited) dilution in the environmental compartment sufficiently for $c_{ij} \gg c_{j,geog}$. For $c_{ij} = c_{j,geog}$, there is apparently no dilution. Hence, as a rule of thumb, in cases where $c_{ij}/c_{j,geog} < 10$, Equations 2.99 and 2.100 have to be replaced by more complex terms that cover the total range $c_{j,geog} < c_{ij} < c = 1$ (g/g). The applicability of Equations 2.99 and 2.100 has to be checked in any case.

$$H^{III}\left(\underline{c}_{ij}, \hat{\underline{m}}_i\right) = -\sum_{i=1}^{k} \hat{\underline{m}}_i \cdot \underline{c}_{ij} \cdot \mathrm{ld}(\underline{c}_{ij}) \tag{2.98}$$

where c_{ij} is defined as

$$\underline{c}_{ij} = \begin{cases} c_{j,geog,g}/100 \\ c_{j,geog,a}/100 \\ c_{ij} \end{cases} \text{for} \begin{cases} i = 1,\dots,k_g \\ i = k_g+1,\dots,k_g+k_a \\ i = k_g+k_a+1,\dots,k \end{cases} \text{for} \left.\begin{cases} \text{gaseous} \\ \text{aqueous} \\ \text{solid} \end{cases}\right\} \text{outputs} \tag{2.99}$$

where
k = number of total output goods
k_g = number of gaseous output goods
k_a = number of aqueous output goods
g = gaseous
a = aqueous

and where $\hat{m}_i$ is defined as

$$\hat{\underline{m}}_i = \begin{cases} \dfrac{\hat{X}_{ij}}{c_{j,geog,g}} \cdot 100 \\[2ex] \dfrac{\hat{X}_{ij}}{c_{j,geog,a}} \cdot 100 \\[2ex] \hat{m}_i \end{cases} \tag{2.100}$$

$$\text{for} \begin{cases} i = 1,\dots,k_g \\ i = k_g + 1,\dots,k_g + k_a \\ i = k_g + k_a + 1,\dots,k \end{cases} \text{for} \begin{cases} \text{gaseous} \\ \text{aqueous} \\ \text{solid} \end{cases} \text{outputs}$$

with

$$\hat{X}_{ij} = c_{ij} \cdot \hat{m}_i \quad \text{normalized substance flow of output } i \tag{2.101}$$

The geogenic concentrations in Equations 2.99 and 2.100 can be replaced by more realistic background concentrations. In this case, the influence of the actual surrounding environment is better reflected in the evaluation process of the investigated system. For a simple comparison of options, this is usually not necessary.

Applied to the output of a system, Equation 2.98 quantifies the distribution of substance j. The maximum of H^{III} is reached when all of substance j is directed to the environmental compartment with the lowest geogenic concentration ($c_{j,geog,min}$). For heavy metals, this is usually the atmosphere, since $c_{j,geog,a} > c_{j,geog,g}$. The maximum of H^{III} is given by Equation 2.102.

$$H^{III}_{max.j} = \text{ld}\left(\frac{1}{c_{j,geog,min}} \cdot 100 \right) \tag{2.102}$$

Using Equations 2.94 and 2.98 through 2.102, the relative statistical entropy $RSE_{j,O} = H^{III}_{rel}$ of the output (index O, measure Y in Figure 2.68) can be calculated. In the same way—using Equations 2.94 through 2.97 and 2.102—the $RSE_{j,I} = H^{III}_{rel}$ for the input (index I, measure X in Figure 2.68) of a system is obtained. The difference in the RSE_j between the input and output of a system can be defined as the SCE of the system.

$$SCE_j = \frac{RSE_{j,I} - RSE_{j,O}}{RSE_{j,I}} \cdot 100 \tag{2.103}$$

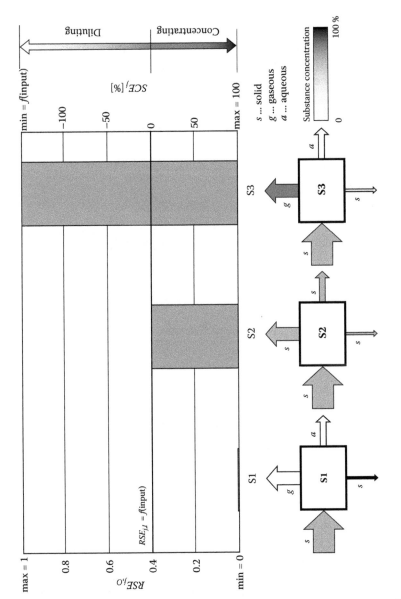

FIGURE 2.71

Relationship between relative statistical entropies RSE_j and substance concentrating efficiency SCE_j. S1 symbolizes a system that achieves maximum concentrating by producing a pure residue containing all of substance j. S2 is an example of a system that does not chemically discriminate between its outputs ($c_i = c_o$). S3 transfers the substance entirely into the atmosphere, which means maximum dilution. (From Rechberger, H. and Brunner, P. H., *Environ. Sci. Technol.*, 36, 809, 2002. With permission.)

The SCE_j is given as percentage and ranges between a negative value, which is a function of the input, and 100%. An SCE_j value of 100% for substance j means that substance j is transferred 100% into one pure output good. $SCE_j = 0$ is the result if RSE_j values of the input and output are identical. This means that the system neither concentrates nor dilutes substance j. Note: This does not imply identical sets of mass flows and concentrations in inputs and outputs. The SCE_j equals a minimum if all of substance j is emitted into that environmental compartment that allows for maximum dilution (in general, the atmosphere). These relationships are illustrated in Figure 2.71.

Incineration transfers 92% of cadmium into fly ash, a small fraction of the 2.5% of total MSW input that can be safely stored in appropriate underground

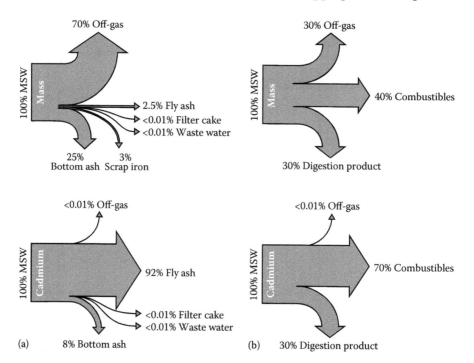

FIGURE 2.72
Partitioning of total mass and cadmium in two waste treatment facilities. (a) MSW incineration. (From Schachermayer, Bauer, Ritter, and Brunner, 1995.) (b) Mechanical–biological treatment of MSW. (From Fehringer, R. et al., (1997), *Auswirkungen unterschiedlicher Szenarien der thermischen Verwertung von Abfällen in Österreich (Project ASTRA)*, Vienna, Austria: Institute for Water Quality, Resource and Waste Management, Technische Universität Wien.) Incineration transfers 92% of cadmium into fly ash, a small fraction of the 2.5% of total MSW input that can be safely stored in appropriate underground disposal facilities. Mechanical–biological treatment does not produce residues with enhanced Cd concentrations with respect to the input, since the partitioning diagrams for mass and cadmium are essentially identical. With regard to management of Cd, solution A is superior to B. Statistical entropy analysis quantifies the advantage: $SCE_A = 42\%$, $SCE_B = 4.2\%$.

disposal facilities. Mechanical–biological treatment does not produce residues with enhanced Cd concentrations with respect to the input, since the partitioning diagrams for mass and cadmium are essentially identical. With regard to management of Cd, solution A is superior to B. Statistical entropy analysis quantifies the advantage: $SCE_A = 42\%$, $SCE_B = 4.2\%$.

2.5.9.4 SEA Applications in MFA

SEA has been applied to MFA results of various waste-treatment facilities and multiprocess systems such as the European copper cycle (see Chapter 3, Section 3.2.2). SEA considers all information of an MFA except the magnitude of stocks. Only inputs into and outputs from a stock are included in SEA. The relevance of a stock has to be assessed by comparison with other geogenic or anthropogenic reference reservoirs (see Figure 2.66). Figure 2.72 gives an example of how SEA quantifies the power of waste-treatment processes to concentrate cadmium.

Classical SFA addresses chemical elements (e.g., heavy metals or nutrients such as nitrogen or phosphorus). However, the chemical speciation of an element is often important to know for assessing impacts. For instance, a simple nitrogen (N) balance of a Waste-to-Energy plant is of limited value regarding environmental pollution because it is important to know whether the nitrogen leaves the plant as molecular nitrogen (N_2), nitrous oxide (NO, NO_2), ammonia (NH_3), or nitrate (NO_3^-), or in some other form. Therefore, Sobańtka and colleagues (Sobańtka, Zessner, and Rechberger, 2012) extended the "classical" SEA to make it applicable to chemical compounds (extended statistical entropy analysis [eSEA]). They applied eSEA to various wastewater treatment plants to determine the efficiency of nitrogen removal, expressed as entropy reduction per amount of costs and energy demand (Sobańtka and Rechberger, 2013; Sobańtka, Pons, Zessner, and Rechberger, 2013a), and they investigated into the efficiency of meat and plant production-based agricultural regions with respect to nitrogen management (Sobańtka, Thaler, Zessner, and Rechberger, 2013b).

3

Case Studies

Looking at the graphic result of a material flow analysis (MFA), it seems easy and straightforward to define the system, collect the data, calculate the results, and draw conclusions. In practice, one does not start with the result but quite often with a badly defined problem that is highly complex and that has to be simplified and well structured first. After the goals of an MFA have been clearly defined, the real art consists of skillfully designing a system of boundaries, processes, flows, and stocks that allows solving a given problem at the least cost. Like in any other art, a precondition for mastering the art is to exercise the basic tools as much as possible. The more experienced a user gets, the easier it becomes to set up an appropriate system in a cost-effective way. An expert skilled in MFA will be able to define a metabolic system in any new field quite efficiently, with only a few alterations of the initial draft. Beginners will often find out that they have to revise their systems several times in order to cope with facts such as incomplete information about the important processes, stocks, and flows within the system; inappropriate systems boundaries; missing, bad, or incompatible data; etc.

MFA is usually a multidisciplinary task. Materials flow through many branches of an economy, and they cross boundaries such as the interfaces anthroposphere–environment or water–air–soil. Hence, it is of prime importance to look for guidance from experts who understand those disciplines that are important for a particular MFA: if regional eutrophication due to poor nutrient management is investigated by MFA, it is necessary to include the knowledge of partners from agriculture, nutrition, sewage treatment, water quality, and hydrology, either by forming a project team or by engaging the experts as consultants when needed. Sometimes, this cooperation leads to new research questions, because the disciplinary research may, so far, not have been directed toward linking their disciplinary knowledge with other fields (cf. Section 3.1.2).

An MFA can be a time-consuming and costly task. This is especially true if an MFA is performed for the first time in a new field, such as a study of regional heavy metal flows (cf. Sections 3.1.1 and 3.4.1). It may well be that the basic data of the region, such as anthropogenic flows and stocks, hydrological data on precipitation, evaporation, and surface and groundwater flows and stocks, have not been assessed before. It should be realized that a minimum amount of information is needed; otherwise, an MFA cannot succeed. Thus, sufficient resources in manpower and funding are required.

It is a distinctly different task to perform an MFA in a particular field for the first time, or to repeat the analysis either for additional materials (e.g., first heavy metals, then nutrients) or for further time periods. The latter two tasks require less effort because the system has been set up and basic data, particularly on the level of goods, have already been collected before. If the costs for an initial MFA seem to be high, it should always be taken into account that the fundamental data can be used for future MFA and similar consecutive studies, such as annual environmental reporting or materials accounting.

The following 18 case studies demonstrate how MFA can be applied for

- Early recognition of beneficial and/or harmful accumulation and depletion of substances in stocks
- Optimization of single processes and of entire metabolic systems
- Policy analysis and policy decisions regarding the three fields of environmental management, resource management, and waste management

In addition, an example of regional materials management (lead) is given in order to show that MFA is especially well suited to address problems related to multiple fields, such as the three described before: the regional lead study by MFA was initially not addressed to any specific problem; it revealed conclusions important for all three fields. The case studies are intended to increase the reader's experience. It is recommended also to read some of the original literature cited in these case studies. Nevertheless, for those who want to *master* the *fine art* of MFA (König, 2002), it will be indispensable to gain additional experience by performing on their own as many MFA studies as possible. Remember that looking at a final graph of an MFA reveals by no means the difficulties even experts encounter when condensing the complex reality of the world into an easily understandable, comprehensive MFA system.

3.1 Environmental Management

Most material flow analysis studies have been undertaken to solve problems related to environmental management. A recent overview of the potential of MFA in this field is given in MAcTEmPo (Brunner et al., 1998). In general, MFA is a tool well suited for

- Early recognition of environmental loadings
- Linking of emissions to sources and vice versa

- Setting of priorities for management measures
- Designing new processes, goods, and systems in view of environmental constraints

As seen in Chapter 2, an MFA is usually the starting point of any life-cycle assessment (LCA) and environmental impact statement (EIS). It is also useful as a base for an environmental management and audit system (EMAS) at the company level (see Section 3.1.4). If a company's financial accounting system is linked to a material input–output flow and stock analysis, it can be efficiently used to measure the company's environmental performance. The following case studies demonstrate that MFA can be used to investigate

- Single-substance issues (e.g., emissions of heavy metals or nutrients)
- Multisubstance problems (e.g., EIS of a coal-fired power plant)

They also show the wide scale of spatial application: a single power plant, a small region of 66 km², and a large watershed such as the entire River Danube basin with 820,000 km² can all be investigated using the same MFA approach.

3.1.1 Case Study 1: Regional Lead Pollution

Heavy metals are important substances for both economic as well as environmental reasons. Because of their physical–chemical properties, they can withstand weathering (zinc coatings of steel), improve the properties of other materials (chromium in steel, cadmium as an additive in polyvinyl chloride [PVC]), or serve to improve the efficiency of energy systems (lead in gasoline, mercury in batteries). Some heavy metals are not essential for the biosphere, but many are toxic for humans, animals, plants, and microorganisms. It is thus important to control the flows and stocks of heavy metals to avoid harmful flows and accumulations and to make the best use of heavy metals as resources.

This case study is taken from RESUB, a comprehensive study on the flows and stocks of 12 elements in a Swiss region (Bunz Valley) of 66 km² and 28,000 inhabitants (Brunner et al., 1990). The purpose was to develop a methodology to assess material flows and stocks within, into, and out of a region in a thorough and integrated way. In addition, the significance of the findings for the management of resources and the environment was to be investigated. There was no given goal in view of environmental management. The case study portrayed in this chapter represents merely a small fraction of the entire RESUB project. Only the flows and stocks of lead relevant to environmental management are discussed. The implications of these flows and stocks for resource management are examined in Section 3.4.1. The detailed procedure described next confirms that an MFA is a multidisciplinary task that requires knowledge, information, and support from many fields.

3.1.1.1 Procedures

In a first step, the region is defined according to Figure 3.1. For the spatial boundary, the administrative boundary of the region "Bunz Valley" is chosen because, by chance, this border coincides well with the hydrological boundary. (This is often the case in mountainous or hilly regions, where the watershed serves well to delineate an administrative boundary.) Because water flow is fundamental for many material flows, it is important that a reliable regional water balance be established. If the spatial boundary does not coincide with the hydrological boundary, it may be difficult to establish a water balance. Hence, it is often crucial to find a good compromise between regional boundaries that match the administrative region, thus allowing

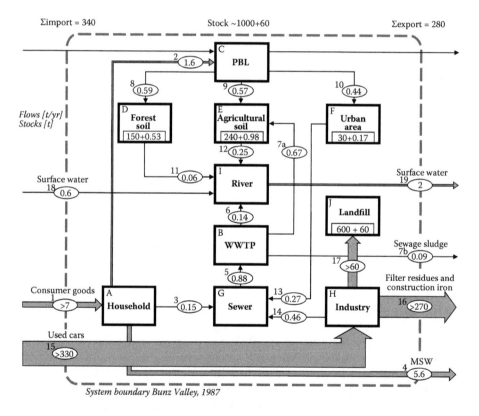

FIGURE 3.1

Results of the MFA of lead flows (t/year) and stocks (t) through the Bunz Valley. (From Brunner, P. H. et al., Industrial metabolism at the regional and local level: A case study on a Swiss region, in *Industrial Metabolism—Restructuring for Sustainable Development*, Ayres, R. U. and Simonis, U. E., Eds., United Nations University Press, Tokyo, 1994. With permission. Brunner, P. H. et al., *RESUB—Der regionale Stoffhaushalt im Unteren Bünztal, Die Entwicklung einer Methodik zur Erfassung des regionalen Stoffhaushaltes*, 1990.)

the use of data collected by the regional administration, and hydrological boundaries that yield a consistent water balance.

As a boundary in time, a period of 1 year is selected because existing data about the anthroposphere (e.g., tax revenues, population data, and fuel consumption) and the environment (e.g., data on precipitation, surface water flow, and concentrations in soil and groundwater) show that a sampling period of 1 year is representative for the region during the period of 1985 to 1990.

In this chapter, each process is labeled with a letter and each flow with a number. These letters and numbers help to identify the corresponding processes and flows in Figure 3.1, Tables 3.1 through 3.10, and the calculations at the end of this chapter. The system is defined by the following 10 processes and 20 flows of goods.

3.1.1.1.1 Private Households

The process *private households* (PHH) summarizes the flows and stocks of materials through 9300 private households of the region. Import goods (1) relevant for lead comprise leaded gasoline (in the process of being phased out) and consumer goods such as lead in stabilizers, caps topping wine bottles, etc. Output goods are exhaust gas from cars (2), sewage (3), and municipal solid waste (MSW) (4).

The lead flows through private households are calculated as follows. Lead input in consumer goods is calculated based on the flows of sewage and MSW. This is a major shortcoming, since neither stocks nor flows of construction materials and appliances in private households are taken into account. To measure flows of lead in such goods is an extremely laborious and costly task. Thus, for this study, it is assumed that all lead that enters households leaves them within 1 year. This hypothesis is incorrect, since it does not account for the lead stock in households. Nevertheless, it is estimated that this error is of little relevance for the conclusions and the overall lead balance of the region.

Figures for lead in sewage are calculated as follows. The number of inhabitants (capita) connected to the sewer system (percentage of the regional population) is multiplied by per capita lead-emission factors (g/capita/year) determined elsewhere in similar regions.

Lead in MSW is similarly calculated. The number of inhabitants (capita) times MSW generation rate (kg/capita/year) times lead concentration in MSW (g/kg) yields the lead flow in MSW. MSW generation rate is available from regional waste-management companies. Lead concentration in MSW is taken from measurements of the residues of waste incineration (see Section 3.3.1).

It is assumed that all lead emitted by car exhausts stems from leaded gasoline. Fuel consumed for room heating contains less than 0.05 t of lead (calculated as the amount of fuel consumed times lead concentration in fuel) and is not taken into further consideration. Lead in car exhaust is calculated as

TABLE 3.1

Calculation of Lead Flows through Process *Private Household*

Flow No.	Operator	Description	Units	Value
Stock				
		Initial value	kg	Not considered
		Rate of change	kg Pb/year	Not considered
Inputs				
1		Consumer goods and leaded gasoline (balanced):		
		Exhaust gas (2)	kg Pb/year	1596
	+	Sewage (3)	kg Pb/year	151
	+	MSW (4)	kg Pb/year	5600
	=	Total lead flow	kg Pb/year	7347
Outputs				
2		Exhaust gas:		
		Number of cars	cars	14,000
	×	Mileage	km/car/year	15,000
	×	Consumption of gasoline	L/km	0.08
	×	Lead content of gasoline	mg Pb/L	95
	=	Total lead flow	mg Pb/year	1.6×10^9
	=		kg Pb/year	1596
3		Household sewage:		
		Number of inhabitants	capita	28,000
	×	Connected to sewer system	–	1
	×	Lead emission per capita	g Pb/capita/year	5.4
	=	Total lead flow	g Pb/year	1.51×10^5
	=		kg Pb/year	151
4		Municipal solid waste (MSW):		
		Number of inhabitants	capita	28,000
	×	MSW generation rate	kg/capita/year	400
	×	Lead concentration in MSW	g Pb/kg MSW	0.5
	=	Total lead flow	g Pb/year	5.6×10^6
			kg Pb/year	5600

Note: Process A in Figure 3.1.

follows. The number of cars licensed in the region is multiplied by the average mileage (in km/year) of a car (taken from national statistics), the average consumption of gasoline per kilometer (l/km, from car manufacturers' statistics), and the mean lead content of the gasoline (mg/L, from gasoline producers and federal statistics). The results are cross-checked by figures from regional traffic monitoring and a model that takes into account the road

TABLE 3.2

Calculation of Lead Flows through Process *Wastewater Treatment Plant*

Flow No.	Operator	Description	Units	Value 1	Value 2	Total Value
Stock						
		Initial value	kg	Not considered		
		Rate of change	kg Pb/year	−16		
Inputs						
5		WWTP input:		TP 1	TP 2 + 3	TP 1 + 2 + 3
		Wastewater flow	L/year	6.1×10^9	2.26×10^9	8.36×10^9
	×	Lead concentration	µg Pb/L	121.7	59.2	
	=	Total lead flow	µg Pb/year	7.42×10^{11}	1.34×10^{11}	
	=		kg Pb/year	742	134	876
Outputs						
6		WWTP output:		TP 1	TP 2 + 3	TP 1 + 2 + 3
		Purified water flow	L/year	6.1×10^9	2.26×10^9	8.36×10^9
	×	Lead concentration	µg Pb/L	20.7	6.3	
	=	Total lead flow	µg Pb/year	1.26×10^{11}	1.42×10^{10}	
	=		kg Pb/year	126	14	140
7a		Sewage sludge (used):		TP 1	TP 2 + 3	TP 1 + 2 + 3
		Sludge flow	kg dry/year	8.06×10^5	2.14×10^5	1.02×10^6
	×	Lead concentration	mg Pb/kg dry	875	216	
	×	Used inside of the region	%	92	36	
	=	Total lead flow	mg Pb/year	6.49×10^8	1.66×10^7	
	=		kg Pb/year	649	17	665[a]
7b		Sewage sludge (exported):		TP 1	TP 2 + 3	TP 1 + 2 + 3
		Sludge flow	kg dry/year	8.06×10^5	2.14×10^5	1.02×10^6
	×	Lead concentration	mg Pb/kg dry	875	216	

(Continued)

TABLE 3.2 (CONTINUED)

Calculation of Lead Flows through Process *Wastewater Treatment Plant*

Flow No.	Operator	Description	Units	Value 1	Value 2	Total Value
	×	Exported out of the region	%	8	64	
	=	Total lead flow	mg Pb/year	5.64×10^7	2.96×10^7	
	=		kg Pb/year	56	30	86

Note: Process B in Figure 3.1.

a Total value does not equal the sum of values 1 and 2, due to rounded values.

TABLE 3.3

Calculation of Lead Flows through Process *Planetary Boundary Layer*

Flow No.	Operator	Description	Units	Value 1	Value 2	Value 3
Stock						
		Initial value	kg	Not considered		
		Rate of change	kg Pb/year	0		
Inputs						
2		Exhaust gas[a]:				
		Total lead flow	kg Pb/year	1596		
Outputs						
8, 9, 10		Deposition:		Forest	Agriculture	Urban
		Deposition ratio	–	3	1	5
		Deposition rate	kg Pb/ha/year	0.294	0.098	0.490
	×	Surface	ha	2000	3700	900
	+	Additional lead (roads)	kg Pb/year	–	200	–
		Total lead flow	kg Pb/year	588	563	441

Note: Process C in Figure 3.1.

a Data from PHH (see Table 3.1).

network (fractions of highways, urban roads, roads outside of settlement areas) and speed-dependent emission factors (figures for lead concentration in gasoline are kept constant). Lead emissions of trucks are considered to be small and are neglected because, in this region, trucks are operated on diesel only, and diesel does not contain significant amounts of lead.

Figures about the total input into households are rounded because they do not include lead-containing goods that contribute to the stock and thus

TABLE 3.4

Calculation of Lead Flows through Process *Forest Soil*

Flow No.	Operator	Description	Units	Value
Stock				
		Initial value	kg	150,000
		Rate of change	kg Pb/year	529
Inputs				
8		Deposition[a]:		
		Total lead flow	kg Pb/year	588
Outputs				
11		Runoff:		
		Deposition	kg Pb/year	588
	×	Runoff factor	–	0.1
	=	Total lead flow	kg Pb/year	59

Note: Process D in Figure 3.1.
[a] Data from PBL (see Table 3.3).

TABLE 3.5

Calculation of Lead Flows through Process *Agricultural Soil*

Flow No.	Operator	Description	Units	Value
Stock				
		Initial value	kg	240,000
		Rate of change	kg Pb/year	982
Inputs				
9		Deposition:		
	+	From PBL	kg Pb/year	563
	+	From WWTP	kg Pb/year	665
	=	Total lead flow	kg Pb/year	1228
Outputs				
12		Runoff:		
		Deposition	kg Pb/year	1228
	×	Runoff factor	–	0.2
	=	Total lead flow	kg Pb/year	246

Note: Process E in Figure 3.1.

have little effect on accuracy. For the overall results and conclusions, this accuracy is sufficient. If it turns out that, for the conclusions, the difference between the calculated and the rounded value is decisive, a more thorough investigation into the lead flows through private households would become necessary.

For detailed information about the calculation, see Table 3.1.

TABLE 3.6

Calculation of Lead Flows through Process *Urban Areas*

Flow No.	Operator	Description	Units	Value 1	Value 2	Total Value
Stock						
		Initial value	kg	30,000		
		Rate of change	kg Pb/year	176		
Inputs						
10		Deposition[a]:				
		Total lead flow	kg Pb/year	441		
Outputs						
13		Runoff:		Buildings	Green	Buildings + green
		Deposition	kg Pb/year	221	221	
	×	Runoff factor	–	1.00	0.20	
	=	Total lead flow	kg Pb/year	221	44	265

Note: Process F in Figure 3.1.
[a] Data from PBL (see Table 3.3).

TABLE 3.7

Calculation of Lead Flows through Process *Sewer*

Flow No.	Operator	Description	Units	Value
Stock				
		Initial value	kg	n.d.
		Rate of change	kg Pb/year	0
Inputs				
3		Household sewage[a]:		
		Total lead flow	kg Pb/year	151
13		Urban area runoff[b]:		
		Total lead flow	kg Pb/year	265
14		Industry sewage (balanced):		
		WWTP input (5)	kg Pb/year	876
	–	PHH sewage (3)	kg Pb/year	151
	–	UA runoff (13)	kg Pb/year	265
	=	Total lead flow	kg Pb/year	460
Outputs				
5		WWTP input[c]:		
		Total lead flow	kg Pb/year	876

Note: n.d. = not determined. Process G in Figure 3.1.
[a] Data from PHH (see Table 3.1).
[b] Data from UA (see Table 3.6).
[c] Data from WWTP (see Table 3.2).

TABLE 3.8

Calculation of Lead Flows through Process *Industry*

Flow No.	Operator	Description	Units	Value
Stock				
		Initial value	kg	Not considered
		Rate of change	kg Pb/year	0
Inputs				
15a		Used cars:		
		Number of used cars	cars/year	120,000
	×	Lead per car (excl. battery)	kg Pb/car	2.5
	=	Total lead flow	kg Pb/year	300,000
15b		Scrap metal:		
		Scrap metal	kg/year	6.50×10^7
	×	Lead content	kg Pb/kg	0.0005
	=	Total lead flow	kg Pb/year	32,500
15		Industry input:		
		Used cars (15a)	kg Pb/year	300,000
	+	Scrap metal (15b)	kg Pb/year	32,500
	=	Total lead flow	kg Pb/year	332,500
Outputs				
16a		Construction iron:		
		Construction iron	kg/year	1.45×10^8
	×	Lead content	kg Pb/kg	0.0005
	=	Total lead flow	kg Pb/year	72,500
16b		Filter residues:		
		Filter residues	kg/year	1.50×10^7
	×	Lead content	kg Pb/kg	0.0133
	=	Total lead flow	kg Pb/year	200,000
16		Industry output:		
		Construction iron (16a)	kg Pb/year	72,500
	+	Filter residues (16b)	kg Pb/year	200,000
	=	Total lead flow	kg Pb/year	272,500
14		Industry sewage[a]:		
		Total lead flow	kg Pb/year	460
17		Automotive shredder residues (balanced):		
		Industry input (15)	kg Pb/year	332,500
	−	Industry output (16)	kg Pb/year	272,500
	−	Industry sewage (14)	kg Pb/year	460
	=	Total lead flow	kg Pb/year	59,540

Note: Process H in Figure 3.1.

[a] Data from Sewer (see Table 3.7).

TABLE 3.9

Calculation of Lead Flows through Process *Landfill*

Flow No.	Description	Units	Value
Stock			
	Initial value	kg	600,000
	Rate of change	kg Pb/year	59,540
Inputs			
17	ASR[a]:		
	Total lead flow	kg Pb/year	59,540

Note: Process J in Figure 3.1.
[a] Data from Industry (see Table 3.8).

3.1.1.1.2 Wastewater Treatment Plant

In the process wastewater treatment plant (WWTP), wastewater (5) is treated, resulting in cleaned wastewater (6), sewage sludge (7a and b), off-gas, and small amounts of sieving residues and sandy sediments. Due to the chemical species of lead in wastewater, off-gas is of no quantitative relevance for this heavy metal. Preliminary sampling and chemical analysis of the concentrations in sievings and sediments show that the amount of lead in these fractions is small. Hence, most lead leaves the WWTP in sewage sludge and purified wastewater. The flows of wastewater, purified wastewater, and sludge are measured during 1 year (m³/year) and are sampled and analyzed for lead (g/m³). The flow of wastewater is determined by a venturi device at the inflow of the WWTP. Samples are taken continuously from wastewater and purified wastewater by a so-called Q/s sampler that samples proportional to the water flow. The flow of sewage sludge is measured as the total volume of sludge transferred to sludge transport vehicles during 1 year. Samples are taken whenever sludge is transferred to transport vehicles. There are three WWTPs in the region: one large plant and two small plants. Only the large plant and one of the small plants are included in the measuring campaign. For the third plant, the same material flows and balances are anticipated as for the other small treatment plant. The data of the three plants are summarized as a single WWTP process.

For detailed information about the calculation, see Table 3.2.

3.1.1.1.3 Planetary Boundary Layer

The process *planetary boundary layer* (PBL) denotes the lowest layer of the atmosphere. It is about 500 m high and is well suited as a "distribution" process for the RESUB case study. In regional studies, it is usually not possible to measure a material balance of the PBL, because it is a daunting analytical and modeling task. However, it is possible to make certain assumptions and simplifications that allow using the PBL as a suitable process to account for flows from the atmosphere to the soil and vice versa.

TABLE 3.10

Calculation of Lead Flows through Process *River*

Flow No.	Operator	Description	Units	Value 1	Value 2	Total Value
Stock						
		Initial value	kg	Not considered		
		Rate of change	kg Pb/year	–948		
Inputs						
18		Surface water (import):		Holzbach	Bunz	Sum
		Water flow	L/year	3.69×10^9	3.19×10^{10}	3.56×10^{10}
	×	Lead concentration	µg Pb/L	4.6	18.4	
	=	Total lead flow	µg/year	1.70×10^{10}	5.87×10^{11}	
			kg Pb/year	17	587	604
11		Forest runoff[a]:				
		Total lead flow	kg Pb/year	59		
12		Agricultural runoff[b]:				
		Total lead flow	kg Pb/year	246		
6		WWTP output[c]:				
		Total lead flow	kg Pb/year	140		
Outputs						
19		Surface water (export):				
		Water flow	L/year	6.70×10^{10}		
	×	Lead concentration	µg Pb/L	29.8		
	=	Total lead flow	µg/year	2.00×10^{12}		
	=		kg Pb/year	1997		

Note: Process I in Figure 3.1.
[a] Data from Forest Soil (see Table 3.4).
[b] Data from Agricultural Soil (see Table 3.5).
[c] Data from WWTP (see Table 3.2).

Because the case study region is surrounded by regions that have basically the same metabolic characteristics, it is reasonable to suppose that the emissions into air are similar for all surrounding regions. Thus, it can be assumed that imported lead corresponds to exported lead, i.e., that the amount of lead imported and deposited in the region equals the amount of domestic lead exported and deposited outside of the region. Note that the total flow of lead through the PBL (not given in Figure 3.1) is about three to four times larger than the flow of lead deposited. The flows from the PBL to the soil consist of wet and dry depositions to forest (8), agricultural land (9), and urban (10) soils. The lead flows to the soil are determined by two methods (Beer, 1990). First, based on the assumption of uniform metabolism of all neighboring regions, the total regional emission to PBL is divided among the land areas, taking into account differences in vegetation and surface. Second, preliminary measurements at 11 sampling stations throughout the region show little significant differences for long measuring periods. Therefore, for a period of 1 year, wet and dry deposition of lead is measured only at two sampling stations in the region. Based on the wet and dry deposition results, and on models given by Beer (1990), lead flows are calculated for the corresponding soils. The deposition on urban soils takes into account that most lead is emitted in the proximity of roads; thus, the load per hectare of urban soils is comparatively larger than that of agricultural and forest soils. Results from the two approaches agree fairly well. The method based on actual measurements of dry and wet depositions yields values about 30% higher than the distribution of the total emissions over the region.

For detailed information about the calculation, see Table 3.3.

3.1.1.1.4 *Lead Flows and Stocks in Soils*

The region consists of 3700 ha soil used for agriculture, 2000 ha forest soil, and 900 ha settlement area. The area actually covered with buildings, roads, and other constructions is much smaller. The hydrological balance reveals the water flow to and from the forest soil and the agricultural soil. Precipitation (measured by continuous automatic rain measurement) minus evaporation, estimated with various models and finally calculated according to Primault (1962), yields the net water flow to the soil. This amount of water is divided among the fractions of agriculture and forest soils, taking into account differences in evapotranspiration of forests and agricultural crops. The water reaching the soil is divided into surface runoff and interflow (both reaching surface waters) and the fraction seeping to groundwater. The concentrations of lead in soil leachate are estimated based on another research project on heavy metal mobility in soils (Udluft, 1981). Erosion is approximated according to von Steiger and Baccini (1990). Both assessments are individually tailored for forest and agricultural soils. The calculations show that 10% of the deposited lead on forest soil (11), 20% of agricultural soil (12), and up to 60% of built-up areas (13) can be found in the runoff. This lead is transported to receiving waters.

For detailed information about the calculation, see Tables 3.4 through 3.6.

3.1.1.1.5 Sewer System

The mixed sewer system receives wastewater from private households (3), urban surface runoff (13), and industry (14), and transports the resulting sewage (5) to the WWTP. Lead in wastewater from industry can be estimated by balancing the process *sewer*, taking into account lead flows in wastewater from private households, in surface runoff from urban areas, and in the resulting sewage.

For detailed information about the calculation, see Table 3.7.

3.1.1.1.6 Industry

The process *industry* proves to be a real challenge. Despite the region's small size and low population, there are 1300 companies with 11,000 employees active in the region. The main task is to find within this large number those companies that play an important role in the regional lead flow. As a first step, all sectors except the production sector are removed from the list. Of the remaining 323 companies, those with less than 20 employees are eliminated. The remaining 102 businesses are included in the investigation, which consists of an interview and a questionnaire about the material flows and stocks of each company. Of these, 61 companies cooperate actively and supply comprehensive data about their material turnover. It appears that only a few are handling lead-containing goods. A car shredder and an adjacent iron smelter using shredded cars to produce construction rods are dominating the process *industry*. Hence, the inputs into the process *industry* are used cars (15). The outputs comprise, on the one hand, construction rods (16a) and filter residues (16b) of the smelter that are exported. On the other hand, the car shredder produces organic shredder residues (17) consisting of plastics, textiles, and biomass (wood, paper, leather, and hair) mixed with residual metals of every kind. These so-called automotive shredder residues (ASRs) are landfilled within the region.

The lead flows through *industry* are assessed as follows. Input in *industry* is calculated as the number of used cars times the concentration of lead in a car. This yields a minimal figure. It may well be that other goods are shredded and treated in the shredder as well. (Note that this uncertainty is not important for the final conclusion. It would be the same even if double the amount of lead were used in *industry*.) The number of cars processed by the shredder is supplied by the shredder operator. Figures for lead concentration in used cars are found in the literature or can be received from car manufacturers. The smelter operator supplies figures about the amount of construction steel produced, the amount of filter residue exported, as well as the lead concentration in the steel and the filter residue. The latter figure is confirmed by local authorities, who periodically monitor emissions of the smelter. Note that there is no emission flow given for the smelter in Figure 3.1. This is due to the excellent air-pollution control (APC) device of the smelter, which keeps annual emissions at a level that is orders of magnitude below lead emissions from gasoline. Thus, the smelter is of no relevance for lead emissions in the region anymore. The amount of lead in industry sewage is similar to other regional lead flows.

For detailed information about the calculation, see Tables 3.8 and 3.9.

3.1.1.1.7 Surface Waters

The process *river* consists of a river flowing through the region and a small tributary originating predominantly within the region. Groundwater entering or leaving the region does not play an important role. The process *river* receives water from the river inflow (10), the agricultural soil (11), the forest soil (12), and the WWTP (8). The river outflow (13) leaves the region. The direct lead flow from settlement areas to surface waters has not been taken into account. First, the region has a mixed sewer system, and most urban runoff is collected and treated in the WWTP. Second, the settlement area is small (<10%) in comparison with the agricultural and forest soils; hence, neglecting this flow may be justified. The surface water flow has been determined in the course of a complete water balance that is measured for the comprehensive RESUB project. Existing measuring stations continuously record the flow of river water in and out of the region. Since these stations are not located at the exact systems boundaries, the differences are compensated for by taking into account the area contributing water to the river. The river water is sampled continuously at the same stations with Q/s samplers. It turns out that these state-of-the-art samplers are well suited to collect dissolved substances and suspended particles, but they cannot catch aliquots of large particles. In the course of a rainstorm, when the river transports larger particles, debris, and chunks of biomass, the samplers are not working appropriately. In addition to the limits in sampling technology, there are practical problems. On several occasions, large water flows during heavy storms have destroyed or carried off the sampling equipment. Hence, it is advisable to have short sampling periods (e.g., 1 week). In case of invalid sampling, the missing data cover a smaller fraction of the total measuring period and thus are of less weight. Given these shortcomings, values for lead flows in the river must be regarded as minimum flows.

For detailed information about the calculation, see Table 3.10.

3.1.1.2 Results

In this section, conclusions are drawn regarding the use of MFA for environmental management. In particular, the results are used to show how MFA serves to provide early recognition of environmental hazards, how it can be used to establish priorities for environmental measures, and how it can be used for efficient environmental monitoring. In Section 3.4.1, the same case study is further used to point out the potential for regional materials management and resource conservation. The numerical results of the MFA analysis of lead in the region are given in Figure 3.1.

3.1.1.2.1 *Early Recognition of Environmental Hazards*

The difference between lead import and lead export amounts to approximately 60 t/year. Hence, lead is accumulated in the region. The existing stock of lead totals about 1000 t. A "doubling time" for the lead stock of 17 years can be calculated. In other words, if the regional flows of lead remain the same for the next 100 years, the stock will have increased from 1000 to 7000 t! (Note that according to Chapter 1, Section 1.4.5, there is no indication yet that lead flows will decrease; on the contrary, based on past developments, they are likely to increase further.) Without the present study, this buildup of lead occurs unnoticed. As shown in Chapter 1, Section 1.4.5, such accumulations of substances are a rule for all urban regions. What makes this case study special is the huge extent of the accumulation. About one-sixth of the lead imported stays "forever" within the region. Thus, it is highly important to investigate the fate of potentially toxic lead in the region. Does the accumulation in the soil result in an increase of lead in plants up to a level of concern for animal or human food? Will there be a steady increase in lead flows from the soil to the surface water? When will lead concentrations reach a level that endangers the standards for surface water or drinking water? What about the concentration of lead in dust; will it increase, too?

While MFA is helpful in identifying the problem and formulating relevant questions, the questions cannot be answered by simple MFA alone. It is necessary to engage specific experts, e.g., in the field of metal transfer between soil and plants, between soil and surface water and groundwater, and between soil and air. The merit of MFA is the ability to identify a future environmental problem that has been neither on the agenda nor even known before the study has begun.

From an environmental point of view, the largest flow, stock, and accumulation in stock are caused by the lead imported in used cars, shredded, and landfilled as ASR. The landfill is by far the main regional "accumulator" for lead. Assuming similar practice over the past 10 years, it can be estimated that approximately 600 t of lead is buried in the landfill. This is due to the fact that the separation of lead by the car shredder is incomplete. Some elemental lead as well as lead compounds in plastic additives are transferred to the ASR. The hypothesis that ASR may contribute to the pollution of the regional hydrosphere is discussed next.

Lead is accumulated in the soil, too. The doubling periods are between 170 (urban soil) and 280 (agricultural soil) years. If the use of lead continues in the same way, standards for lead concentrations in soils will be exceeded in the future. It is a matter of soil-protection strategy (and, in a wider perspective, environmental protection) whether such a slow approach to a limit needs to be controlled or not, and if so, when. The "filling up" strategy raises the question of what options future generations will have when they inherit the "full" soil. If soil inputs and outputs can be kept in equilibrium, the concentrations in the soil will remain constant. The case-study region will come close to this condition if leaded gasoline is phased out. (Note that since the residence time of aerosol-borne lead in the atmosphere is several days, which facilitates the transport of lead over long distances, this measure will be effective only if neighboring regions adopt the same strategy.) Lead concentration increases fastest in urban soils. MFA results suggest that persons eating food grown in urban areas, such as home gardeners, consume the highest amount of heavy metals such as lead. Hence, material balances and analysis of urban soils and gardens are needed in order to protect consumers of homegrown products in cities.

3.1.1.2.2 *Establish Priorities for Environmental Measures*

The landfill is the most important stock of lead, so it must be investigated and controlled first. Based on the following calculation, it can be hypothesized that lead is leaching into the groundwater and surface waters. The balance of the process *river* reveals a deficit of 0.95 t/year. Compared with the lead flows from soils and WWTP, this is a large figure. The most likely stock that can lead to such a large flow is the landfill stock. At present, the fate of lead in the landfill is not known. Since much of the ASR has been landfilled without bottom and top liners, it may be that lead is leaching into groundwater and surface water. Even if the landfills are constructed according to the state of the art, with impermeable bottom and top liners, it still has to be expected that the liners will become permeable over the long run (>100 years), thus polluting ground and surface waters for long time periods (Baccini and Lichtensteiger, 1989).

The following issues are crucial for possible mobilization and emission of lead from the landfill: the interaction of the landfill body with the surroundings (atmosphere, precipitation), the transformations of ASR within the landfill due to biochemical and chemical reactions, and the chemical speciation of the lead in ASR. Investigations into these aspects are specialized tasks that cannot be performed by MFA. Such investigations require in-depth analysis by experts in the field of transformation and leaching processes in soils and landfills. It is not possible within the case study to follow up this hypothesis (e.g., by collecting leachates and analyzing it for lead). In any case, it is of first priority to ensure that the ASR is landfilled in a manner that ensures long-term immobilization of heavy metals. If disposing of raw ASR leads to leaching of heavy metals and significant water pollution, pretreatment of ASR before landfilling will be mandatory.

The second largest regional flow of lead is due to MSW, which is exported from the region and incinerated. This flow is an order of magnitude larger than the lead flow in sewage. Thus, compost from MSW is less suited for application to land than sewage sludge because it will overload the soil with lead in a comparatively short time. For this region, it is recommended that sewage sludge not be applied to soils as well, since it loads the soil with additional lead. It is clear that a decision to use compost or sludge in agriculture cannot be based on lead alone. The approach taken here is exemplary and also has to be applied to a full range of substances such as other heavy metals, nutrients, and organic substances.

As mentioned previously, MSW is incinerated outside the region. In order to protect the soil from the large flow of lead in MSW, thermal treatment of MSW has to be combined with efficient APC. If incineration transfers less than 0.01% of lead into air, the resulting increase of lead in soil will be below 1% in 8000 years (assuming uniform deposition within the region). State-of-the-art MSW incinerators exhibit such transfer coefficients (TCs) for lead to the atmosphere. Regarding the smelter, the MFA supports the conclusion that the emissions to air are of no priority ever since the furnace has been equipped with a high-efficiency fabric filter system, reducing the lead emissions to less than 50 kg/year.

3.1.1.2.3 Environmental Monitoring

Once an MFA of a region is established, many opportunities for monitoring substance flows and stocks arise.

MFA can replace soil monitoring programs. Such programs are costly and are limited in their forecasting capabilities. If statistically significant changes in soil concentrations are to be detected by traditional soil monitoring, then either (1) very intensive sampling programs with large numbers of samples or (2) sampling over long time periods are required. Because the funds for such intense sampling are not usually available, it takes decades until significant changes in the soil become visible. However, with a single measuring campaign, MFA can predict how the soil concentration will evolve over time. The results indicate whether there is a danger before high concentrations are reached. If inputs to the soil are changed, e.g., through the addition of sewage sludge or a ban on leaded gasoline, the effect of such measures can be evaluated by an MFA before they are implemented. In contrast, traditional soil monitoring would take years to confirm accumulation or depletion of soil pollutants in a statistically significant way.

Combining MFA with the analysis of sewage sludge allows monitoring of the process *industry*. For example, before the smelter was equipped with high-efficiency fabric filters, a wet scrubbing system removed metals from the off-gas stream. By accident, it happened that some of the lead-loaded scrubber effluent reached the sewer system and severely contaminated the activated sludge in the treatment plant. While wastewater has a short residence time in the treatment plant, sewage sludge in a digester or storage tank represents a "memory" of several weeks. Thus, sludge samples from the digester can show a prolonged increase in lead concentrations. In combination with concentration data for other metals, it may be possible to identify the source of pollution by assigning metal "fingerprints" (concentration ratios) of sewage sludge to those of scrubber liquid.

Likewise, the combination of MFA and monitoring of MSW incineration residue allows one to assess the flow of lead, as well as other substances, through private households (see Section 3.3.1).

Finally, MFA facilitates an additional type of monitoring. If there is no information available about a process, it may be possible to estimate the missing material flows by mass-balancing the adjacent processes. In the aforementioned case study, data about substance flows through the shredder are not available. Rough data about lead in new cars from the last decade, combined with data supplied by the smelter on lead in the two outputs (filter dust and construction rods), allows the flow of lead through the car shredder to be assessed without analyzing the shredder itself.

3.1.1.3 Basic Data for Calculation of Lead Flows and Stocks

In the calculations presented in Tables 3.1 through 3.10, each process is labeled with a letter and each flow with a number. These letters and numbers help to identify the corresponding processes and flows in Figure 3.1. The description of each process is structured as follows:

- Name of the process
- Lead stock inside the process
- Rate of change of the lead stock
- Name of input flows (including a list of quantities that are used to calculate the lead flows)
- Name of output flows (including a list of quantities that are used to calculate the lead flows)

3.1.2 Case Study 2: Regional Phosphorous Management

Nutrients such as nitrogen, phosphorous, potassium, and carbon are essential for the biosphere. They are the key factors controlling growth and enabling species and populations to develop or causing them to vanish. They are especially crucial for the production of food for humans and animals. Because of limitations inherent to the soil–plant system, not all nutrients delivered to the soil can be taken up by plants (Scheffer, 1989). Thus, agricultural losses of nutrients are common, and they cannot be avoided. Yet, they can be reduced by farming practices that are directed toward minimizing losses to the environment. Nutrients in surface waters enhance the growth of algae (eutrophication). As a consequence, the oxygen content in surface water is reduced due to the increased plankton mass, mass death, and decomposition of organisms. As the oxygen concentration decreases, fish and other organisms find it increasingly difficult to survive. Due to transformations in soil and groundwater, nitrogen can also be lost as NO_x or NH_3 to the atmosphere, contributing to the formation of tropospheric ozone and particulate matter, respectively. Hence, the control of nutrients is of prime importance for the management of resources as well as of the environment.

Case studies 2 and 3 both relate to nutrient pollution. The difference between the two is the scale: a small region of 66 km^2 and 28,000 inhabitants in case study 2 versus the entire River Danube with a watershed of 820,000 km^2 and 85 million inhabitants in case study 3. It is noteworthy that for both scales, the same MFA approach can be taken. Nevertheless, there are focal differences in the task of balancing nutrients on these two scales. The challenge on the large scale is to put together a team (often international) that uses the same approach along the entire stream, allowing true comparison and combination of the individual results. In addition to the present case study, another case study on P is presented in Section 3.5 about regional materials management. In this case study, the challenge of accounting for varying P flows and stocks over longer time periods is discussed, too.

3.1.2.1 Procedures

Like the lead example in Section 3.1.1, case study 1 is a part of the comprehensive RESUB project; it focuses on flows and stocks of phosphorous (P). The procedure is the same as for lead, with some small changes due to the way phosphorous is used. The systems boundaries in space and time are identical. Only agricultural soil is taken into account, since the flow of P on forest and urban soils is comparatively small. Two additional processes for animal breeding and plant production are introduced. Hence, again, 10 processes and 19 flows of goods are taken into account (Figure 3.2).

As a first step, the water balance is estimated. Water is important for the flow of phosphorus because P can be transported both in a dissolved state (leaching) and as a particle (runoff and erosion). Hence, a comprehensive water balance for the region is needed (Figure 3.3). To minimize the costs for an annual water balance, the relevant hydrological flows and processes must be identified by a systems analysis (Figure 3.4).

By a provisional semiquantitative water balance, the main water flows and stocks are identified in order to set priorities for the following costly assessment and measurement program. The main purpose is to achieve sufficient accuracy with the least number of expensive measurements. A potential problem for water balancing is the mismatch between the regional (administrative) and hydrological boundaries. In this study, the two definitions of the region coincide well. The small deviations are compensated for assuming the same net precipitation for areas within and outside the administrative region. Determining the flows and stocks of groundwater is a necessary but usually difficult and resource-consuming task. It is therefore beyond the possibility of most regional MFAs. If groundwater data are not available, and if there are major groundwater inflows, outflows, or changes in stock, a hydrological balance might not be possible. In such cases, MFA has to be limited to a specific regional problem not related to the hydrosphere, or it fails altogether. Data for evapotranspiration can be calculated using various formulas (according to Penman, 1948 or Primault, 1962) and regional data on climate and vegetation. The path of water from precipitation to groundwater and surface water can only roughly be assessed, too. In the present study,

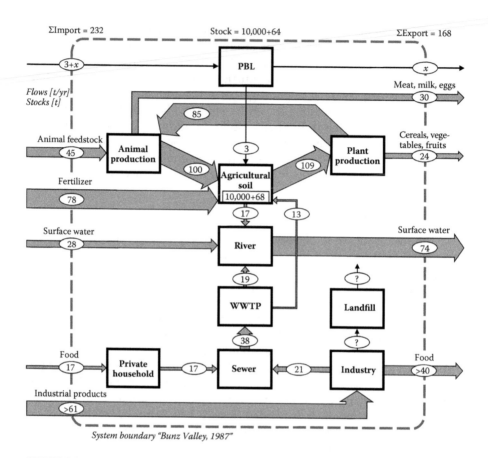

FIGURE 3.2
Regional phosphorous flows and stocks. (From Brunner, P. H. et al., Industrial metabolism at the regional and local level: A case study on a Swiss region, in *Industrial Metabolism—Restructuring for Sustainable Development*, Ayres, R. U. and Simonis, U. E., Eds., United Nations University Press, New York, 1994. With permission. From Brunner, P. H. et al., *RESUB—Der regionale Stoffhaushalt im Unteren Bünztal, Die Entwicklung einer Methodik zur Erfassung des regionalen Stoffhaushaltes.*, 1990.)

there is sufficient information about groundwater outflow from the region to neighboring regions.

The following equation is used for the hydrological balance:

$$\text{Precipitation} + \text{surface water import} + \text{groundwater import}$$
$$+ \text{drinking water import} = \text{evapotranspiration}$$
$$+ \text{surface water export} + \text{groundwater export}$$
$$+ \text{drinking water export} + \text{change in stock}$$

The flows and stocks of water in eight goods, listed in Table 3.11, are measured for a period of 1 year. Samples are taken for the same time period for

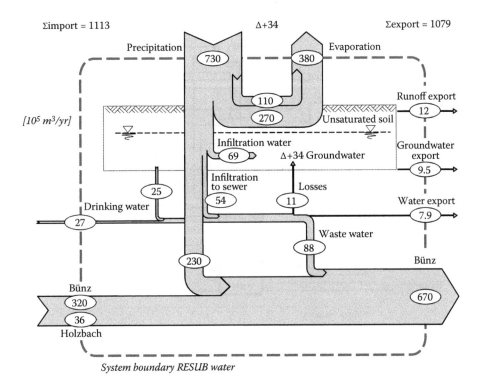

FIGURE 3.3
Results of regional water balance: while the river passes the region, the flow of surface water is doubled by the net precipitation input (precipitation minus evapotranspiration). Bunz and Holzbach are two rivers in the valley. (From Henseler, G. et al., *Vom Wasser*, 78, 91, 1992. With permission.)

most of these goods. Since drinking water is produced from groundwater, it is assumed that drinking water and groundwater have the same concentrations. Measurement and sampling methods, frequencies, and locations are given in Table 3.11 and Figure 3.4. For more information about establishing regional water balances, refer to Henseler, Scheidegger, and Brunner (1992) After analyzing the water balance, the next step is to measure the flows and stocks of phosphorus. For each good investigated, the flow is multiplied by the concentration of P within that good to determine the phosphorus fluxes. In the following, it will be explained how the data presented in Figure 3.2 are assessed.

3.1.2.1.1 Private Household

The flow of food-derived P into private households is established using data about household food consumption (BAS, 1987) and the nutrient content of food (Lentner, 1981). Phosphorus in household detergents and cleaners is not taken into account, since federal legislation banned P for these purposes.

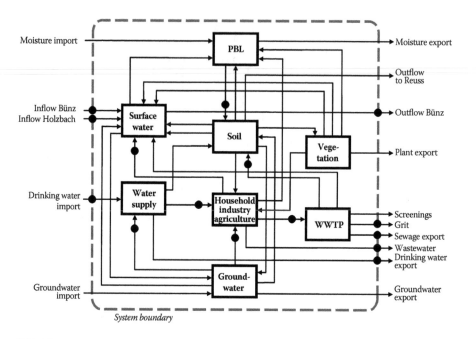

FIGURE 3.4

Determination of regional water balance. (• = flow measurements and sampling points). (From Henseler, G. et al., *Vom Wasser*, 78, 91, 1992. With permission.)

A rough estimation of other P flows showed that they are so small that they do not have to be taken into account (<1% of total regional flow, <10% of flow through private households). The P output of households is not measured but instead is calculated according to Figure 3.5 and the conservation of mass. Of the P entering private households, 90% is assumed to leave by means of wastewater and the remaining 10% by MSW. MSW is not considered further, since it is treated in an MSW incinerator outside the region. It is worth mentioning that composting of MSW can hardly be justified on the grounds of nutrient conservation, since its maximum contribution to regional P management is marginal and about 1% of the total nutrient use in agriculture.

3.1.2.1.2 River

The good *surface water* is flowing in and out of the region at rates of 35 × 10⁶ and 67 × 10⁶ m³/year, respectively (Figure 3.3). The P concentrations in the input and output of the region are measured (0.8 mg/L and 1.1 mg/L, respectively) and multiplied by the corresponding water flow, resulting in 28 and 74 t P/year, respectively (Figure 3.2). Some phosphorus flows such as those in precipitation, drinking water import and export, evapotranspiration, and groundwater are less than 1% of the total regional phosphorus flow. Therefore, they are not taken into account for the phosphorus balance.

TABLE 3.11

Measuring and Sampling Procedure to Establish a Regional Water Balance

Good	Number of Measuring Stations	Method of Flow Measurement	Measuring Period	Method of Substance Sampling	Sampling Period
Precipitation	3	Rain gauge	1 year (365 × 24 h)	Composite sample	27 × 2 weeks
Surface water	3	River gauge	1 year (365 × 24 h)	Composite sample	27 × 2 weeks
Wastewater	2	Venturi	1 year (365 × 24 h)	Composite sample	27 × 2 weeks
Sewage sludge	2	Container	1 year	Composite sample	10 per year
Sievings from WWT[a]	1	Balance	1 year	Grab sample	3 per year
Sand from WWT[b]	1	Volume	1 year	Grab sample	3 per year
Drinking water	9	Meter	1 year	Grab sample	9 per year
Groundwater	5	Water table	1 year	No sampling[c]	–

Source: Henseler, G. et al., *Vom Wasser, 78,* 91–116, 1992.

[a] Screenings from wastewater pretreatment.
[b] Sediment of wastewater pretreatment.
[c] Groundwater identical to drinking water.

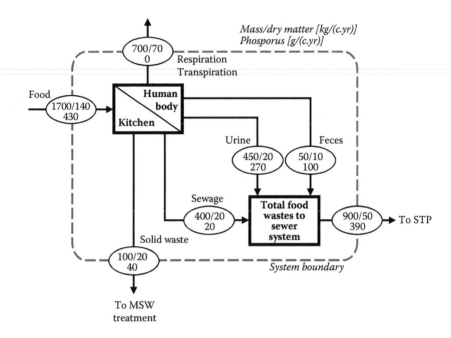

FIGURE 3.5
Flow of food, food dry matter, and phosphorous contained in food through private house-holds. STP: sewage treatment plant. (With kind permission from Springer Science+Business Media: *Metabolism of the Anthroposphere*, 1991, Baccini, P. and Brunner, P. H.)

3.1.2.1.3 Sewer System

The flows of P through the sewer system are calculated using data from household outputs and measured WWTP inputs. The figure for P in "industry" wastewater is calculated as the difference between WWTP input and household wastewater (38 − 17 = 21 t P/year). In the WWTP, the output to the surface waters is measured by multiplying the volume of treated wastewater by the concentration of P measured in 52 biweekly samples of treated wastewater (19 t P/year). P contained in sludge and applied in agriculture (13 t P/year) is measured in metering the total flow of sludge when transferred to transport vehicles, and samples of P are taken for analysis during this transfer.

3.1.2.1.4 Industry

Two aspects are important for the process *industry*: food is stored temporarily in a large stock of interregional importance, and polyphosphates are used in some amounts in regional chemical and other companies. The P contained in food leaves the region unchanged as an export good. P in polyphosphates is transferred to the sewer and is a large source for P in the WWTP.

3.1.2.1.5 Landfill

The process *landfill* is not investigated, although it is possible that phosphorous-containing wastes (biomass, detergents) have been landfilled in the past and that P is leaching to groundwater and surface water.

3.1.2.1.6 Agriculture

Three types of agricultural production systems—*animal breeding, crop raising*, and *miscellaneous*—are defined based on their different managerial characteristics. These agricultural practices are investigated under three processes: *animal production* (production of animals and dairy produce), *plant production* (production of wheat, corn, vegetables, etc.), and *agricultural soil*, as shown in Figure 3.2. For each production system, the use of mineral fertilizer, manure, animal products, and harvested goods are measured per unit of agricultural area and monitored for 2 years. Phosphorus contents of all goods are determined analytically to estimate the annual entry of phosphorus to the soil. All data are double-checked against the values taken from agricultural information sources. Input through manure and fertilizer and output through harvest are then extrapolated, taking the values of the three production systems described previously and considering actual farming practice in the region (e.g., number of animals, amount of produce, area for crop production, etc.). Flows of goods in the process *animal production*, such as animals, fodder, and dairy products, are checked through field accounts. Figures for sewage sludge are collected from WWTPs, and those for deposition, erosion, and runoff are taken from the literature. Figure 3.2 displays the amount of P in the output goods, namely, harvested plants like cereals, vegetables, and fruits (export of 24 t P/year), and animal feedstock cycled within the region (85 t P/year).

The flow of P to plant production including agricultural soil (X) is calculated as follows:

$$X = \text{fertilizer} + \text{manure} + \text{atmospheric deposition}$$
$$+ \text{sewage sludge} - (\text{animal feed produced} + \text{cereals, vegetables, fruits})$$
$$= 194 - 109 = 85 \text{ t P/year}$$

The amount of P stored in the agricultural soil (S) is calculated as follows:

$$S = X - (\text{erosion} + \text{runoff and leaching to surface and groundwater})$$
$$= 85 - 17 = 68 \text{ t P/year}$$

The groundwater inflow into the region is close to zero, and the groundwater outflow from the region is small. Therefore, it has been assumed that all P running off and leaching from soils is eventually reaching the regional river.

3.1.2.2 Results

The results of this case study include the regional water balance as well as the P flows and stocks of the region. The water balance is summarized in Figure 3.3. There are two large imports (precipitation and river inflow) and exports (river outflow and evapotranspiration) of water. The main water flow through the region consists of humidity contained in air; this flow has not been considered here because it is not relevant for the P case study. On its way through the region, the surface water flow (river water) is doubled by the input of net precipitation (precipitation minus evapotranspiration). Of the surface water produced within the region, 28% is treated wastewaters from one major and two small sewage treatment plants. Thus, the ratio of wastewater to surface water flow is relatively high, and accordingly, the regional potential for dilution of wastewaters is rather small. Hence, efficient wastewater treatment is highly important for the quality of the river water.

An increase in groundwater stock of ≈10% of net precipitation is observed during the assessment campaign. It reveals that the year of measurement is a rather "wet" year; it distinctly deviates from the 10-year average of the hydrological balance, which shows a tendency toward decreasing groundwater stock.

The results of the analysis of flows and stocks of phosphorus are presented in Figure 3.2. As in the case of lead, the P imports (232 t/year) outweigh the P exports (168 t/year) by far, resulting in an accumulation of 64 t of P/year. The main sink for P is the soil. It contains already 10,000 t, and 68 t/year is added. (Note: the difference between 68 and 64 t/year given for regional accumulation stems from the uncertainty of the processes *WWTP* and *sewer* that are not in balance.) In MFA, it is often the case that inputs, outputs, and changes in stocks of processes do not match, and hence, uncertainties remain (see Chapter 2, Section 2.3). The largest amount of P is imported for agricultural activities (45 t/year of fodder for animals and 78 t/year of fertilizer for plant production). By manure (100 t/year), fertilizer (78 t/year), sewage sludge (13 t/year), and atmospheric deposition (3 t/year), 194 t/year of P is applied to the regional soil. Plants take up 109 t/year, and 17 t/year passes to the surface waters by leaching and erosion. A regional silo for food holds large amounts of P that are continuously replenished, accounting for a P flow of 40 t/year. The use of P for industrial water treatment amounts to another 21 t/year. Possibly, more P is used but not accounted for in *industry*.

3.1.2.2.1 Environmental Protection

Besides the unknown amount of P in landfills, there are two main issues concerning P management in the region. First, P is accumulating in the soil (+68 t/c, corresponding to an increase of the P stock in the soil of +0.68% per annum) and eroding/leaching to the surface waters (17 t/year). Second, P is directly discharged with purified sewage into the receiving waters. The load of P in the river increases from 28 t/year at the inflow to 74 t/year at the outflow. The flow of river water is doubled by the addition of regional net precipitation. Hence, the P concentration in the river increases by ≈30%. If all upstream and downstream regions contribute to the P load in surface waters in the same way, eutrophication is likely to take place in downstream lakes and reservoirs. Thus, the maximum allowable P load to the river must be assessed by also taking into account the potential and limitations for P dilution of the surface waters outside of the region.

3.1.2.2.2 Early Recognition and Monitoring

The MFA of P facilitates early recognition of P accumulation in the soil before it actually happens. This is important for water-pollution control. If the load of P into the river needs to be limited, there are two theoretical options (the uncontrolled landfills are not discussed here as a potential source because they have not been investigated):

1. The removal efficiency for P in the sewage treatment plant WWTP can be increased from about 30–50% to >90% in a relatively short time (months).
2. The flow of P from the soil to the surface waters can be reduced.

The second option does not allow quick reduction of P flows: for a given agricultural practice, the amount of P eroded is mainly a function of the P stock in the soil. Hence, it is necessary to either change agricultural practices or to reduce the P stock in the soil, which takes a long period of time (decades to centuries). MFA makes it possible to forecast accumulation (as well as depletion) of P in the soil long before it actually happens. Taking into account the current amount of P in the soil (≈10,000 t) and the annual accumulation of 68 t/year, it can be assessed that the P concentration in soils will be doubled in about one and a half centuries, if present agricultural practice is maintained. This will lead to a large increase in eroded P, offsetting the reduction of P in the river due to improved elimination of P in sewage treatment.

Direct soil monitoring yields results with large standard deviations. Thus, even an intensive soil sampling and analysis campaign will not identify P accumulation within a few years, because mean values will not be significantly different within one decade. MFA provides timely predictions of the change in soil stocks with one single measuring campaign of soil concentrations and P inputs into the soil. Of course, if agricultural practice is changed, the data and calculations have to be adapted to the new situation, too.

3.1.2.2.3 Priorities

Comparing the various flows of P to the soil, it becomes clear that in this region, sewage sludge is, comparatively, a small source of P, supplying less than 10% of the total soil input. Thus, in terms of resource conservation, the application of sewage sludge on farmland is of little importance and is of low priority. The dominant flows are due to the cycling of a large stock of P within agricultural production by the soil–plant–animal–soil system. The ratio "input to product output" of the different processes is noteworthy: The process *animal* (production of animals and diary produce) consumes 130 t P/ year to produce 30 t P/year; *plant production including soil* requires 194 t P/year to produce 109 t P/year in plants harvested. This translates to efficiencies of 23% for the use of P in *animal production* and 56% for *plant production including soil*. Clearly, if P becomes a limited resource, priorities are either to increase the efficiency of P in animal farming or to shift the dietary intake toward less meat and more vegetarian foods.

It is also noteworthy that composting of household garbage is insignificant and of low priority regarding the P flow within the region. Assuming that less than 20% of food bought by private households is discarded as MSW (80% being eaten, eventually transformed to urine and feces, and collected with sewage), composting of separately collected garbage would supply only about 3 t P/year for agricultural production, equaling ≈2% of total agricultural input.

3.1.3 Case Study 3: Nutrient Pollution in Large Watersheds

Case study 3 is part of an in-depth investigation into water-quality management of the entire Danube Basin comprising 11 different riparian countries. It is described comprehensively in the report "Nutrient Balances for Danube Countries" (Somlyódy et al., 1997).

The case study is included here to demonstrate the following: (1) The application of MFA is independent of scale; hence, the same methodology can be applied to small (farm) and very large (international watershed) systems. Nevertheless, there are clear differences in the focus and procedures according to the scale. In general, multinational MFAs on large scales like an investigation into a transnational watershed require the joint effort of several research groups from each of the participating countries.

(2) The results of a large-scale study directed toward decision making in environmental protection can have different consequences for the partners engaged. While one country may not be a large factor for the pollution of the watershed, another may turn out to be a major contributor. Hence, if a common level of water protection is established and corresponding remediation measures are taken, the financial consequences may be quite severe for the latter and only marginal for the former. It is thus of utmost importance to use the same, appropriate and uniform methodology that is accepted by all partners. It is necessary to acquire an adequate, comprehensive and compatible data set for each country using equal definitions. Otherwise, if terms are not equal, data and results of the different riparian states cannot be compared. This holds true for flows and stocks of nutrients from agriculture, industry and trade, private households, water and wastewater management, and waste management alike. And it is essential to use a uniform methodology

in collecting, calculating, and evaluating the data, too. MFA represents such a methodology well suited for multinational teams collaborating in delicate situations. For transnational application of MFA, capacity building and know-how transfer is important to ensure that all groups and participants are applying the same methodology. If MFA is standardized in the future, multinational analysis of metabolic systems will be much facilitated.

High nutrient loads are recognized as one of the most severe problems of the River Danube, the Danube Delta, and the "final sink" Black Sea. The ecosystem of the Danube Delta is severely endangered, and a large part of the Black Sea is critically eutrophic (Mee, 1992) The main objective of this study is to prepare a basis for decisions regarding the protection of the water quality of the Danube, its delta, and the Black Sea. In particular, the goal is to use MFA to establish reliable and uniform information about sources, flows, stocks, and sinks of phosphorous and nitrogen in the Danube Basin (Brunner and Lampert, 1997; Somlyódy, Brunner, and Kroiß, 1999). The main difference between Case Studies 2 and 3 is the scale: Instead of an area of 66 km² and a population of 28,000 inhabitants (case study 2), the "Danube" case study covers an area more than 1000 times larger (820,000 km²) and includes 12 countries with a population more than 3000 times larger (85 million inhabitants). Despite the large difference in scale, the same MFA methodology is applied.

Key questions of this case study are as follows: what are the main sources of nutrients, and what measures are appropriate to reduce the nutrient flows to environmentally acceptable levels? Traditionally, emission inventories and ambient water-quality measurements are used to answer these questions. As a novel approach, comprehensive material flow analysis is applied to the entire watershed. The main advantage is that all nutrient-related processes in the region are looked at uniformly, and the total inputs, outputs, and stocks are investigated. Nutrient flows are tracked from their very beginning (fertilizer, animal feedstock, agricultural production) to the consumer (private households), to waste management, to surface water and groundwater, and finally to the River Danube. Since the balance principle is applied to all processes, cross-checking of flows and stocks becomes possible at many points within the system investigated.

3.1.3.1 Procedures

As mentioned before, one of the main tasks when exploring such a large system is to set up a broad international group that learns and uses the same MFA methodology. Ten national teams from Austria, Bulgaria, Czech Republic, Germany, Hungary, Moldavia, Romania, Slovakia, Slovenia, and Ukraine, each consisting of several experts, are participating in the study. In a first step, the common MFA methodology as well as water-quality goals and principles are established. The system boundaries are defined in space and time. The least number of processes is selected that allows full description of all necessary nutrient flows and stocks and still does not result in excessive work. To facilitate assembly of the individual data and results, the system is defined uniformly for all teams (see Figure 3.6).

Next, data are collected to balance each of the processes of Figure 3.6. Existing measurements, regional statistics, literature data, expert advice, and sometimes additional measurements are used to assemble a data set as comprehensive as possible. For example, for the process *agriculture including soils*, this means finding information about all process inputs such as mineral fertilizer, atmospheric deposition, nitrogen fixation, sewage sludge, compost, seedlings, and process outputs such as crops harvested, animal products, eroded

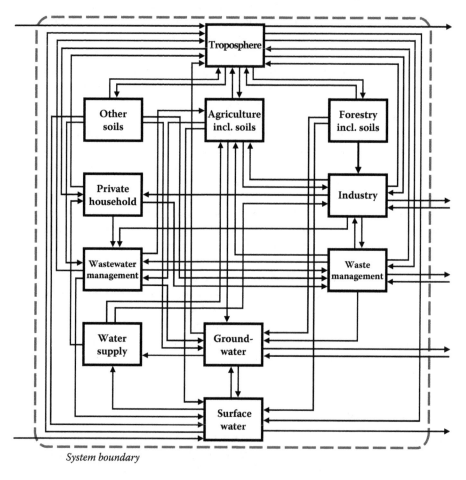

System boundary

FIGURE 3.6
System definition for nutrient balancing in the Danube River watershed. The same system is used for all national balances and for the balance of the total catchment area. (From Somlyódy, L. et al., *Nutrient Balances for Danube Countries*. Final Report Project EU/AR/102A/91, Service Contract 95–0614.00, PHARE Environmental Program for the DanubeRiver Basin ZZ 9111/0102. Vienna, Austria: Consortium TU Vienna Institute for Water Quality and Waste Management, and TU Budapest Department of Water and Waste Water Engineering, 1997.)

soil, gaseous losses, leachate, and percolation. Manure is recycled within the process and thus can be looked at as both an output and an input at the same time if no export or import of manure takes place. Stocks comprise nutrients in the soil, stored manure, animal and vegetable biomass, and stockpiles of fertilizer. The procedure is similar to the one described for case study 2 in Section 3.1.2. All other processes of Figure 3.6 are balanced similarly.

Some processes are not easy to balance: erosion from forest and agricultural soils in alpine areas can only be roughly estimated. Denitrification in natural systems (e.g., soil, aquifer) is not well known. The fate of intermediate stocks in the River Danube and in the soil over time is not sufficiently understood yet. Data about the efficiency of wastewater treatment and about corresponding nutrient removal are not available in all of the Eastern European countries. During the time of centrally planned economies, much information on agriculture, water quality management, and waste management was collected and stored on a large scale. However, since the transition of these economies to a free-market economy, much less information is available. In part, this is because it is too costly to gather comprehensive data. On the other hand, the price to access existing information increased dramatically after the economic transition.

It is important that all partners exchange information during collection of data. They also must make sure that they use compatible figures. For instance, it is likely that the balance of a cow (a process in MFA terminology) is similar in most countries of the Danube Basin, and thus that the input and output figures are comparable for most teams. If there are differences, like the significant variation between the nutrient metabolism of a Ukrainian and an Austrian cow, explanations must be available. Often, the balance principle brings such differences to light and allows cross-checking and verifying such differences. Thus, the balance principle can be highly valuable in the negotiation process within a group comprising teams from many countries. It ensures transparency, enables data verification, and results in acceptance of each other's results.

3.1.3.2 Results

A large-scale multinational MFA proves to be a time- and resource-consuming task. It takes time until all know-how is transferred, incorporated, and well applied in practice. It takes even more time to find all the necessary data. The exchange of information, iterations, and adaptations of the individual work of the different participating groups again takes time. It may be that a partner is not able to perform its task and that a new team has to be engaged in the middle of the project. Given these factors, it is clear that a large-scale MFA cannot be undertaken in a couple of weeks. It is likely to take 1 year or more to complete such a comprehensive and challenging task.

The Danube case study produces a lot of data and many results that can be used to support decisions regarding water quality management. For results regarding wastewater management and water pollution control, see Zessner, Fenz, and Kroiss (1998). The results presented here identify the most important sources and pathways of nutrients in the Danube River basin. The purpose of this presentation is to demonstrate how a very large data set can be compressed to present the relevant key results. In Figure 3.7, the system given in Figure 3.6 is transformed and presented in a format that identifies the major imports and exports of nutrients into the surface waters of the Danube catchment area. This format still shows the same 11 processes, but it centers on the surface waters. Imports and exports are quantified, and conclusions regarding the importance of all flows can be taken according to their mass flow. Note that mass flows alone do not permit one to evaluate the effects of nutrients in surface waters. It is necessary first to transform nutrient flows into nutrient concentrations by dividing nutrient flows through water flows.

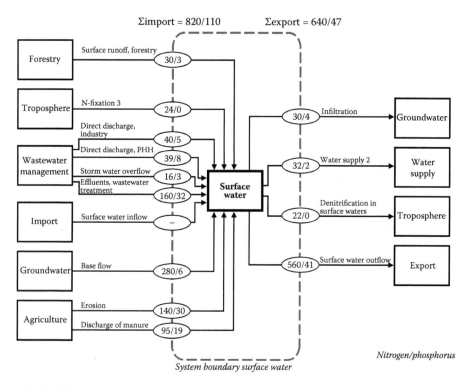

FIGURE 3.7

Nitrogen and phosphorous flows in surface waters of the Danube River basin in 1992, kt/year. The system shown in Figure 3.6 is transformed to Figure 3.7 in order to present the main inputs and outputs of the surface waters. This allows identifying the importance of each process as a source of nutrients for the Danube River. (From Somlyódy, L. et al., *Nutrient Balances for Danube Countries*. Final Report Project EU/AR/102A/91, Service Contract 95–0614.00, PHARE Environmental Program for the DanubeRiver Basin ZZ 9111/0102. Vienna, Austria: Consortium TU Vienna Institute for Water Quality and Waste Management, and TU Budapest Department of Water and Waste Water Engineering, 1997.)

The same case study data are even more condensed in Figure 3.8. It becomes clear that the more aggregated the data are, the easier it is to get a message across: Figure 3.8 clearly identifies agriculture as the dominant source of nutrients in the Danube Basin. In addition, it suggests the hypothesis that improving existing WWTPs is probably more important for reducing emissions than connecting all households and industries to sewers. This hypothesis of course has to be verified with data about the fraction of people and companies connected to sewer systems, and on nutrient removal in wastewater treatment within the catchment area. In addition, the costs of upgrading WWTP and of connecting households and industries to sewers need to be known. The advantage of an MFA approach is that such hypotheses can be set up, making further investigations more straightforward.

Another way of presenting data is displayed in Tables 3.12 and 3.13, which combine results about sources (agriculture, household, industry, etc.) with

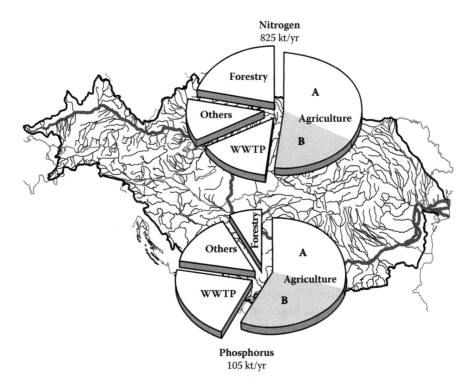

FIGURE 3.8

Sources of nutrients in the catchment area of the Danube River in 1992, kt/year. The charts show clearly the importance of agriculture for P and N emissions. Erosion and leaching from agricultural fields dominates all sources (A). Direct discharges and discharges via treatment plants of animal wastes are the second most important path of nutrients to the Danube (B). The direct inflows from private households and industry (others) are smaller than the effluents from WWTPs. Diffusive inputs from forestry are small for P but more significant for N. (From Brunner, P. H. and Lampert, Ch., *EAWAG News*, 43E, June 1997, pp. 15–17. With permission.)

TABLE 3.12

Sources and Pathways of Nitrogen in the Danube River (1992)

N, %	Agriculture	Private Household	Industry	Others	Total
Erosion/runoff	17	0	0	4	21
Direct discharges	12	4	6	2	24
Base flow	17	4	0	13	35
Sewage treatment plant	6	10	5	0	20
Total	51	19	10	19	100

Source: Brunner, P. H., and Lampert, C. (1997). "Nährstoffe im Donauraum, Quellen und letzte Senken" ("The Flow of Nutrients in the Danube River Basin"). *EAWAG News, 6* (43D+E+F), 15–17.

Note: Total input equals 100%. Base flow represents flows to the Danube via groundwater.

TABLE 3.13

Sources and Pathways of Phosphorous in the Danube River (1992)

P, %	Agriculture	Private Household	Industry	Others	Total
Erosion/runoff	28	0	0	3	31
Direct discharges	18	6	6	3	33
Base flow	2	2	0	2	6
Sewage treatment plant	9	14	7	0	30
Total	57	22	13	8	100

Source: Brunner, P. H., and Lampert, C. (1997). "Nährstoffe im Donauraum, Quellen und letzte Senken" ("The Flow of Nutrients in the Danube River Basin"). *EAWAG News, 6* (43D+E+F), 15–17.

Note: Total input equals 100%. Base flow stands for flows to the Danube via groundwater.

results about pathways. Hence, they are well suited to serve as a basis to set priorities for decisions regarding nutrient emission reductions. The following conclusions can be drawn: Agriculture is the main source of nutrient inputs into the River Danube. Erosion and runoff are the main pathway of nutrients from agriculture to surface waters for both P and N. The direct inputs of liquid manure are high and amount to about 12% of total N and 20% of total P loads. Private households are the second largest source of nutrients, contributing around 20% of total N and P. Approximately 10% of both N and P originates from industry. MFA yields the following results regarding pathways: for surface waters, the main nitrogen load (35%) is due to exfiltration of groundwater. Erosion/runoff; direct inputs from agriculture, households, and industry; and effluents from WWTP each contribute about 20–25% of the total load to the Danube River. About 60% of N and 40% of P originate from nonpoint sources. Retention (sedimentation and denitrification) amounts to 15% of N and close to 50% for P.

The detailed results of all teams (presented in Somlyódy et al., 1997) make it possible to identify the nutrient contribution of each country. Almost half of the nutrient input into the Danube catchment area comes from Romania, while Austria, Germany, and Hungary together contribute about one-third of the total load. An interesting and not yet resolved question concerns the allocation of the dilution potential of the River Danube to the riparian countries: what is the amount of nutrients a country may release to the Danube? Assuming that the carrying capacity for nutrients of the Danube basin is known, there are several ways to answer this question. A per capita load limit favors countries with large populations; it can be justified on the grounds that every human being has a similar metabolism and thus should have an equal share. This method of allocation neglects the fact that some countries are better suited for agriculture than others and, because of their agricultural activities, will have a larger nutrient turnover. A per-area load limit favors large countries but does not consider population density. A per-net-precipitation limit takes into account the regional dilution of nutrients: if regional nutrient emissions are heavily diluted by a large amount of net precipitation (precipitation minus evapotranspiration), the resulting concentration in the River Danube may still be low and below carrying capacity. However, this argument does not hold for the Black Sea, where the total flow is important.

The River Danube, like many of the large river systems in the world, has become an important path for wastes such as nutrients from countries within its catchment area. The main question is how future loads will develop. At present, eastern European countries are experiencing a low standard of living. It can be assumed that the per capita turnover will rise rapidly in the future and that population will grow again, too. Both factors will increase total nutrient turnover as well as waste generation. If no actions are taken, the capacity of the River Danube, the delta, and the final sink Black Sea will be overloaded, with serious ecological and economic consequences. It is not the "classical" resource problem (lack of nutrients) that will limit the development of the region. Rather, it will be the lack of appropriate sinks that restricts progress. Strategies to limit nutrient loads need to be discussed and developed. Crucial issues will be type of agriculture; population density, lifestyle, and consumption; and standards and enforcement for emissions from industry, households, and WWTP. In any case, transregional and international agreements will be required to solve the allocation problem. As proven by this case study, MFA can play a major role in supporting policy decisions to protect the River Danube, the delta, and the Black Sea.

3.1.4 Case Study 4: Support Tool for EISs

In an environmental impact assessment (EIA), potential impacts of a project such as a new plant (power plant, municipal incinerator) or system (road, harbor) on the environment are identified, quantified, evaluated, predicted, and monitored. In the event that a significant impact is acknowledged, more detailed studies have to be carried out, finally leading to the preparation of an EIS. Meanwhile, more than half of the nations around the world require an EIA for certain projects. The US National Environmental Policy Act of 1969 (NEPA) provided one of the earliest sets of EIA requirements. Many countries followed and modeled their requirements after NEPA (Canter, 1996). In Europe in the 1970s, some federal environmental legislation began to require mandatory EIAs. However, the final breakthrough was not achieved until the European Directive 85/337/EEC on the "assessment of the effects of certain public and private projects on the environment" became effective (European Commission, 1985). MFA that is based on a sound methodological framework is considered a useful tool to support both EIA and EIS (Brunner and Baccini, 1992).

In the case study SYSTOK (Schachermayer, Rechberger, Maderner, and Brunner, 1995), the impact of electricity production from coal on the local and regional environment is investigated. For this purpose, a three-process system comprising coal mining, the coal-fired power plant, and landfilling of the ash is defined (Figure 3.9). The contribution of this system to the anthropogenic and geogenic metabolism of the region is determined (Figure 3.10). The case study focuses on particular technologies of mining, power generation, APC, and landfilling. It is clear that the results cannot be generalized to other coal-fired power plants. If the technology is changed, if the coal composition and heating value is different, or if the landfill leaks to the groundwater, the impact on the environment will also be different. The

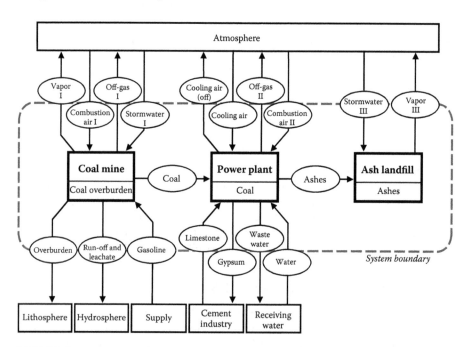

FIGURE 3.9
Systems definition of electricity production in a coal-fired power plant including the processes *coal mining* and *ash landfilling*. The figure includes all exports and imports that are necessary to establish an EIA and EIS.

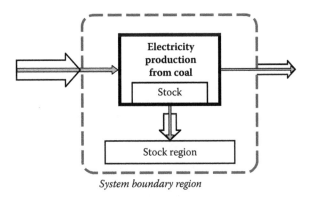

FIGURE 3.10
The contribution of a coal-fired power plant to a region's metabolism. The system consists of the three processes, *coal mining, power generation,* and *landfilling of the ash* that can be summarized in a single process, *electricity production from coal.*

reason for including this case study is to show that MFA serves well as a base for EIS and EIA. MFA can be applied independently of the technology used or the input composition.

The findings of SYSTOK serve as a basis for the operator to optimize the plant and to prepare an EIS. One of the features of SYSTOK is that a system, in this case *electricity production from coal,* is to be embedded comprehensively into a region. However, the definition of the region is not clear a priori, so appropriate approaches for the demarcation of spatial and temporal systems boundaries have to be developed. SYSTOK exemplifies how a comparatively simple MFA can lead to valuable new conclusions regarding system boundaries.

3.1.4.1 Description of the Power Plant and Its Periphery

All coal used in the power plant (1 million t/year) is produced in a nearby open-pit coal mine with a surface area of 2 km². Temporary interruptions in mining or power generation are buffered by interim coal storage of 2 million tons, located at the premises of the power plant. In order to gain 1 ton of coal, an average of 6.7 tons of overburden have to be removed. Of this mining waste, 70% is transported to locations outside of the mine, and the rest is filled back into the mine. For extraction, transportation (trucks and conveyors), and processing of the coal and overburden, 1400 t/year of gasoline and 25 million kWh/year of electricity are needed. The reservoir of coal still in the mine is estimated to be 11 million tons.

Brown coal has a low heating value of 8.4 to 13 MJ/kg and bears much noncombustible ash, making transportation expensive. Thus, the power plant is situated near the coal mine in order to avoid long transport distances. Because of high specific costs of power generation at this plant, it operates only during peak demand. The average time of operation is 4000 h/year, and the maximum electric capacity is 330 MW. The coal (C) is pulverized to coal dust and injected into the combustion chamber at a feed rate of 300 t C/h at full load. The average coal consumption (ca. 265 t C/h) is somewhat lower due to periods of partial loading. Approximately 33 kg/t C of wet bottom ash (40% water content) is removed from the water basin that serves as an air seal toward the combustion chamber. The power plant is equipped with an electrostatic precipitator (ESP) to collect particulates (ESP ash) with an efficiency of 99.85%. The ratio of dry bottom ash to dry ESP residue (or fly ash) is ca. 1:9. The ESP residue is humidified (20% water content) and, together with bottom ash, transported by conveyor belts to the landfill. Sulfur dioxide (SO_2) is removed from the flue gas in a wet scrubber using limestone ($CaCO_3$, 20 kg/t C). Absorption of SO_2 occurs with an efficiency of more than 90%, producing ca. 37 kg of gypsum per t C (water content is 12%). Finally, nitrogen oxides (NO_x) are reduced to molecular nitrogen (N_2) in a catalyst by injecting ca. 1.2 kg of ammonia solution (NH_3, 33%) per t C. Other chemicals used include hydrochloric acid (HCl, 0.02 kg/t C, 33%) and sodium hydroxide (NaOH, 0.007 kg/t C, 50%) for conditioning of feeder and perspiration water as well as chemicals used to stabilize water hardness, inhibit corrosion, etc. (<0.002 kg/t C). Combustion requires 4000 kg/t C of air, resulting in ca. 5000 kg/t C of off-gas. Water is used to feed the steam cycle (20 kg/t C) and for cooling (2300 kg/t C). Waste heat is dissipated in a cooling tower having an air-exchange rate of 8500 kg/t C. Part of the cooling water (25%) is discharged to surface water; the rest evaporates in the cooling tower.

The landfill for the ashes consists of a natural vale made of quartzite with a surface of ca. 0.35 km². This basin, which is a former mine area of coal, has a low permeability for water. Hence, precipitation (ca. 1000 l/m²) creates a lake in the landfill. The water is used during dry periods to wet the ashes and is supposed to finally evaporate. No leachate leaves the ash landfill as long as the landfill is maintained by the operator.

3.1.4.2 System Definition

The system displayed in Figure 3.9 is divided further into three processes: *coal mining*, *power plant* including APC, and ash *landfill*. For EIA and EIS, a black-box approach is appropriate. It is not necessary to take into account more detailed subprocesses that would require much additional information. The substances are selected based on knowledge about combustion processes and the main product "coal." Carbon is the priority substance in any combustion process, since the content of organic carbon in ashes and flue gas is a measure of combustion efficiency and of the formation of organic pollutants. Past experience with coal-fired power plants has shown that they are major sources of emissions of SO_2, NO_x, HCl, and heavy metals such as arsenic and selenium (Greenberg, Zoller, and Gordon, 1978). Modern legislation (Clean Air Act) sets stringent standards for these emissions. The procedure to select substances relevant for EIAs is given in Table 3.14. The substance

TABLE 3.14

Determination of Relevant Substances for the Analysis of a Coal-Fired Power Plant by Relating Average Concentrations of Selected Substances in Coal and Ash to Concentrations in the Earth's Crust

Substance	Concentration in Coal (A), mg/kg	Concentration in Ash (B), mg/kg	Concentration in Earth's Crust (C), mg/kg	Ratio A/C	Ratio B/C
Arsenic	12	64	1	12	64
Lead	6	37	13	0.5	2.8
Cadmium	0.1	0.64	0.2	0.5	3.2
Chromium	30	170	100	0.3	1.7
Copper	13	84	55	0.24	1.5
Selenium	0.9	2.2	0.05	18	44
Zinc	27	190	70	0.4	2.7
Nickel	27	130	75	0.36	1.7
Mercury	0.3	0.45	0.08	3.8	5.6
Sulfur	6500	3000	260	25	12
Chlorine	1000	–	130	7.7	–
Nitrogen	12,000	–	20	600	–

concentrations in coal are compared with concentrations in the Earth's crust. The following six elements are significantly more concentrated in coal than in the crust: As, Se, Hg, S, Cl, and N. The relevant goods are determined by mass-balancing each process and by assuming that goods inducing a substance flow <1% of the total throughput of a substance can be neglected. This requires an iterative approach among the steps *establishing substance balances*, *selection of goods*, and *establishing total mass balances*. For a first step, the premises of mining, power station, and landfilling are selected as a spatial system boundary. Materials balances are determined for an average operational year (temporal system boundary). Figure 3.9 shows the system defined.

3.1.4.3 Results of Mass Flows and Substance Balances

The mass flows of all goods are listed in Table 3.15. Except for mining, where it is not known whether and how much runoff and leachate is draining to surface water and groundwater, all processes are mass-balanced. The main quantitative features of the system are as follows: (1) The coal mine will be exhausted in about 10 years. (2) Compared with the emissions of the power station, the off-gas from mining is negligible. (3) The power plant is actually a giant fan moving huge amounts of air. The mass of air required for cooling dominates the mass flows of the system, exceeding combustion airflow by more than an order of magnitude. Additionally, the power plant consumes more than 1 t of water per ton of coal. (4) The main flow of solid waste is generated during mining (overburden). The power plant produces a large net "hole," since coal is extracted and the volume of backfilled ash is much smaller.

TABLE 3.15

Mass Balance for Goods of the System *Electricity Production by a Coal-Fired Power Plant*, 1000 t/Year

Input		Output	
Process Coal Mining			
Combustion air I	29	Off-gas I	31
Storm water I	2000	Vapor	2000–n.d.
Gasoline	1.4	Overburden	4700
		Runoff and leachate	n.d.
		Coal	1000
Total	2000		7700
Stock (coal)	11,000		1000
Stock (total)	145,000[a]	Stock change	–5700[a]
Process Power Plant			
Coal	1000	Ashes	280
Cooling air (input)	85,000	Cooling air (output)	87,000
Combustion air II	4000	Off-gas II	5000
Water	2500	Wastewater	760
Limestone	20	Gypsum	390
Total	93,000		93,000
Stock (coal)	2000	Stock change	0
Ash Landfill			
Storm water III	350	Vapor III	350
Ashes	280		
Total	630		350
Stock (ashes)	4100	Stock change	+280

Note: Values are rounded; n.d. = not determined.
[a] Estimated, includes coal and overburden.

The substance balances as displayed in Figure 3.11 provide an overview of the qualitative behavior of the system. These balances are calculated based on data for substance concentrations in overburden ("soil" in Table 3.19), gasoline, and coal (Table 3.14), and the TCs for the power plant that have been measured on site (see Table 3.16).

A comparison of sulfur concentration in coal and in the Earth's crust shows that a large amount of sulfur is extracted from the crust via coal. The power plant transfers sulfur quite efficiently into the product gypsum (86%). Before desulfurization became a part of the APC system of the plant, this sulfur was emitted, too, resulting in a sulfur transfer to the atmosphere of >90%.

The foremost flow of mercury is associated with the good "overburden" or mining waste. During combustion, the atmophilic mercury is evaporated and leaves the plant evenly distributed between ESP residue and off-gas. A small part is precipitated with gypsum. Electricity production in coal-fired power plants extracts significant amounts of mercury from the Earth's crust and disperses

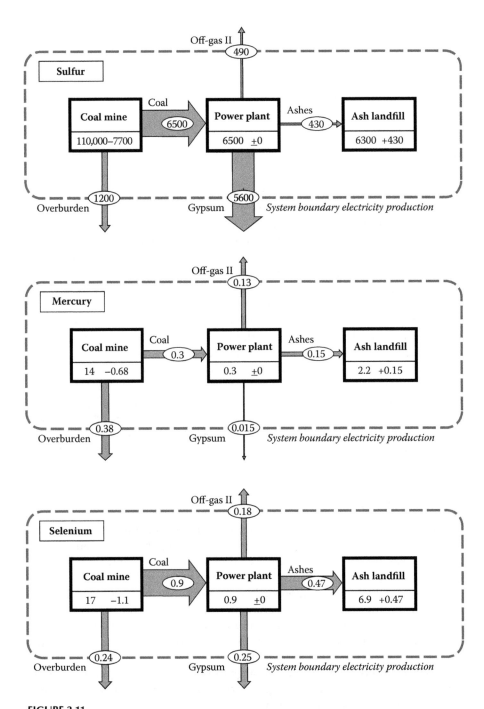

FIGURE 3.11
Substance balances for sulfur, mercury, and selenium for the system *electricity production* (flows are in t/year and stocks in t). The stock in the process *coal mine* includes coal and overburden.

TABLE 3.16

Partitioning of Selected Substances in a Coal-Fired Power Plant, % of Input

Substance	ESP Fly Ash	Bottom Ash	Gypsum	Off-Gas II
Carbon	2	1.3	<0.05	97
Sulfur	5.7	0.93	86	7.6
Mercury	50	0	5	45
Arsenic	99	0.4	0.4	<0.1
Selenium	52	0.6	28	20

a substantial amount via the stack. So far, 2.2 t of mercury has been deposited in the ash landfill and more than 10 t in the overburden deposit. It is interesting to compare these stocks with other mercury stocks. The consumption of mercury has been assessed to range between 0.66 g/capita/year in Stockholm and 1.5 g/capita/year in the United States. For stock, 10 g/capita has been determined in Stockholm (Jasinski, 1995; Bergbäck, Johansson, and Mohlander, 2001). This means that the landfill contains the same amount of mercury as is stored in buildings, infrastructure, and long-lasting goods associated with a region of 220,000 inhabitants. In the overburden deposit, a mercury stock corresponding to more than 1 million persons is contained. While this mercury is dispersed over a large region (2000 km²), the landfill's mercury is located in a comparatively small area (0.35 km²) and therefore is easier to control.

About half the amount of selenium that enters the power plant is transferred to the ashes and landfilled, and 20% is emitted into the atmosphere. The relevance of this path for the environment will be discussed in the following sections. Note that gypsum holds about 30% of the selenium contained in coal.

The relation between the power plant and the surrounding region is discussed in the following sections.

3.1.4.4 Definition of Regions of Impact

The mine, power plant, and ash landfill have several impacts. In the first place, they supply power to consumers. Thus, a region can be defined in a product-related way. Second, they create jobs and income and thus serve an economic region. Third, the mine, power plant, and landfill are situated in an administratively defined region such as a community or a province. Fourth, they have an impact on the environment. Depending on the "conveyor belt" that transports an emission, substances released by the coal-fired power plant may affect a small or large area. The ash landfill has only a local impact during the transfer of the ash to the landfill site. Sulfur and mercury emitted by off-gases are distributed over a large (global) area through atmospheric transport. Thus, the size of the region is determined by the distance an emitted substance travels from the plant and by the effect of this substance on the environment. In each of these four regions of impact, there are specific problems, benefits, and stakeholders with regard to the power plant.

In the SYSTOK study, three regions of impact are defined.

3.1.4.4.1 Product-Related Region

The power plant supplies electricity to a certain area. In principle, this area is defined by the amount of electricity the plant supplies, by the demand per customer, and by the population (or customer) density. Due to the liberalization of the electricity market, this area can only be defined as a virtual region, since customers may be served far away from the plant. The product region is changing constantly in response to the market situation. Thus, for SYSTOK, the size of this virtual region is calculated by the average electricity production of the plant, the average consumption per capita, and the national population density.

$$A_{PR} = \frac{P \cdot h \cdot f}{e \cdot \rho_P} = 2100 \text{ km}^2$$

where
P = output of the power plant (330 MW)
h = operating hours per year (4000 h)
f = factor considering partial-load operation (0.85)
e = specific demand for electricity in Austria (including private households, industry, service, administration, traffic, agriculture; 5.6 MW·h/capita/year)
ρ_P = population density in Austria (95 capita/km²)

3.1.4.4.2 Administratively Defined Region

The region is defined by the borders of the administrative unit, i.e., the district that represents the legislative and administrative authority for the plant operator. The advantage of this definition is twofold:

1. The region as a spatial unit is well accepted and known and is governed by an authority supervising the plant.
2. Data are usually collected on the level of administrative regions, thus facilitating the allocation of data.

3.1.4.4.3 Region Defined by Potential Environmental Impacts

As mentioned before, this area is different for particulate, gaseous, and aqueous emissions, and it is also substance specific. For gaseous emissions, dispersion models help to determine the region. Criteria for selecting the boundaries may be

1. Concentration limit (ambient standard) for a substance ($c_{crit} = c_{lim}$)
2. A fraction of the concentration limit, since the limit should not be used up by the power plant alone ($c_{crit} = c_{lim}/10$)

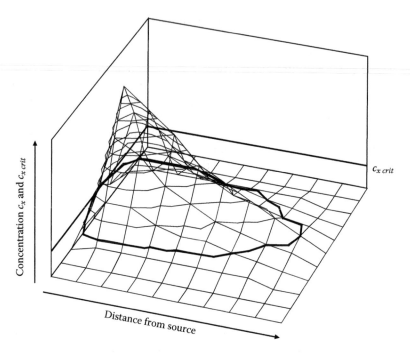

FIGURE 3.12
Application of a dispersion model to determine the border of a substance-specific, environ-mentally relevant region. The regional boundary with regard to substance x is defined as the area within $c_x > c_{x\ crit}$.

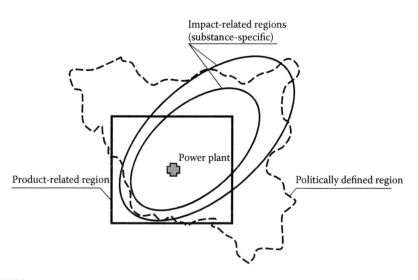

FIGURE 3.13
Regions defined according to three criteria—*consumers of electricity* (product-related region), *administrative designation*, and *environmental impacts*—overlap but are not identical.

TABLE 3.17

Sizes of Differently Defined Regions for a 330 MW Coal-Fired Power Plant

Region Definition	Size, km^2
Administrative (district)	678
Product (electricity)	2100
Environmental (example SO_2)	620

Note: The boundary for the environmentally defined region of SO_2 is determined by $c_{crit} = c_{geog} \times 1.1$, where $c_{geog} = 10$ mg/m^3.

3. A proviso that the power plant does not change the current ambient concentrations in a significant way ($c_{crit} = c_{background} \times 1.1$)

4. A proviso that the power plant does not change the geogenic (or "natural," without present anthropogenic influences) concentrations in a significant way ($c_{crit} = c_{geog} \times 1.1$; see Figure 3.12)

Figure 3.13 exemplifies the differently defined regions, and Table 3.17 presents sizes of regions calculated according to different definitions for SYSTOK.

3.1.4.5 Comprehensive Regional Significance

In addition to the procedure given in Figure 3.12, there are other means of determining the relevance of power produced, emissions, and wastes of the coal mine, power plant, and landfill for the region. In Figure 3.14, emissions from the power plant are compared with the total regional emissions of various air pollutants. As a basis for comparison, the product-related region is chosen. The emissions of the region (R_i) are assessed by the following equation:

$$R_i = X_i - P_i \cdot \frac{A_{PR}}{A_{AU}} + P_i$$

with X_i as the average emissions of an appropriate administrative unit (AU) for which data are available (state, district). P_i stands for the emissions of the power plant, and A_{PR} and A_{AU} are the respective areas of product-related region and administrative unit.

The power plant is responsible for about half of the region's CO_2 emissions. Removal of SO_2 and catalytic reduction of NO_x significantly reduced the plant's contribution to the regional emissions. A further decrease in particulates and NO_x will result in modest improvements of air quality only (<10%). For CO, the power plant's emissions are not relevant at all.

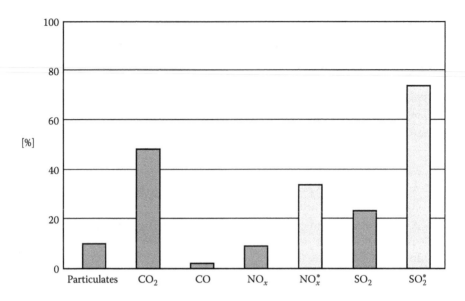

FIGURE 3.14
Contribution of the power plant to the total emissions of the product-related region (=100%).
* Indicates emissions before introduction of advanced air-pollution control.

The throughput of heavy metals by the power plant is set in relation to the product-related region, too. Table 3.18 shows the annual flows of selected metals and nonmetals through the power plant. Comparing these flows to the corresponding total flows through the region is time consuming. Much information, which is usually not available, is required. Therefore, only the materials flows through private households are taken into account. The contributions of industry and the service sector are not considered. Since the consumption of heavy metals in private households is not well known either, the amount of metals in MSW, which is available from measurements (see Section 3.3.1), is taken as a reference. Table 3.18 gives the ratio of substance flows via coal and flows via MSW on a mass-per-capita and year basis. For comparison, the flows of carbon and sulfur in fossil fuels are also shown. The coal-fired power plant is responsible for a high turnover of arsenic, selenium, and sulfur when compared with the generation of MSW in private households.

The relevance of heavy metal emissions can be assessed by the "anthropogenic versus geogenic flows" approach described in Chapter 2, Section 2.5.8. Applied to SYSTOK, the following question has to be answered: does the power plant change substance concentrations in any of the environmental compartments? A simplified model is used to identify the effect of power plant emissions on the soil concentrations in the product-related region of 2100 km². By deposition, metals such as lead and cadmium may be accumulated in the top 30 cm layer of soil, the soil depth that is

TABLE 3.18

Comparison of Substance Flows Induced by a Coal-Fired Power Plant
and by MSW in the Product-Related Region

Substance	Flow via Coal, g/Capita/Year	Flow via MSW, g/Capita/Year	Coal/MSW
Arsenic	18	0.8	21
Copper	19	100	0.2
Lead	9	170	0.1
Cadmium	0.15	2.3	0.1
Mercury	0.45	0.4	1.2
Selenium	1.3	0.2	8
Chromium	45	53	0.9
Nickel	40	18	2.3
Zinc	40	230	0.2
Carbon	420,000	930,000.0[a]	0.4
Sulfur	10,500	2000.0[a]	5

[a] Figures for carbon and sulfur include the contribution by fossil fuels, which is much larger than MSW.

turned over by plowing. Assuming an average soil density of 1.5 kg/m^3, the regional compartment *soil* holds a mass of 2100×10^6 m^2 × 0.3 m × 1.5 kg/m^3 = 950×10^6 kg.

The cumulative emissions of the power plant can be estimated based on the total coal throughput of 33 million t, the mean substance concentrations in coal, and the TCs for off-gas.

$$E_i t = 33 \times 10^6 \text{ t} \times c_i \text{ mg/kg} \times 10^{-6} \times TC$$

Table 3.19 shows that only emissions of selenium and mercury are of relevance. Note that the model draws on two major simplifications. First, it is understood that the product-related region is identical to the substance-specific impact-related regions. Second, it is assumed that deposition is evenly distributed over the region. Concerning the first simplification, one has to consider that particulate removal takes place in an efficient ESP. Hence, the emitted particulates are small, most with diameters <2 μm. Such particles (aerosols) have a long residence time in the atmosphere (≈1 week) and do not sediment in the vicinity of the power plant. They are washed out of the atmosphere by precipitation. Thus, the average frequency of rainfall determines the travel distance of these particles. In Central Europe, this is around 1 week and results in a significantly larger region than the product-related region (at least 10 times larger). Hence, the product-related region overestimates substance accumulation in the soil by more than one order of magnitude.

TABLE 3.19

Accumulation of Metals in Soils due to Emissions of a Coal-Fired Power Plant

Substance	Coal, mg/kg	TC	Emission, t/τ^a	Soil[c], mg/kg	Soil Reservoir[b], t	Enrichment, %/τ^a
As	12	0.001	0.4	8	7600	0
Pb	6	0.005	0.99	25	23,800	0
Cd	0.1	0.041	0.14	0.2	190	0.1
Cr	30	0.001	0.99	40	38,000	0
Cu	13	0.014	6	15	14,300	0
Se	0.9	0.2	5.9	0.1	95	6.3
Zn	27	0.009	8	30	28,500	0
Ni	27	0.003	2.7	20	19,000	0
Hg	0.3	0.45	4.5	0.2	190	2.3

[a] τ = total time of operation (ca. 33 years).
[b] Soil reservoir is calculated based on the product-related region.
[c] Scheffer, 1989.

The relevance of the second simplification can be assessed when dispersion models for particulates are analyzed. In most cases, the ratio between maximum concentration and mean concentration is less than 10. This means that the assumption of substances being evenly distributed over the region underestimates the actual accumulation by a maximum factor of 10. Considering both simplifications, it can be concluded that the chosen model rather overestimates the enrichment in soils caused by the power plant, and the emissions can be rated as not relevant with the exception of selenium and mercury, where more detailed investigations are necessary.

3.1.4.6 Conclusions

The case study shows that the power generation based on coal is relevant for enhanced flows of arsenic, selenium, mercury, sulfur, and carbon within the region's metabolism. The landfill represents a considerable reservoir for certain metals within the region. It can be considered as a point source that is comparatively easy to control. On the other hand, in the event of insufficient immobilization or leaching of the containment, the landfill can substantially contribute to the pollution of the region's environment. The landfill requires constant water management. Solutions have to be developed for the future, when the power plant is not in operation anymore and when funds are no longer available for landfill aftercare. Since leaching will be a constant threat, it is necessary to investigate whether immobilization of filter ash is more economic than aftercare of the landfill for several thousand years.

The emissions of the power plant are of little relevance for the region, with the exceptions of CO_2 (greenhouse gas) and, of minor importance, SO_2. The comparatively low retention capacity for mercury and selenium should be a focus for future improvement efforts of the operators after more detailed investigations and measurements. Generally, the study shows that the power plant actually does not pose a severe burden to the region's environment. The operators use these results for their EIS and for communication with concerned local people.

PROBLEMS—SECTION 3.1

Problem 3.1:

Assess the effects of the following measures on the regional lead flows and stocks given in Figure 3.1. Show quantitatively and discuss the following effects of reductions in lead concentrations in soil, surface waters, and landfill: (a) ban on leaded gasoline, (b) ban on application of sewage sludge to land, and (c) construction of a new MSW incinerator in the region with an air pollution control efficiency for lead of 99.99% treating the waste of 280,000 persons from the Bunz Valley and neighboring regions.

Problem 3.2:

Consider a region of 2500 km^2 and 1 million inhabitants. Only one river flows through this region. At the inflow, the river has a flow rate of 1 billion m^3/year, and the concentration of phosphorous (P) is 0.01 mg/L. The river discharges into a lake that represents a reservoir of 2.8 billion m^3; residence time of water in the lake is 1 year. (Precipitation, evaporation, etc., are not considered; assume that the river is unchanged when flowing through the lake.) The anthroposphere of the region comprises the following processes: agriculture, food industry, private households, composting of biomass wastes from private households, and wastewater treatment. Per capita consumption of P for nutrition is 0.4 kg/capita/year; 20% of this demand is supplied by food industry within the region. For cleaning purposes, 1 kg/capita/year of P is used, with 70% contained in detergents for textiles; all detergents are imported. Assume that 90% of total nutritional P and 100% of total detergent-based P are directed to the WWTP. The transfer coefficient (TC) for P into sewage sludge is 0.85. The remaining 10% of nutritional P is contained in biomass waste from households that is composted without loss of P and applied to the soil. The stock of P is assessed at ca. 380,000 t. Agriculture imports 2400 t of P in fertilizers and 1200 t of P in animal feed; 80% of this P flow goes to the soil as manure, dung, and residues from harvesting. The balance is input to the regional food industry. The TC for P into food production wastes is 0.6. Food products that are not consumed within the region are exported. Approximately 1% of P that is annually applied to soil escapes to the surface water (river) as a result of erosion.

1. Draw a qualitative flowchart for the region described (system boundary, flows, processes).

2. Quantify the P flows and stocks (t/year and t) of the system. What is the accumulation of P in the soil?

3. Assume that P in detergents for textiles is phased out. Is this measure sufficient to prevent eutrophication (limit for eutrophication = 0.03 mg/l) of the lake?

4. What further measures do you suggest to prevent eutrophication?

Problem 3.3:

Discuss and quantify the reaction time of different measures to control phosphorous flows in the Danube River basin. *Reaction time* is defined as the time span in days, weeks, months, years, etc., between the decision to take an action and a measurable effect in the Danube River. Note that reaction time also includes planning and implementation (construction, startup).

1. Reduction of phosphorous fertilizer input to soils by a resource tax

2. Connecting 95% of all private households to sewer systems

3. Increasing the removal efficiency for P in sewage treatment from 50% to >80%

4. Banning direct discharges from agriculture

5. Assessment of reaction time if P is banned in all detergents (assuming that one-third of the P flow through private households in the Danube basin stems from phosphorus-containing detergents)

Draw a general conclusion regarding the reduction of P flows to the river Danube when you evaluate the effectiveness of the measures discussed.

Problem 3.4:

Compare the materials turnover of a coal-fired power plant and an MSW incinerator. The feed rate for coal is 300 t/h; for an MSW, it is 30 t/h. Select the substances As, Pb, Cd, Cu, Se, Zn, Hg, S, Cl, and N. Substance concentrations for coal are given in Table 3.19; for MSW, see Table 3.20. Discuss your findings with respect to air pollution control.

The solutions to the problems are given on the website http://www.MFA -handbook.info.

TABLE 3.20

Example of Average Substance Concentrations in MSW, mg/kg

As	Pb	Cd	Cu	Se	Zn	Hg	S	Cl	N
10	500	10	1000	1	1200	1	3000	7000	5000

3.2 Resource Conservation

The main advantage of the application of MFA for resource conservation is the comprehensive information about sources, flows, and sinks of materials. This makes it possible to set priorities in resource conservation, to recognize early the benefit of material accumulations (e.g., in urban stocks), and to design new processes and systems for better control and management of resources. In this chapter, two groups of substances (nutrients and metals) and two groups of goods (plastic materials and construction materials) are discussed in view of resource conservation.

3.2.1 Case Study 5: Nutrient Management

Nutrients are essential resources for the biosphere. Life without nitrogen and phosphorus is not possible. The atmosphere represents an unlimited reservoir for nitrogen. The industrial transformation of N_2 to chemical compounds such as ammonium and nitrate that can be taken up by plants requires energy. Hence, the amount of nitrogen available within the anthroposphere is limited mainly by energy supply. In contrast, phosphorus is taken from concentrated phosphate minerals that are limited in extent. It is assessed that at present consumption rates, concentrated phosphate deposits might be used up in about 100 years (Steen, 1998). Thus, in order to conserve energy and resources, both nitrogen and phosphorus have to be managed with care.

The purpose of the following case study is to show how MFA can be used to set priorities in resource conservation. Measures are analyzed in view of their effectiveness regarding nutrient recycling. The total flows of phosphorus and nitrogen are investigated. The entire activity to nourish is analyzed from agriculture to food processing to private households. Losses and wastes are identified and quantified along the process chain. Because MFA of nutrients is discussed in several chapters of this book (see Sections 3.1.2, 3.1.3, and 3.5.2), the main emphasis is on the interpretation of the results. The procedure for establishing nutrient flow analysis of the activity *to nourish* is not given in detail. For further information, see Baccini and Brunner (2012).

3.2.1.1 Procedures

The activity to nourish is investigated on a national level. A system comprising five processes for the production (cultivating, harvesting), industrial processing (including distribution), household processing, and consumption (including digestion) of food is defined and investigated (Figure 2.8). For each process, information about inputs and outputs is obtained from available sources, including national statistics about import, export, and production of fertilizer, agricultural produce, and food; agricultural information databases about the use of fertilizer and production of agricultural products; reports from food-processing companies, wholesale companies, and distributors of food; medical literature about human consumption and excretion of nutrients; and databases about concentrations and loadings of nutrients in wastewater, municipal solid wastes, and compost.

It is important to start with reliable data about the structure of the national agricultural sector: what are the main agricultural products; how are they produced; what is the nutrient input required for the production; and how large is the amount of nutrients actually harvested? Internal cycles of agriculture are to be investigated, such as the soil–plant–animal–manure–soil nutrient cycle. Figures for total losses of nutrients in agricultural practice are usually not available. Farmers use different definitions for wastes and losses. Shortfalls have to be calculated as the difference between total input and total output of the agricultural sector. The same method can be applied to the processes *industrial processing and distribution* to calculate or cross-check figures for losses, wastes, and wastewaters.

Using the sources pointed out previously, the process *household* can be balanced as shown in Figure 2.8. The average amount of food consumed per capita and per year is taken from national statistics. Note that if such statistics are based on bookkeeping of individual households, they usually do not contain out-of-house consumption; in such cases, it will be necessary to increase the figure for food consumption by 20–30%. Waste analysis data yield the amount of food residues in MSW. If such data are not available, it can be assumed that 5–10% of food purchased is discarded with MSW. Information about kitchen wastewater is taken from studies about sewage production in households. It can also be estimated that about 20–25% of food entering a household is discarded via the kitchen sink. Note that cooking water may contain considerable amounts of dry matter and (dissolved) salts. Of course, the partitioning of food in households between MSW, wastewater, and human consumption is a function of cultural aspects, too: in societies that are traditionally scarce in resources, the amount of kitchen wastes is considerably smaller. If grinders are installed in kitchen sinks, the food fraction in wastewater will be higher. If fast food plays a major nutritional role, food wastes in kitchens will be smaller because most food entering households has already been processed. In such cases, packaging wastes may be larger.

Data about respiration, urine, and feces is found in the medical literature on human metabolism. This information contains figures about N and P in food, urine, and feces, too. It is important to cross-check all data. The output of agricultural production can be compared with the input into the food industry, the output of the food industry to the consumption of the total population, and the output of the total population to the input into wastewater treatment and waste management. If the data for balancing the individual processes have been collected independently for each process, the redundancy of such cross-checking will be high, and the accuracy of the total nutrient balance can be improved significantly.

3.2.1.2 Results

To demonstrate the relevant results, the five processes in Figure 2.8 are combined into the three processes presented in Figures 3.15 and 3.16. Food-related

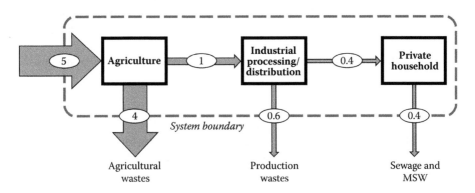

FIGURE 3.15
Phosphorus flow through the activity *to nourish*, kg capita^{-1} year^{-1}.

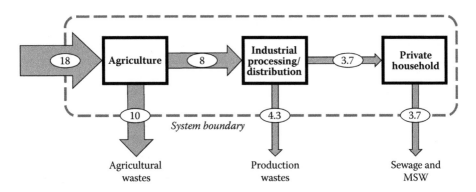

FIGURE 3.16
Nitrogen flow through the activity *to nourish*, kg capita^{-1} year^{-1}.

flows of P and N through agriculture, industrial processing and distribution, and consumption are displayed on a per capita basis. In view of resource conservation, agriculture is the most important process, where 80% of P and close to 60% of N are lost during agricultural production. Losses are flows to groundwater, surface water, and air for N, and erosion/surface runoff and accumulation in soils for P. In order to optimize nutrient management, agricultural practice has to be changed as a first priority. Since nutrients are still comparatively cheap, there is no economic incentive yet for such a change. It seems timely to investigate how new or other technologies can make better use of nutrients in agriculture. While the primary objective today is to prevent nutrient losses in order to protect the environment, it is likely that within a century, resource scarcity of phosphorus may become a driving force for changes in agriculture.

A key factor for nutrient losses in agriculture is consumer lifestyle. During the change from a resource-scarce society to an affluent society, the dietary tradition usually changes from low meat consumption to a diet that is rich in animal protein. The production of meat and poultry requires a much larger nutrient turnover than cereals and vegetables. Hence, the shift in dietary habits causes an increase in nutrient losses, too.

Losses of nutrients in industrial processing and distribution are much smaller than in agriculture, similar to those in households. The main difference between industrial processing/distribution and households is the number of sources: There are about 1000 times more households. Thus, from a reuse point of view, it is much more efficient to collect and recycle wastes from industrial sources than from consumers. The results summarized in Table 3.21 demonstrate clearly the limited contribution of individual households to the overall nutrient flows.

TABLE 3.21

Partitioning of Food-Derived P and N in Private Households

	P, g capita^{-1} year^{-1}	P,^{a} %	P,^{b} %	N, g capita^{-1} year^{-1}	N,^{a} %	N,^{b} %
Food input	430	100	8.6	3700	100	20.5
Output						
MSW	40	9	0.8	300	8	1.7
Kitchen wastewater	20	5	0.4	200	6	1.1
Respiration	0	0	0.0	110	3	0.6
Urine	270	63	5.4	2600	70	14.4
Feces	100	23	2.0	490	13	2.7
Total food-related output	430	100	8.6	3700	100	20.5

Source: With kind permission from Springer Science+Business Media: *Metabolism of the Anthroposphere (1st Edition)*, 1991, Baccini, P., and Brunner, P. H.

[a] Percent of food nutrient input into household.

[b] Percent of total nutrient import into activity *to nourish* from Figures 3.15 and 3.16.

If all food-derived nutrients from households are recycled, less than 10% of P and about 20% of N requirements of agriculture can be satisfied. Table 3.21 also shows the contribution of each household output to nutrient conservation, thus serving as a basis for decisions regarding nutrient conservation and waste management. Composting of MSW is an inefficient measure to recycle nutrients. If all MSW were turned into compost, the contribution to agriculture would only be 1–2%. The fraction of nutrients in wastewater from households is about 10 times larger. Thus, the priority in nutrient recycling should be on wastewater and not on solid waste.

Another interesting fact is revealed by MFA and presented in Table 3.21: the amount of nutrients in urine is three (P) to five (N) times larger than in feces. This opens up new possibilities. Separate collection of urine could allow more than half of all nutrients entering a household to be accumulated in a relatively pure, concentrated, and homogeneous form. Several concepts have been proposed to manage this so-called anthropogenic nutrient solution (ANS) (Larsen et al., 2001). They are all based on a new type of toilet that is designed to collect urine separately from feces. The sewer system would be used after midnight to collect ANS stored in households during the daytime, thus permitting specific treatment and recycling of N, P, and K. Or ANS could be stored in households for longer time periods and collected separately with mobile collection systems. In any case, in order to prepare a fertilizer of high value, hazards such as endocrine substances and pharmaceuticals would have to be removed before ANS could be applied in agriculture.

Note that MFA of the activity *to nourish* is the basis for identifying the relevant nutrient flows and for developing alternative scenarios. In order to test the feasibility of the scenarios, technological, economic, and social aspects have to be investigated. New ways of managing urine and feces will only be successful if the same or greater convenience for the consumer is guaranteed.

3.2.2 Case Study 6: Copper Management

In Chapter 1, Section 1.4.5.1, it has been documented that modern economies are characterized by unprecedented material growth. Consumption of metals has increased while metal prices have decreased due to more efficient mining and refining technologies (Metallgesellschaft Aktiengesellschaft, 1993). Up to 80–90% of all resources consumed by mankind have been used in the second half of the twentieth century (Figure 3.17).

Within the anthropogenic metabolism, heavy metals are comparatively unimportant from a *mass* point of view, since they represent less than 10% of all inorganic goods (excluding water) consumed (Baccini and Bader, 1996). However, heavy metals play an important role in the production and manufacture of many goods. They can improve the quality and function of goods and are often crucial in extending the lifetime and range of application of goods. Their importance is based on their specific chemical and

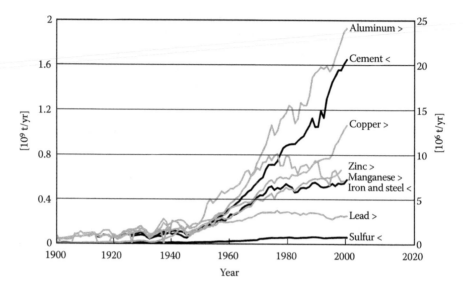

FIGURE 3.17
World use of selected resources, t/year. About 80–90% have been used since 1950. (From Kelly, T. D., and Matos, G. R., Historical statistics for mineral and material commodities in the United States (2016 version): *U.S. Geological Survey Data Series 140*, 2014. Retrieved from http://minerals .usgs.gov/minerals/pubs/historical-statistics/.)

physical properties, e.g., corrosion resistance, electrical conductivity, ductility, strength, heat conductivity, brightness, etc.

In 1972, the Club of Rome was among the first to point out the scarcity of resources in the book *The Limits to Growth* (Meadows, Meadows, Randers, and Behrens, 1972). Meadows et al. predicted that resources such as copper will be depleted within a short time of only a few decades. Prognoses about the depletion time (the number of years left until a resource is exhausted) of metals have been constantly revised and extended as a result of newly found reserves and advanced exploitation technologies. For certain metals essential for modern technology—lead, zinc, copper, molybdenum, manganese, etc.—some authors expect shortages within the next several decades (Kesler, 1994). There is controversy about whether this limitation will restrict future growth (for more information, see Becker-Boost and Fiala, 2001). Up to now, some but not all functions of metals can be mimicked by other materials.

Current metal management cannot be considered sustainable. During and after use, large fractions of metals are lost as emissions and wastes. Consequently, in many areas, concentrations of metals in soils as well as in surface water and groundwater are increasing. As discussed in Chapter 1, Section 1.4.5.2, human-induced flows of many metals surpass natural flows. Figure 1.7 displays the example of cadmium (Baccini and Brunner, 2012). While geogenic processes mobilize roughly 5.4 kt/year of cadmium, human activities extract about 17 kt/year from the Earth's crust. Comparatively large anthropogenic emissions into the atmosphere are causing a significant accumulation of cadmium in the soil. Global emissions of cadmium should be reduced by an order of magnitude to achieve similar deposition rates as those determined for natural cadmium deposition. On a regional basis, the reduction goal should be even higher. Since most of the anthropogenic activities are concentrated in the Northern Hemisphere, the cadmium flows in this region have to be reduced further in order to protect the environment properly. The stock of anthropogenic cadmium grows by 3–4% per year. It needs to be managed, disposed of, and recycled carefully in order to avoid short- and long-term environmental impacts.

Heavy metals are limited valuable resources, but they are also potential environmental pollutants. New strategies and methods are needed for the management of heavy metals. A first prerequisite for efficient resource management is appropriate information about the use, location, and fate of these substances in the anthroposphere (Landner and Lindeström, 1999). Based on such information, measures to control heavy metals in view of resource optimization and environmental protection have to be designed. This case study discusses sustainable management of copper using information about copper flows and stocks in Europe as determined by Spatari and colleagues (2002).

3.2.2.1 Procedures

The copper household is evaluated by statistical entropy analysis (SEA). In Chapter 2, Section 2.5.9, the SEA method was introduced for single-process systems. In this case study, a system consisting of multiple processes is analyzed, requiring additional definitions and procedures. SEA can be directly applied to copper databases with no further data collection and little computational effort. The procedure has been developed and described by Rechberger and Graedel (1999).

3.2.2.1.1 Terms and Definitions

A set of *material flows* consists of a finite number of material flows. The *distribution* of a substance represents the partitioning of a substance among a defined set of materials. The distribution (or distribution pattern) is described by any two of the three properties $\dot{M}_i$, $\dot{X}_i$, c_i for all materials of the set (see Figure 3.18).

3.2.2.1.2 Calculations

The following equations are used to calculate the statistical entropy H of a set of solid materials. If gaseous and aqueous flows (emissions) are also to be considered, more complex equations such as given in Chapter 2, Section 2.5.9.3, have to be applied. The system analyzed in this section contains solid

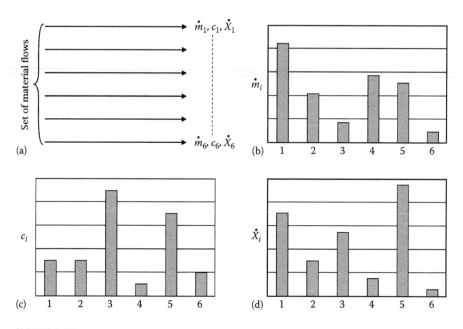

FIGURE 3.18
(a) Exemplary set of six material flows. (b) Mass flows of the set, mass/time. (c) Concentrations of the substance in the material flows, mass/mass. (d) Distribution of the substance among material flows (fraction).

materials/copper flows only. The number of materials in the set is k, and the flow rates $(\dot{m}_1, \ldots, \dot{m}_k)$ and substance concentrations $(c_1, \ldots, c_k)$ are known.

$$\dot{X}_i = \dot{m}_i \cdot c_i \tag{3.1}$$

$$\tilde{m}_i = \frac{\dot{m}_i}{\displaystyle\sum_{i=1}^{k} \dot{X}_i} \tag{3.2}$$

$$H(c_i, \tilde{m}_i) = -\sum_{i=1}^{k} \tilde{m}_i \cdot c_i \cdot \mathrm{ld}(c_i) \geq 0 \tag{3.3}$$

The concentrations in Equations 3.1 and 3.3 are expressed on a mass-per-mass basis in equivalent units (e.g., $g_{substance}/g_{product}$ or $kg_{substance}/kg_{product}$, etc.), so that $c_i \leq 1$. If other units are used (e.g., %, mg/kg), Equation 3.3 must be replaced by a corresponding function (Rechberger, 1999). The variable $\tilde{m}_i$ represents standardized mass fractions of a material set. If the c_i and $\tilde{m}_i$ are

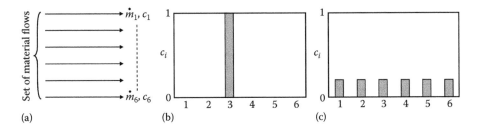

FIGURE 3.19
(a) A set of material flows representing the distribution of a substance defined by the couple $(\dot{m}_i, c_i)$. (b) If the substance is only contained in one material flow, the statistical entropy H is 0. If the substance is equally distributed among the material flows, H reaches the maximum. (c) Any other distribution yields an H value between 0 and max. (From Rechberger, H. and Graedel, T. E., *Ecol. Econ.*, 42, 59, 2002. With permission.)

calculated as described, the extreme values for H are found for the following distributions (see Figure 3.19):

1. The substance is only contained in one of the k material flows ($i = b$) and appears in pure form $\Sigma \dot{X}_i = \dot{X}_b = \dot{m}_b$. Such a material set represents the substance in its highest possible concentration. The statistical entropy H of such a distribution is 0, which is also a minimum, since H is a positive definite function for $c_i \leq 1$ (Figure 3.19b).

2. The other extreme is when all material flows have the same concentration ($c_1 = c_2 = ... = c_k$). Such a material set represents the substance in its highest possible diluted form. For such a distribution, the statistical entropy is a maximum. Any other possible distribution produces an H value between these extremes (Figure 3.19c).

The maximum of H is expressed as

$$H_{max} = \text{ld}\left(\sum_{i=1}^{k} \tilde{m}_i\right) \tag{3.4}$$

Finally, the relative statistical entropy (RSE) is defined as

$$RSE \equiv H/H_{max} \tag{3.5}$$

A material flow system usually comprises several processes that are often organized in process chains. Figure 3.20 displays such a system comprising four processes (P) linked by 10 material flows (F), including one loop (recycling flow F9).

The procedure for evaluating a system by SEA depends on the structure of the system. For the system investigated in this chapter, the statistical entropy development can be calculated as described in the following two sections.

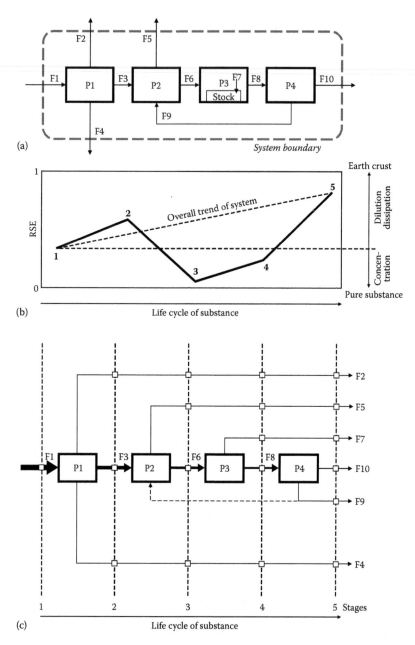

FIGURE 3.20
(a) Basic structure of a system made up of a process chain including one recycling loop.
(b) Allocation of the system's material flows to five stages. For example, stage 3 is represented and
defined by flows F2, F5, F6, and F4. Stages 2 to 5 represent the transformations of the input (stage
1) caused by processes 1 to 4. (c) The partitioning of the investigated substance in each stage corresponds to a relative statistical entropy (RSE) value between maximal concentration (0) and maximal dilution (1). (From Rechberger, H. and Graedel, T. E., *Ecol. Econ.*, 42, 59, 2002. With permission.)

3.2.2.1.2.1 Determination of Number and Formation of Stages If the number of processes in the system is n_P, then the number of stages is $n_S = n_P + 1$, where the stage index $j = 1, 2, ..., n_S$. The system as a whole can be seen as a process that transfers the input step by step, with each step designated as a *stage*. Stages are represented by a set of material flows (see Figure 3.20b). The first stage is defined by the input into the first process of the process chain. The following stages are defined by the outputs of processes 1 to n_P. So stage j ($j > 1$) receives (1) the outputs of process $j - 1$ and (2) all outputs of preceding processes that are not transformed by the system (export flows and flows into a stock). Flows out of a stock are treated as input flows into the process. Flows into a stock are regarded as output flows of the process (see process P3, flow F7 in Figure 3.20a). This means that the stock is actually treated as an independent external process. However, for the sake of clarity, stocks are presented as smaller boxes within process boxes (see Chapter 2, Figure 2.1). Finally, recycling flows are treated as export flows. The allocation of material flows to stages is displayed in Figure 3.20b. The diagram shows how substance flows through the system become increasingly branched from stage to stage, resulting in different distribution patterns of substances.

3.2.2.1.2.2 Modification of Basic Data and Calculation of RSE for Each Stage The basic data, flow rates of materials, and substance concentrations ($\dot{m}_i$, c_i) of the investigated system are determined by MFA. Normalized mass fractions $\tilde{m}_i$ are derived using Equations 3.1 and 3.2. Application of Equation 3.3 to the couple (c_i, $\tilde{m}_i$) or to each stage yields the statistical entropy H for that stage. H_{max} is a function of the total normalized mass flow represented by a stage (see Equation 3.4). This normalized mass flow grows with subsequent stages if the concentrations of the materials decrease, since $\Sigma c_i \times \tilde{m}_1 = 1$ (combine Equations 3.1 and 3.2). One can assume maximum entropy when materials of a stage have the same concentration as the Earth's crust (c_{EC}) for the substance under study. H_{max} is then given by

$$H_{max} = \text{ld}\left(\frac{1}{c_{EC}}\right) \qquad (3.6)$$

The reason for this definition of H_{max} is related to resource utilization. If, for example, copper is used to produce a good that has a copper concentration of 0.06 g/kg (the average copper content of the Earth's crust (Krauskopf, 1967), this product has the same resource potential for copper as the average crustal rock. Thus, a stage with entropy $H = H_{max}$ defines a point at which enhanced copper resources no longer exist. Using Equations 3.5 and 3.6, the RSE for each stage can be calculated. Figure 3.20c demonstrates that a system as a whole can be either concentrating, "neutral" (balanced), or diluting, depending on whether the RSE for the final stage is lower than, equal to, or higher than for the first stage.

3.2.2.1.3 Copper Data and Copper System of Study

Figure 3.21 illustrates copper flows and stocks in Europe in 1994, developed as part of a comprehensive project carried out at the Center for Industrial Ecology, Yale University. For a discussion of the quality, accuracy, and reliability of the data, see Graedel et al. (2002) and Spatari et al. (2002). Evaluating copper management practices on the basis of these data poses a challenge. At present, Europe is an *open system* for copper and depends heavily on imports. The total copper import (2000 kt/year) is more than three times higher than the domestic copper production from ore (≈590 kt/year; ore minus tailings and slag). Large amounts of production residues result from the use of copper, but with the present system boundaries, they are located outside of the system and therefore are not considered in an evaluation of European copper management. For a *true* evaluation, exports of goods containing copper and imports such as old scrap have to be taken into account, too. Thus, it is necessary to define a virtual *autonomous system* that (1) is independent of import and export of copper products and wastes and (2) incorporates all external flows into the system. Hence, in this virtual system, the copper necessary to support domestic demand is produced entirely within the system, depleting resources and producing residues. The estimated data for this supply-independent scenario are given in parentheses in Figure 3.21,

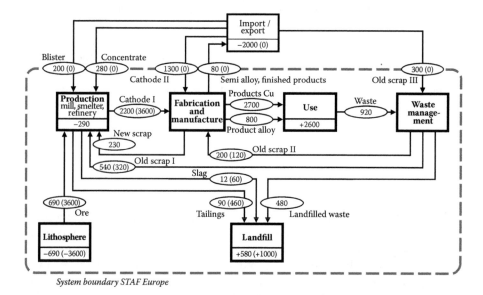

System boundary STAF Europe

FIGURE 3.21
Copper flows and stocks for Europe in 1994 (values rounded, kt/year). The values given in parentheses represent a virtual and autonomous copper system with the same consumption level but no copper imports and exports. (From Rechberger, H. and Graedel, T. E., *Ecol. Econ.,* 42, 59, 2002. With permission.)

which represents a closed system that includes all material flows relevant for today's copper management.

Table 3.22 gives the data that are used to calculate the entropy trends. The flow-rates for copper ($\dot{X}_i$) are from Spatari et al. (2002). The concentrations for copper (c_i) and their ranges are either from literature references or best estimates. The ranges provide the basis for assessing the uncertainty of the final entropy trends. The flow rates for materials ($\dot{m}_i$) are calculated using Equation 3.7:

$$\dot{m}_i = \dot{X}_i/c_i \times 100 \tag{3.7}$$

TABLE 3.22

Data on Material Flows of European Copper Management

Material	Material Flow ($\dot{m}_i$), kt/Year	Copper Concentration (c_i), g/100 g	Copper Flow ($\dot{X}_i$),* kt/Year
Ore	69,000	1 (0.3–3)	690
Concentrate	930	25 (20–35)	280
Blister	205	98 (96–99)	200
Cathode I	2200	100	2200
Flow out of stock (production)	290	100	290
Cathode II	1300	100	1300
Tailings	90,000	0.1 (0.1–0.75[a])	90
Slag	1700	0.7 (0.3[b]–0.7)	12
New scrap	260	90 (80–99)[c]	230
Old scrap I	680	80 (20–99)	540
Old scrap II	250	80 (20–99)	200
Old scrap III	380	80 (20–99)	300
Semialloy and finished products	110	70 (7–80)[c]	80
Products (pure Cu)	27,000	10 (1–50)[c]	2700
Products (Cu alloy)	11,000	7 (1–40)[c]	800
Flow into stock (use)	1,200,000	0.2 (0.1–0.3)[c]	2600
Wastes	460,000	0.2 (0.1–0.3)[c]	920
Landfilled wastes	460,000	0.10[d]	480

Source: Rechberger, H. and Graedel, T. E., *Ecological Economics*, 42, 59, 2002. With permission. Data from DKI Deutsches Kupferinstitut, *Kupfer, Vorkommen, Gewinnung, Eigenschaften, Verarbeitung, Verwendung Informationsdruck.* Duesseldorf: Deutsches Kupferinstitut, 1997; Zeltner, C. et al., *Regional Environmental Change,* 1 (1), 31–46, 1999; Gordon, R. B., *Resources, Conservation and Recycling,* 36 (2), 87–106, 2002.

Note: Values are rounded.

[a] Higher value for period around 1900.

[b] Lower value for period around 1925.

[c] Informed estimate.

[d] Calculated by mass balance on waste management process.

*Spatari, Bertram, Fuse, Graedel, and Rechberger (2002).

3.2.2.2 Results

3.2.2.2.1 RSE of Copper Management and of Alternative Systems—
Status Quo and Virtual Supply: Independent Europe

The entropy trends are calculated using Equations 3.3 to 3.7, the data given in Table 3.22, and the appropriate flowcharts. Figure 3.22 shows the trend of the RSE along the life cycle of copper for two systems: (1) the status quo of 1994 and (2) the supply-independent Europe (both displayed in Figure 3.21). The assignment of material flows to stages is illustrated in Figure 3.23.

Both systems behave similarly, with the production process reducing the RSE from stage 1 to 2, since ore (copper content = 1 g/100 g) is refined to plain copper (content > 99.9 g/100 g). Note that the RSE for stage 2 is not 0, since mining ores and the smelting concentrates produce residues (tailings and slag). The more efficient a production process is (efficiency being measured by its ability to transform copper-containing material), the more closely the RSE of stage 2 approaches 0, meaning that the total amount of copper appears in increasingly purer form. Note: For the reduction of the RSE from stage 1 to 2, external energy (crushing ores, smelting concentrate, etc.) is required. The impact on the RSE induced by this energy supply is not considered within the system, since the energy supply is outside the system boundary. Whether or not the exclusion of the energy source has an impact

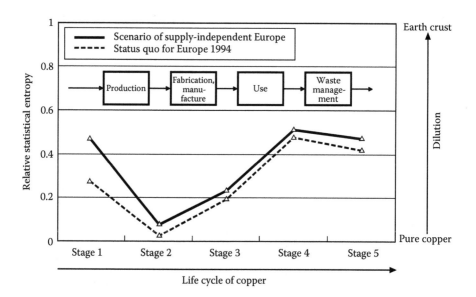

FIGURE 3.22
Change of the relative statistical entropy along the life cycle of copper for the status quo in Europe in 1994 (open system) and for a virtual, supply-independent Europe (closed or autonomous system). The shapes of the trends are identical, but the overall performances (differences between stages 1 and 5) of the systems are different. (From Rechberger, H. and Graedel, T. E., *Ecol. Econ.*, 42, 59, 2002. With permission.)

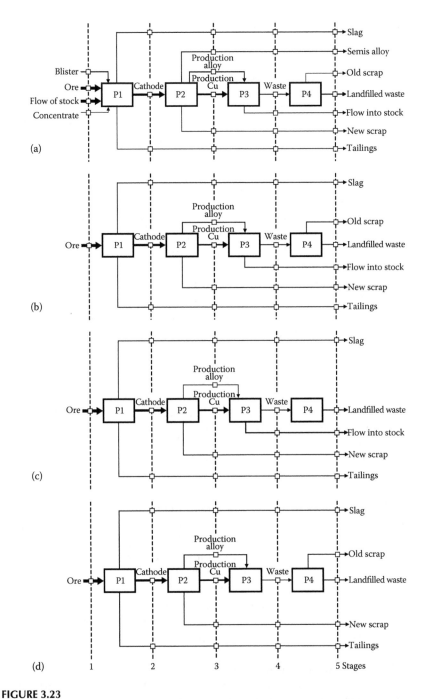

FIGURE 3.23
Assignment of material flows to stages. (a) Status quo, (b) supply-independent system, (c) supply-independent system without recycling, and (d) supply-independent system in steady state. (From Rechberger, H. and Graedel, T. E., *Ecol. Econ.*, 42, 59, 2002. With permission.)

on the RSE development depends on the kind of energy source (coal, oil, and hydropower) used. However, in this chapter, the system boundaries are drawn as described by Spatari et al. (2002).

Producing semiproducts and consumer goods from refined copper increases the RSE from stage 2 to 3 because of the dilution of copper that occurs in manufacturing processes. It is obvious that dilution takes place when copper alloys are produced. Similarly, installing copper products into consumer goods (e.g., wiring in an automobile) or incorporating copper goods into the built infrastructure (transition from stage 3 to 4, e.g., copper tubing for heating systems) "dilutes" copper as well. In general, the degree of dilution of copper in this stage is not well known. Information about location, concentration, and specification is a sine qua non condition for future management and optimization of copper. For a first hypothesis, it is sufficient to assume that the mean concentration of copper in the stock is the same as the mean concentration in the residues that leave the stock. This concentration level can be determined from copper concentrations and relevant waste generation rates such as municipal solid waste, construction and demolition debris, scrap metal, electrical and electronic wastes, end-of-life vehicles, etc. (Bertram, Rechberger, Spatari, and Graedel, 2002). During the transition from stage 4 to 5, the entropy decreases, since waste collection and treatment separate copper from the waste stream and concentrate it for recycling purposes. The "V" shape of the entropy trend—the result of entropy reduction in the production (refining) process and entropy increase in the consumption process (see Figure 3.22)—was described qualitatively, e.g., by O'Rourke, Connelly, and Koshland (1996), Ayres and Nair (1984), Stumm and Davis (1974), and Georgescu-Roegen (1971).

The differences in the entropy trends between the status quo and the supply-independent system are noteworthy. First of all, the status quo system starts at a lower entropy level, since concentrated copper is imported in goods. The differences in stage 2 are due to the increased ore production in the supply-independent system, resulting in larger amounts of production residues, which are accounted for in stage 2. In stages 3 and 4, the difference between the status quo and the supply-independent system remains rather constant, since the metabolism for both scenarios does not differ significantly in these stages. The effectiveness of waste management is lower in the supply-independent system, as there is no old scrap imported and the recycling rate is therefore lower. In the following, only the supply-independent system and some variations of it are discussed, since it comprises all processes and flows relevant for European copper management and includes external effects within Europe's hinterland.

The overall performance of a system can be quantified by the difference between the RSEs for the first and the final stages. In this case,

$$\Delta RSE_{total} = \Delta RSE_{15} = [(RSE_5 - RSE_1)/RSE_1] \times 100 \qquad (3.8)$$

where $\Delta RSE_{total} > 0$ means that the investigated substance is diluted and/ or dissipated during its transit through the system. From a resource conservation and environmental protection point of view, such an increase is a drawback. If maintained indefinitely, such management practice will result in long-term problems. In contrast, scenarios with high recycling rates, advanced waste management, and nondissipative metal use show decreasing RSE trends ($\Delta RSE_{total} \leq 0\%$). Low entropy values at the end of the life cycle mean that (1) only small amounts of the resource have been converted to low concentrations of copper in products (e.g., as an additive in paint) or dissipated (in the case where emissions are considered) and (2) large parts of the resource appear in concentrated (e.g., copper in brass) or even pure form (e.g., copper pipes). Wastes that are disposed of in landfills should preferably have Earth-crust characteristics or should be transformed into such quality before landfilling (Baccini, 1989). Earth-crust-like materials are in equilibrium with the environment, and their exergy approaches 0 (Ayres and Martinas, 1994; Ruth, 1995; Ayres, 1998). Thus, waste management systems must produce (1) highly concentrated products with high exergy that are not in equilibrium with the surrounding environment and (2) residues with Earth-crust-like quality. Low- or zero-exergy wastes can easily be produced by dilution, e.g., by emitting large amounts of off-gases with small concentrations in high stacks, or by mixing hazardous wastes with cement, thus impeding future recycling of the resource. A low RSE value for a stage thus means that both highly concentrated (high-exergy) and low-contamination (low-exergy) products are generated.

3.2.2.2.2 Recycling in Supply-Independent Europe

The relevance of recycling on the entropy trend is investigated using Figure 3.24. Numbers in parentheses show the supply-independent system without any recycling of old and new scrap. Compared with the supply-independent scenario displayed in Figure 3.21, this results in a higher demand for ore (+63%) and larger requirements for landfills for production and consumption wastes (+220%).

The entropy trend for the nonrecycling scenario is given in Figure 3.25. All RSEs are higher, showing the effects of not recycling production residues (new scrap) in stages 2 and 3 and the zero contribution of waste management in stage 5. The resulting $\Delta RSE_{15} = +28\%$ indicates a bad management strategy. At present, the overall recycling rate for old scrap is about 40%. Some countries within the European Union achieve rates up to 60% (Bertram, Rechberger, Spatari, and Graedel, 2002). Assuming that in the future, all countries will achieve this high rate, ΔRSE_{15} would be reduced from –1% to –4% (recycling rate of 90%: $\Delta RSE_{15} = –11\%$) for the supply-independent system. This shows that the impact of today's waste management on the overall performance of the system is limited. The reason is that the copper flow entering waste management is comparatively small.

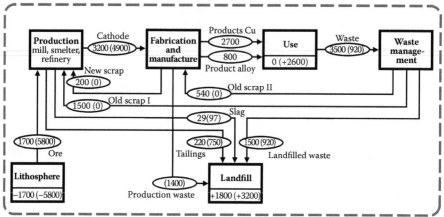

System boundary STAF Europe

FIGURE 3.24
Copper flows and stocks of a supply-independent Europe with no accumulation of copper in the process *use*, kt/year (steady-state scenario). Values in parentheses stand for a scenario without copper recycling. (From Rechberger, H. and Graedel, T. E., *Ecol. Econ.*, 42, 59, 2002. With permission.)

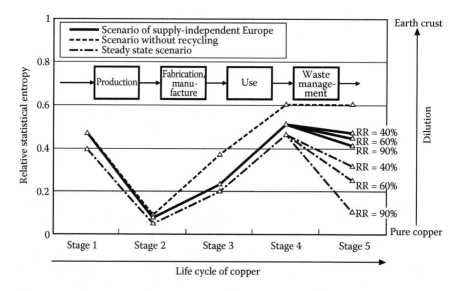

FIGURE 3.25
Comparison of the effect of different scenarios on the relative statistical entropy along the life cycle of copper: scenario of supply-independent Europe versus scenario of no recycling and scenario of steady state producing no stocks. The assessment shows that waste management and recycling can play a crucial role in future resource use. (From Rechberger, H. and Graedel, T. E., *Ecol. Econ.*, 42, 59, 2002. With permission.)

3.2.2.2.3 Supply-Independent Europe in Steady State

Figure 3.24 also gives the flows for a steady-state scenario in which the demand for consumer goods is still the same as in the status quo, but the output equals the input of the stock in the process *use*. This scenario may occur in the future when, due to the limited lifetime, large amounts of materials turn into wastes (Brunner and Rechberger, 2001). Assuming a recycling rate of 60% results in $\Delta RSE_{15} = -47\%$. This shows that in the future, waste management will be decisive for the overall management of copper. A recycling rate of 90% will result in $\Delta RSE_{15} = -77\%$. Such a high recycling rate cannot be achieved with today's design of goods and systems. Also, better information bases on the whereabouts of copper flows and, especially, stocks are needed. If the design process is improved, if necessary information is provided, and if advanced waste management technology is employed, future management of copper can result in declining RSE rates, contributing to sustainable metal management.

3.2.2.2.4 Uncertainty and Sensitivity

The uncertainty of the data (material flow rates $\dot{m}_i$ and substance concentrations c_i) and the accuracy of the results are fundamental pieces of information for the evaluation process. In most cases, data availability constrains the application of statistical tools to describe materials management systems. Statistics on material flows do not customarily provide information on reliability and uncertainty, such as a standard deviation or confidence interval. Sometimes, substance concentration ranges can be determined by a literature survey. In Figure 3.26, upper and lower limits of the RSE are presented for the supply-independent scenario.

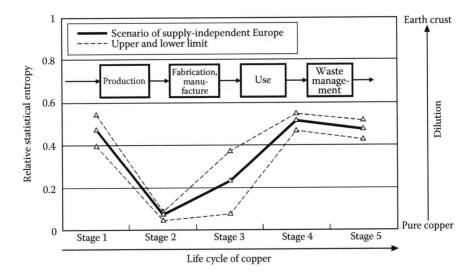

FIGURE 3.26

Variance of relative statistical entropy based on estimated ranges of basic data. (From Rechberger, H. and Graedel, T. E., *Ecol. Econ.*, 42, 59, 2002. With permission.)

These limits are calculated using the estimated ranges for copper concentrations as given in Table 3.22. Thus, the limits are not statistically derived but estimated. Since the ranges in Table 3.22 have been chosen deliberately to be broad, the possibility that the actual RSE trend lies within these limits is high. This is despite the fact that the uncertainty in the material flow rates is not considered. The range for ΔRSE_{15} lies between −23% and 28% (mean, −1%), sufficient for a first assessment. The uncertainties for the different stages vary considerably. The range for stage 1 is due to the range of the copper content in ores (0.5–2%). The range for stage 2 is quite small, meaning that the RSE for this stage is determined with good accuracy. The largest uncertainty is found for stage 3, since the average copper concentrations of many goods are poorly known. The uncertainties for stages 4 and 5 are lower, with a range similar to that for stage 1.

The result emphasizes the hypothesis that the stock in use has the potential to serve as a future resource for copper. Both stages 4 and 5 show the same entropy level for 1 t of copper. When calculating the RSE, the stock is characterized by the estimated average concentration of copper in the stock, meaning that the copper is evenly distributed and maximally diluted in the stock. This can be regarded as a *worst-case* assumption. Having more information about the actual distribution of copper in the stock would result in lower RSE values for stage 4. Provided that this information can be used for the design and optimization of waste management, the high recycling rates necessary to achieve $\Delta RSE_{15} < -70\%$ should be feasible.

3.2.2.3 Conclusions

Contemporary copper management is characterized by changes in the distribution pattern of copper, covering about 50% of the range between complete dilution and complete concentration. Copper flows and stocks through the (extended) European economy are more or less balanced due to recycling of new and old scrap and the small fraction of dissipative use of copper in goods. It is confirmed that the stock of copper currently in use has the potential for a future secondary resource. This can be even further improved by appropriate design for recycling of copper-containing goods. Provided that waste management is adapted to recycle and treat the large amounts of residues resulting from the aging stock, copper can be managed in a nearly sustainable way. Thus, this case study exemplifies how nonrenewable resources can be managed in order to conserve resources and protect the environment.

3.2.3 Case Study 7: Construction Waste Management

Construction materials are important materials for the anthropogenic metabolism. They are the matrix materials for the structure of buildings, roads, and networks and represent the largest anthropogenic turnover of solid materials (see Table 3.23). They have a long residence time in the anthroposphere and thus are a legacy for future generations. On one hand, they are

TABLE 3.23

Per Capita Use of Construction Materials in Vienna from 1880 to 2000

Period, Decade	Per Capita Use of Construction Materials, m^3 Capita^{-1} Year^{-1}
1880–1890	0.8
1890–1900	0.4
1900–1910	0.1
1910–1920	0.1
1920–1930	0.1
1930–1940	1.4
1940–1950	0.1
1950–1960	0.1
1960–1970	0.1
1970–1980	2.4
1980–1990	3.3
1990–2000	4.3

Source: Fischer, T. (1999). *Zur Untersuchung verschiedener methodischer Ansätze zur Bestimmung entnommener mineralischer Baurohstoffmengen am Beispiel des Aufbaus von Wien (Diploma Thesis).* Technische Universität Wien, 1999.

a resource for future use; on the other hand, they can be a source of future emissions and environmental loadings. An example of reuse would be recycling of road surface materials, which is widely practiced in many countries. Examples of emissions are polychlorinated biphenyls (PCBs) in joint fillers and paints, and chlorinated and fluorinated carbohydrates (CFCs) in insulation materials and foams. Hence, construction materials have to be managed with care in view of both resource conservation and environmental protection. A main future task will be to design constructions in a way that allows the separation of construction materials after the lifetime of a building, with the main fraction being reused for new construction, leaving only a small fraction for disposal via incineration in landfills. (Incineration will be necessary to mineralize and concentrate hazardous substances that are required to ensure long residence times of, for example, plastic materials.)

In this case study, construction materials are discussed in view of resource conservation. Both *volume* and *mass* are considered as resources. The purpose is twofold. First, it is shown that MFA can be used to address *volume-related* resource problems, too. Also, some of the difficulties of bringing construction wastes back into a consumption cycle are explained. Second, two technologies for producing recycling materials from construction wastes are compared by means of MFA.

3.2.3.1 The "Hole" Problem

Excavation of construction materials from a quarry or mine usually results in a hole in the ground. Since construction materials are used to create buildings with residence times of several decades, it takes some 30 to 50 years before these holes can be filled up with construction debris. In a growing economy, the input of construction materials into the anthroposphere at a given time is much larger than the output. Thus, as long as the building stock of a city expands, the volume of holes in the vicinity of the city expands as well.

In Figures 3.27 and 3.28, the total and per capita use of construction materials in Vienna is given for the time span from 1880 to 2000. The extraction of construction materials varies much from decade to decade. The effect of an economic crisis, such as the Great Depression of the 1930s and the postwar periods, on construction activities is evident. If accumulated over the time period of 120 years, a total hole of 207 million m³ results (Figure 3.28). This corresponds to about 140 m³/capita for today's population (1.5 million inhabitants).

It is interesting to note that the holes created by the needs of a prosperous, growing city of the 1990s are much larger than the volume of all wastes available for landfilling. In Figure 3.29, Lahner (1994) presents a construction material balance established for Austria. The input of construction materials exceeds the output of construction wastes by nearly an order of magnitude. Besides the hole problem discussed here, another important implication arises from input >> output: the amount of construction wastes available for recycling is small when compared with the total need for construction materials. Thus, even if all wastes were recycled, they would replace only a small fraction of primary materials. It may be difficult to create a market for a product

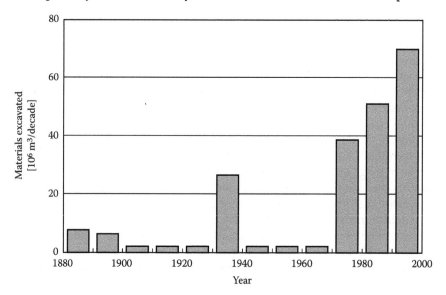

FIGURE 3.27
Construction material excavated from the ground and built into Vienna from 1880 to 2000, m³ per decade. (From Fischer, T., *Zur Untersuchung verschiedener methodischer Ansätze zur Bestimmung entnommener mineralischer Baurohstoffmengen am Beispiel des Aufbaus von Wien (Diploma Thesis)*. Technische Universität Wien, 1999.)

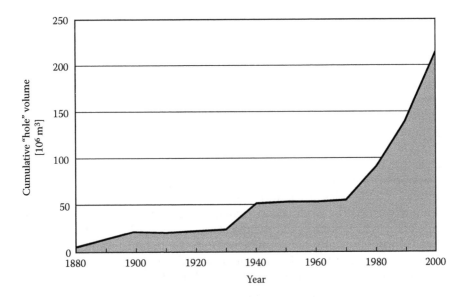

FIGURE 3.28
Cumulative "hole" volume in the vicinity of Vienna due to excavation of construction materials between 1880 and 2000. (From Fischer, T., *Zur Untersuchung verschiedener methodischer Ansätze zur Bestimmung entnommener mineralischer Baurohstoffmengen am Beispiel des Aufbaus von Wien (Diploma Thesis).* Technische Universität Wien, 1999.)

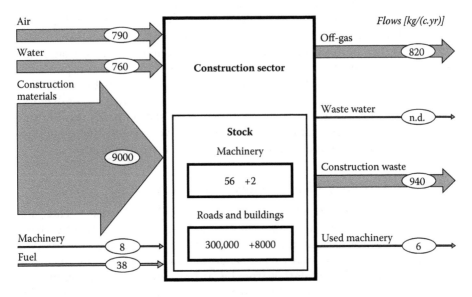

FIGURE 3.29
Materials used for construction in Austria (1995), kg capita^{-1} year^{-1}. The input of construction materials into a growing economy is much larger than the output of construction wastes. (From Lahner, T., *Müll Magazin, 7,* 9, 1994. With permission.)

with such a small market share, especially if there is uncertainty with respect to the quality of the new and as-yet-unknown material and if there is only a small advantage in price. For successful introduction of recycling materials, it is necessary to establish technical and environmental standards, to develop technologies that produce sufficiently high quality at a competitive price, and to persuade consumers of the usefulness and advantages of the new product.

In the case of Vienna, the total waste (MSW, construction waste, etc.) generated annually for disposal during the 1990s was about 600,000 tons measuring 800,000 m³ (or 400 kg/capita at 0.53 m³/capita). Wastes that are recycled are not included in this figure. This is approximately eight times less than the annual consumption of construction materials (4.3 m³ capita⁻¹ year⁻¹). Thus, it is not possible to fill the holes of Vienna by landfilling all wastes. Note that the actual volume of wastes to be landfilled in Vienna is considerably smaller due to waste incineration, which reduces the volume of municipal wastes by a factor of 10.

Landfilling is usually not a problem from the point of view of quantity (volume or mass); rather, it is an issue of quality (substance concentrations). The wastes that are to be disposed of in landfills do not have the same composition as the original materials taken from these sites. Thus, the interaction of water, air, and microorganisms with the waste material is likely to differ from the original material, resulting in emissions that can pollute groundwater and the vicinity of the landfill. On the other hand, the native material has been interacting with the environment for geological time periods. Except for mining and ore areas, the substance flows from such native sites are usually small ("background flows and concentrations") and not polluting.

The conclusion of the "hole balance" problem is as follows: Growing cities create holes; hence, "hole management" is important and necessary. These void spaces can be used for various purposes, such as for recreation or for waste disposal. If they are used as landfill space, qualitative aspects are of prime importance and have to be observed first. Wastes to be filled in such holes need to have stonelike properties. They require mineralization (e.g., incineration with after treatment), and they should be in equilibrium with water and the environment. The new objective of waste treatment thus becomes the production of immobile stones from waste materials.

3.2.3.2 MFA for Comparing Separation Technologies

Construction wastes are the largest fraction of all solid wastes. Thus, for resource conservation, it is important to collect, treat, and recycle these wastes. There are various technologies available to generate construction materials from construction wastes. Their purpose is to separate materials well suited as building materials from hazardous, polluting, or other materials inappropriate for construction purposes. MFA serves as a tool to evaluate the performance of construction waste sorting plants with regard to the composition of the products (e.g., production of clean fractions versus accumulation of pollutants in certain fractions).

In order to design and control construction waste recycling processes, it is necessary to know the composition of the input material that is to be treated in a sorting plant. The composition and quantity of construction wastes depend upon the "deconstruction" process. If a building is broken down by brute force of a bulldozer, the resulting waste is a mixture of all possible substances. If it is selectively dismantled, individual fractions can be collected that represent comparatively uniform materials such as wood, concrete, bricks, plastics, glass, and others. These fractions are better suited for recycling. After crushing, they can be used either for the production of new construction materials or as fuel in industrial boilers, power plants, or cement kilns. Both types of deconstruction yield at least one fraction of mixed construction wastes. While indiscriminate demolition results in mixed construction wastes only, the mixed fraction obtained in selective dismantling is much smaller and comprises mainly nonrecyclables such as plastics, composite materials, and contaminated constituents.

Construction waste sorting plants are designed to handle mixed fractions. The objectives of sorting are twofold: First, sorting should result in clean, high-quality fractions suited for recycling. Second, sorting should yield nonrecyclables that are ready for treatments such as incineration or landfilling. In Figures 3.30 and 3.31, two technologies for construction waste recycling are presented. They differ in the way they separate materials. Plant A (25 t/h) is a dry process, including handpicking of oversize materials, rotating drum for screening, crusher and pulverizer, zigzag air classifier, and dust filters. In plant B (60 t/h), the construction waste is similarly pretreated before it is divided into several fractions by a wet separator. In order to evaluate and compare the performance of the two processes with regard to resource conservation, both plants are investigated by MFA. The results serve as a basis for decisions regarding the choice of technologies for construction waste sorting.

3.2.3.3 Procedures

Since it is not possible to determine the chemical composition of untreated construction wastes by direct analysis, the input material into both plants is weighed only and not analyzed. The composition of the incoming waste is established by sampling and analyzing all products of sorting, and by calculating for each substance the sum of all output flows divided by the mass of construction wastes treated within the measuring period. This procedure is chosen because the sorting plants produce fractions that are more homogeneous in size and composition, and thus, they are easier and less costly to analyze than the original construction waste. The input into both plants is not the same, because the two collection systems that supply construction wastes to plants A and B are also different.

The method of investigation is described in by Schachermayer, Lahner, and Brunner (2000) and Brunner and Stämpfli (1993). Mass balances of input

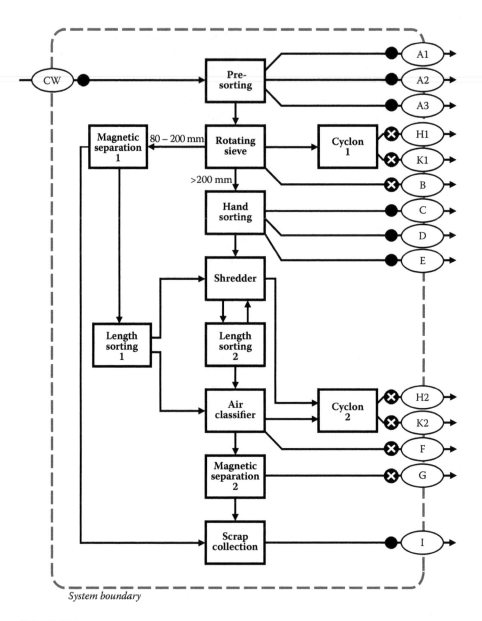

FIGURE 3.30
Construction waste (CW) sorting plant A, dry process. Fraction A1, large pieces of concrete and stones; A2, metals; A3, oversize combustibles; B, <80 mm; C, concrete and stones; D, metals; E, oversize material; F, light fraction; G, heavy fraction; H1 and H2, dust from cyclones 1 and 2; I, scrap iron; K1, off-gas drum and shredder; K2, off-gas air classifier. ●, measurement of mass flow, m³/h and t/h; X, measurement of substance concentration, mg/kg.

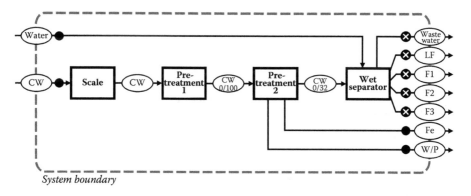

FIGURE 3.31

Construction waste (CW) sorting plant B, wet process. Pretreatment 1: crusher and sieve, 0 to 100 cm; pretreatment 2: hand sorting, magnetic separator, pulverizer, and sieve, 0 to 32 cm; CW 0/100 and CW 0/32: construction waste crushed, pulverized, and sieved by a mesh size of 100 and 32 cm, respectively; wastewater includes settled sludge; LF: light fraction; F1 to F3: construction materials for recycling (F1, 16 to 32 mm; F2, 4 to 16 mm; F3, 0 to 4 mm); Fe, scrap iron; W/P, fraction containing wood and plastics. ●, measurement of mass flows, m³/h and t/h; X, measurement of substance concentrations, mg/kg.

and output goods are performed for time periods between 2 and 9 h. The wet process is analyzed in five short campaigns, the dry process in a comprehensive investigation of 9 h. Samples of all output goods are taken and analyzed for matrix substances (>1 g/kg) and trace substances (<1 g/kg) at hourly intervals. Off-gases and wastewater are sampled according to standard procedures for such materials. The size of solid samples is between 5 and 500 kg. Aliquots of the samples are crushed and pulverized until particles are smaller than 0.2 mm. Metal fractions such as magnetically separated iron are not crushed; their composition is roughly estimated according to the individual components present. Oversize materials of concrete and stones are not analyzed either. For concrete, literature values are taken; composition of stones is assumed to be the same as in other, smaller stone fractions. For fractions that cannot be analyzed due to the lack of pulverized samples, it is tested to see if the overall material balance is sensitive against these assumptions. The fractions not analyzed amount to less than 5% of the total construction waste treated. Since the matrix (bulk) compositions of these fractions are known (e.g., the magnetically separated fraction contains <80% iron), errors in the assumptions proved not to be decisive for the overall mass balance and the transfer coefficients.

3.2.3.4 Results

3.2.3.4.1 Composition of Construction Wastes

As expected, the composition of the construction wastes treated in plant A is not the same as in plant B (see Table 3.24). The material treated in plant A

TABLE 3.24

Composition of Construction Wastes Treated in Dry-Separation Plant A and Wet-Separation Plant B Compared with the Average Composition of the Earth's Crust

Substance	Construction Waste Plant A (Mixed Construction Wastes)	Construction Waste Plant B[a] (Presorted Construction Wastes)	Earth's Crust
Matrix Substances, g/kg			
S	5.8	1.1–2.9	0.3
TC	93	47–79	0.2
TIC	33	35–69	
TOC	60	2–21	
Si	121	100–150	280
Ca	150	120–200	41
Al	9.5	8–15	81
Fe	40	7–20	54
Trace Elements, mg/kg			
Zn	790	24–66	70
Pb	630	3–103	13
Cr	150	13–32	100
Cu	670	8–23	50
Cd	1.0	0.10–0.22	0.1
Hg	0.2	0.05–0.55	0.02

Note: TC, total carbon; TIC, total inorganic carbon; TOC, total organic carbon.

[a] Data for plant B are the result of four sampling campaigns with different input materials; thus, ranges are given.

contains more sulfur (gypsum), organic carbon, and iron than the input into plant B, and the concentration of trace metals is about one order of magnitude higher. Construction wastes treated in plant A are more contaminated and contain less inorganic materials than the product for plant B. The reason for this difference has not been investigated. Possible explanations are as follows.

1. Plant A is located in Switzerland and was analyzed in 1988, while the mass balance of plant B, operating in Austria, was conducted in 1996. During the time period of 8 years, construction waste management experienced swift development. In the 1980s, mixed construction wastes were treated in separation plants, while the 1990s saw a shift toward selective deconstruction and dismantling, resulting in cleaner and more uniform input fractions for such plants.

2. At the time of analysis, Switzerland and Austria had distinctly different legislation and practices in construction waste management. In Switzerland, no legislative framework had been established at the time of analysis. The MFA of the sorting plant is a first investigation into the power of such plants to produce appropriate secondary construction materials. The results are used to establish a new strategy, giving preference to selective deconstruction (see the results that follow). Eight years later in Austria, it was mandatory to separately collect uniform fractions such as wood, metals, plastic, concrete, etc. when a certain mass flow per construction site is exceeded. The cleaner input into plant B indicates that the decision made in Switzerland (selective deconstruction) is appropriate.

3. Most construction waste stems from demolition and not from new construction sites. Due to different economic cycles (Austria was at a low level of economic development after World War II and was slow in recovering), buildings demolished in Switzerland and Austria are of different time periods. Some of the Swiss construction waste resulted from comparatively new buildings that had been constructed only 20 to 40 years ago, while in Austria, the buildings demolished in the 1990s were older.

Thus, the composition of construction wastes in plant A may resemble the construction materials of the 1950s and 1960s, while for plant B, the input most likely stems from prewar periods (1930–1940) and hence has a different composition in trace substances.

Note that the three reasons stated here have not been investigated in detail; they are merely given as possible explanations for different compositions of construction materials. In order to derive significant results about differences in the composition of construction wastes, the analysis would have to be planned from a statistical point of view, which was not intended when the mass balance was conducted in plant A.

In summary, at the time of investigation, plant A was fed by mixed construction wastes as received when indiscriminately demolishing a building. Plant B received construction debris that resulted from more-or-less controlled dismantling and represented a fraction that looked well suited for recycling, where much of the unsuitable material had already been removed at the construction site.

3.2.3.4.2 *Mass Flow of Products of Separation*

The balance of plant A is given in Table 3.25. Dry separation generated 14 different products. Four products are wastes and have no further use (dust from cyclones 1 and 2 and off-gases from the drum, shredder, and air classifier). Some of the remaining 10 fractions are, in part, quite similar. Therefore, they have been rearranged into the five fractions *I, II, III, metals,* and *rest* seen at the bottom of Table 3.25.

- Major fractions
  Fraction I, pieces <80 mm
  Fraction II, light materials
  Fraction III, heavy materials
- Minor fractions
  Scrap iron
  Rest, consisting of useless residues (filter dust and off-gas)

The rationale for this new grouping will become apparent when the chemical compositions of the individual fractions are discussed.

The balance of goods for plant B is given in Table 3.26. A priori, this plant produces fewer fractions. Only two of the seven fractions generated are of major importance. The light fraction only amounts to 5.1 g/100 g, indicating again that the input into plant B contains less organic waste (plastic, paper, light wood, and the like) than plant A. In contrast to plant A, plant B produces

TABLE 3.25

Mass Flow through Construction Waste (CW) Sorting Plant A

Material	Consisting of	Mass Flow, 10^3 kg/Day	Fraction, g/100 g CW
Total input	Construction wastes	225.3	100
Fraction			
A1	Concrete, stones	8.5	3.8
A2	Metals	3.08	1.3
A3	Oversize combustibles	3.75	1.7
B	<80 mm	102	45.3
C	Concrete, stones	4.14	1.8
D	Metals	2.36	1.0
E	Oversize material	0.43	0.2
F	Light fraction	51.4	22.8
G	Heavy fraction	47.7	21.2
H1	Dust cyclone 1	0.16	0.06
H2	Dust cyclone 2	0.10	0.04
I	Iron metals	1.73	0.8
K1	Off-gas drum/shredder	n.d.	n.d.
K2	Off-gas air classifier	n.d.	n.d.
New fractions	I + II + III + metals + rest	225.3	100
I (= B)	<80 mm	102	45.3
II (= F + E + A3)	Light fraction	55.6	24.7
III (= G + C + A1)	Heavy fraction	60.3	26.8
Metals (= A2 + D + E)	Iron	7.13	3.1
Rest (= H + K)	Dust and off-gases	0.26	0.1

Note: n.d., not determined.

TABLE 3.26

Mass Flow through Construction Waste (CW) Sorting Plant B (Presorted Construction Waste)

Material	Consisting of	Mass Flow, 10^3 kg/h	Fraction, g/100 g CW
Total input		370–380	≈500
Presorted CW		75	100
Water		300	400
Total output		200–270	260–360
Wastewater		130–190	170–250
(Wastewater sediment)[a]	(Wastewater sludge from pond)	(2.5–3.7)	(3.3–4.9)
LF	Light fraction	3.8	5.1
F1	Sorting fraction 16–32 mm	15	20
F2	Sorting fraction 4–16 mm	27	36
F3	Sorting fraction 0–4 mm	25	33
Fe	Scrap iron	0.13	0.17
W/P	Wood and plastic fraction	0.05	0.07

Note: The difference between input and output is due to the loss of water when the drenched fractions leave the wet process and are stored and dewatered on site without measuring water losses. It is not possible to quantify this difference.

[a] Wastewater sediment is included in wastewater and is generated in a process outside the system's boundary (sedimentation in a wastewater sludge pond).

a significant amount of fine-grain material of <4 mm particle size. The operator of plant B finds a good market for this material, while plant A's customers are asking for coarser materials. Note that due to waste separation on the construction site, the percentage of the scrap-iron fraction is 20 times smaller for plant B than A.

3.2.3.4.3 Composition of Products of Separation

The compositions of the products of the two construction waste recycling plants are presented in Table 3.27. In both plants, fractions rich in carbonates and silicates and poor in organic carbon are produced. Also, both plants produce light fractions containing approximately 20% of total organic carbon (TOC) and scrap-iron fractions. The difference in chemical composition of the products obtained in the two plants is mainly due to the difference of input materials.

Because of the given input, all fractions of dry separation in plant A exceed concentrations of heavy metals in the Earth's crust. Since construction waste treated in plant B is considerably cleaner, the compositions of the wet products come closer to Earth-crust quality. Nevertheless, concentrations of lead and mercury are above that of the Earth's crust for all fractions analyzed in plant B, too. The fraction most polluted is light fraction II from

TABLE 3.27

Composition of Products from Dry (A) and Wet (B) Construction Waste Separation

Substance	Products of Plant A					Products of Plant B				Wastewater Sludge	Earth's Crust
	I	II	III	G	Iron Metals	F1	F2	F3	LF		
Matrix Elements, g/kg											
Si	160	n.d.	180	170	n.d.	170 ± 10	170 ± 16	190 ± 13	170 ± 8	170	280
Ca	180	91	160	160	n.d.	160 ± 9	160 ± 19	140 ± 18	100 ± 17	160 ± 18	41
Fe	12	16	20	22	800	15 ± 5	16 ± 6	16 ± 5	20 ± 5	20 ± 3	54
TC	62	210	48	47	n.d.	54 ± 4	59 ± 6	59 ± 6	210 ± 90	98 ± 23	0.2
TIC	41	17	38	34	n.d.	53 ± 5	52 ± 10	47 ± 8	22 ± 8	47 ± 6	–
TOC	21	190	9.9	12	n.d.	1.8 ± 1	7 ± 6	11 ± 3	190 ± 95	51 ± 25	–
Al	8.8	8.3	12	12	8.1	15 ± 4	15 ± 5	11 ± 3	21 ± 6	20 ± 3	8.1
S	7.3	5.7	3.9	4.3	n.d.	1.6 ± 0.54	1.3 ± 0.2	1.4 ± 0.2	3.8 ± 0.4	2.4 ± 0.5	0.3
Trace Elements, mg/kg											
Zn	540	1400	170	200	4900	35 ± 8	34 ± 8	48 ± 5	65 ± 9	200 ± 91	70
Cu	47	420	330	410	11,500	16 ± 3	21 ± 6	22 ± 6	30 ± 7	45 ± 4	50
Pb	200	940	930	1200	1800	30 ± 54	16 ± 15	25 ± 10	46 ± 37	75 ± 11	13
Cr	160	90	130	140	760	24 ± 3	25 ± 9	25 ± 10	110 ± 22	41 ± 7	100
Cd	0.7	2.3	0.5	0.6	n.d.	0.12 ± 0.01	0.11 ± 0.005	0.13 ± 0.01	0.2 ± 0.07	0.31 ± 0.08	0.1
Hg	0.2	0.3	0.1	0.1	n.d.	0.11 ± 0.07	0.17 ± 0.08	0.47 ± 0.31	0.7 ± 0.03	3.1 ± 1,7	0.02

Note: n.d., not determined.

dry separation. The material is similar to MSW and exhibits a high content of organic carbon (20%). Thus, this fraction is not suited for recycling as a construction material. Instead, it can be utilized to recover energy from waste in an incinerator equipped with sophisticated air-pollution devices to remove acid gases, particulates, and volatile heavy metals like mercury and cadmium.

The light fraction from plant B is similar to the one from plant A. The main differences are that trace-metal concentrations are smaller in B and that the amount of light fraction that the wet plant B produced per unit of construction waste (5.1 g/100 g CW) is about five times smaller than for the dry plant A (24.7 g/100 g CW). Both differences are due to differences in the input materials for the two plants. Plant B produces a large amount of wastewater containing suspended solids. Most of this wastewater is treated in a sedimentation pond, where a sludge (sediment) is formed and deposited. Contaminant concentration of this sludge is higher than in any other product of plant B, confirming the hypothesis that a lot of heavy metals are present on small particles that are removed and transferred to the water phase during wet separation. A significant amount of less contaminated wastewater is not controlled and is "lost" on the site (the plant stands on a river bank).

3.2.3.4.4 *Partitioning of Metals and Transfer Coefficients*

The main purpose of construction waste sorting is to produce clean secondary construction materials. In chemical terms, sorting must direct hazardous substances contained in construction wastes to those fractions that are not intended for reuse. Preferably, the resulting substance concentration in recycling fractions is close to the concentration of materials used for the primary production of construction materials such as limestone, granite, and gypsum. A second goal is to maximize mass flows of useful and clean fractions. A third goal is to produce separation wastes that are well suited for disposal, by either landfilling or incineration. All of these goals can be achieved if mechanical sorting succeeds in controlling the flow of hazardous substances to certain fractions of sorting. Hence, it is of first importance to know the partitioning of heavy metals among the sorted products.

Table 3.28 lists the transfer coefficients (partitioning coefficients) for the two plants A and B. The results show that neither the dry nor the wet processes achieves the goal of directing the whole array of hazardous substances from recycling fractions to disposal fractions. Transfer coefficients for mass and substances are quite similar for most fractions, showing that *true* enrichment or depletion does not take place. It becomes clear that the superior qualities of the products of plant B are due to the clean input material and not because of a better separation by the wet process. MFA reveals the potential of the two technologies, and the transfer coefficients allow comparison of the separation efficiencies.

TABLE 3.28

Transfer Coefficients *k* of Selected Substances in Construction Waste Sorting Plants A and B, ×10^{-2}

| Substance | Plant A | | | | Plant B[a] | | | | |
	I	II	III	Metals	F1	F2	F3	LF	Sludge
Mass	45	25	27	3	20	36	33	5.1	≈ 4
Si	60	n.d.	40	n.d.	21	38	33	3.9	4.6
Ca	56	15	29	n.d.	23	41	28	2.7	5.0
Fe	14	10	13	63	18	34	27	4.3	5.5
TOC	16	80	4	n.d.	2.2	15	21	46	15
Al	42	21	34	3	23	40	25	6.0	6.8
S	57	24	18	n.d.	20	29	24	9.1	6.7
Zn	31	44	5	20	16	28	31	5.5	21
Cu	3	15	13	69	16	38	31	5.5	10
Pb	14	37	40	9	25	24	30	7.2	14
Cd	29	57	14	n.d.	19	33	30	5.9	12
Hg	43	36	12	n.d.	5.7	16	35	6.8	37

Note: n.d., not determined.

[a] Transfer coefficient k_{Fe} for scrap metals in plant B is 0.11. Transfer coefficient k_S for wastewater in plant B is 0.11. All other ks for wastewater are <0.003.

Transfer coefficients display the partitioning of elements only; they do not yet allow direct comparison of the enrichment or depletion of substances. In Figure 3.32, the quotients *substance concentrations in main fractions over concentration in construction waste* are presented for plant A on a log scale. These quotients are chosen to measure accumulation and depletion. In plant A, the most enriched elements are iron, copper, zinc, and chromium in the metal fraction. Dry sorting successfully concentrates these metals in the metal fraction. Organic carbon, cadmium, mercury, and lead are enriched in the light (combustible) fraction II. Fractions I and III are similar. In both, the matrix substances Si, Ca, and inorganic carbon are slightly enriched, while organic carbon and some heavy metals are modestly depleted. Except for copper in fraction I, all substances are depleted by less than an order of magnitude in fractions I to III. For mixed construction wastes, this order of magnitude is necessary if the process is to produce materials that are similar to the composition of the Earth's crust or to primary construction materials (see Table 3.24). There are no mechanical means yet to appropriately control the flow of all hazardous substances in sorting of mixed construction wastes.

3.2.3.5 Conclusions

Dry separation in plant A successfully concentrates combustible materials in the light fraction and construction-like materials in two other fractions. The processing yields about 70% of potentially useful construction products in

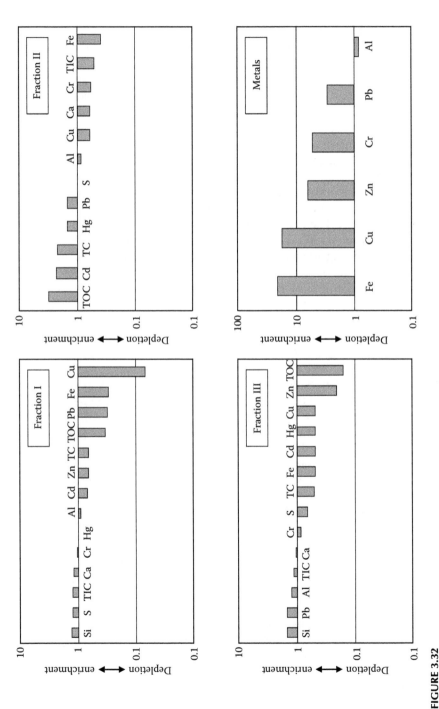

FIGURE 3.32
Enrichment, [concentration of X in fraction I]/[concentration of X in CW], of selected substances in main fractions of CW sorting plant A.

two fractions and about 3% of metals for recycling. The remaining fraction of 25% is not suited for recycling or landfilling; it has to be incinerated. Plant A is not capable of reducing the contaminant level of any fraction significantly. The main disadvantage of all products is the high trace-metal concentration. When the recycling products from plant A are being used for construction, the buildings will contain heavy metals that are significantly above Earth-crust concentrations. When the light fraction is incinerated, sophisticated and expensive air pollution control is required. Thus, it is most important that contaminants be removed by selective dismantling before entering the construction waste recycling plant.

Due to a cleaner input, wet separation in plant B results mainly in two comparatively clean fractions well suited for recycling. Although a few of the heavy-metal concentrations are elevated compared with the Earth's crust, they are (because of the cleaner input) generally of much lower concentration than in plant A. The overall performance of the wet process is similar to that of the dry process. While it is possible to produce a fraction rich in TOC and combustibles, significant accumulation or depletion of hazardous metals in any of the fractions is not observed. As for plant A, the light fraction contains much organic carbon, too, with the content of TOC reaching nearly 20%. Landfilling of a material with such a high TOC requires long after-care periods. Thus, it seems appropriate to utilize the light fraction as a fuel. However, due to the presence of heavy metals such as Hg (see Table 3.27), boilers designed to utilize the light fraction must be equipped with efficient air pollution control devices for atmophilic metals.

Despite the differences between the inputs into the two separation processes, MFA and transfer coefficients allow a comparison of the performance of the two plants. From a recycling point of view, the main differences are the products, with plant A producing gravel substitutes and plant B producing sand and gravel. The regional market situation determines whether sand or gravel is to be preferred. From an environmental point of view, there are no important differences. Because neither plant can sufficiently enrich or deplete hazardous materials, the substance concentrations of the main product fractions are similar to the concentrations of the incoming construction wastes.

The results of the MFA of the two plants support the strategy of selective deconstruction. Neither of the two processes is able to accumulate or deplete significantly (factor 10) hazardous materials in any of the resulting fractions. Once again, it becomes evident that at today's stage of development, mechanical processes are of limited use for the chemical separation of waste materials. Thus, wastes from indiscriminate demolishing of buildings are not well suited to produce recycling materials in construction waste sorting plants. For optimum resource conservation, it is important to separately recover materials during the deconstruction process and to recycle uniform fractions such as bricks, concrete, wood, and metals individually. In most cases, the remaining fraction can be mechanically sorted to recover a combustible fraction. Due to the composition of this fraction, containing plastic materials, paints, tubing, and cables, it is mandatory that energy recovery take place in incinerators equipped with state-of-the-art air pollution control devices suited to remove heavy metals such as mercury.

3.2.4 Case Study 8: Plastic Waste Management

Plastic materials were introduced in the 1930s. Ever since, polymers such as polyvinyl chloride (PVC), polyethylene (PE), polypropylene (PP), and polyamide (e.g., nylon) have shown large growth rates. Today they are among the most important man-made materials for many activities. At present, most plastics are made from fossil fuels that represent nonrenewable carbon sources. The production of plastics accounts for about 5% of the total fossil fuel consumption. They are used in cars, construction, furniture, clothes, packaging materials, and many other applications. Often, they contain additives to improve their properties. In particular, long-living plastic materials such as window frames, floor liners, and car fenders have to be protected from degradation and weathering by ultraviolet light, aggressive chemicals, temperature changes, and the like. Hence, plastic materials are usually mixtures of polymers with stabilizers, softeners, pigments, and fillers.

3.2.4.1 Plastic as Significant Fraction of MSW

Plastics make up between 10% and 15% of the total MSW flow. In addition, industrial and construction wastes are important sources of plastic wastes. Some plastic wastes (in particular from plastic manufacturing) are relatively clean and homogeneous and thus suitable for recycling. Others are mixtures of several goods and substances and hence cannot be recycled. Most plastic materials have a high energy content, and turned to waste, they can be used as a fuel. Due to stabilizers that contain heavy metals (lead, tin, zinc, cadmium, and others) and the chlorine content of some polymers (PVC, polyvinylidene chloride), thus yielding dioxins during incineration, incinerators for plastic wastes generally must be equipped with advanced air pollution equipment.

As shown in Table 3.29, packaging materials are comparatively clean and may be used as a secondary resource. On the other hand, the stock of long-living plastics contains large amounts of hazardous substances that will have to be dealt with in the future. Hence, plastic recycling and waste management needs tailor-made solutions that are appropriate for the individual material and its ingredients.

Figure 3.33 shows the plastic flows and stocks through Austria (Fehringer and Brunner, 1996). The figure was prepared using data from plastic manufacturers, waste management, and other sources. In the following discussion, the focus is on plastic waste management, emphasizing plastic wastes as energy resources and as sources of hazardous materials. In 1992, about

TABLE 3.29

Additives in Plastic Materials in Austria

Material	Total Consumption (1992), 1000 t/Year	Packing Material Consumption (1992), 1000 t/Year	Total Stock (1994), 1000 t
Plastics	1000	250	6700
Softener	14	3	180
Ba/Cd stabilizers	0.25	0.0002	4
Pb stabilizers	1.6	0.002	27
Flame retardants	2	0	34

Source: Fehringer, R., and Brunner, P. H. (1996). Kunststoffflüsse und die Möglichkeiten der Kunststoffverwertung in Österreich. Vienna, Austria: Umweltbundesamt Wien GmbH.

Note: Plastics with short residence times such as packaging materials are comparatively clean. The long-lasting stock in construction, cars, and other applications contains large amounts of hazardous materials such as cadmium, lead, and organotin compounds.

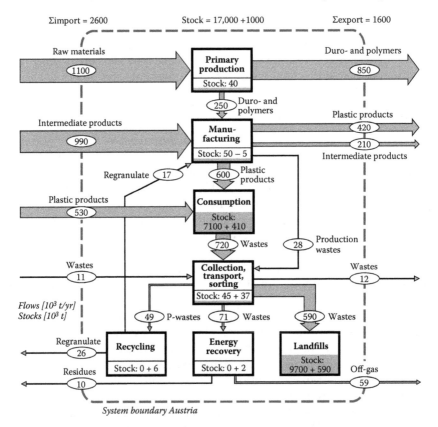

FIGURE 3.33

Plastic flows and stocks in Austria. (From Fehringer, R. and Brunner, P. H., Kunststoffflüsse und die Möglichkeiten der Verwertung von Kunststoffen in Österreich, UBA Monographien Band 80, Umweltbundesamt, Vienna, 1996. With permission.)

8 million Austrian consumers bought roughly 1.1 million tons of plastic materials. A large portion is used in goods with long residence times (floor liners, window frames, car parts, etc.) and thus is incorporated into the "anthropogenic stock." In Figure 3.33, this stock is assigned to the process *consumption*. The rest of the plastic is used for products with short residence times such as packaging materials and other consumer goods. The net flow (input minus output) into the stock of the process *consumption* amounts to 410 kt/year. Of the 720 kt/year of plastic wastes that leave the process *consumption*, 590 kt/year is landfilled, and the rest is either incinerated or recycled. It is interesting to note that the packaging ordinance that was instated in Austria in 1992 does not change much of this situation. Only about 7% (49 kt/year out of 759 kt/year) of all plastic wastes are controlled by the packaging ordinance and are directed toward material recycling. About 71 kt/year is being incinerated together with MSW in MSW incinerators. By far the largest amount of plastic wastes (590 kt/year) was still disposed of in landfills. Hence, much energy was wasted, since 1 t of plastics corresponds roughly to 1 t of fossil fuels. The landfilling of plastic wastes is not only a waste of resources; it also offends the Austrian Waste Management Act (BMUJF, 1990). The goals of this law are directed toward the conservation of resources such as energy and materials, and the law explicitly calls for the minimization of landfill space. Neither of these requirements is observed by plastic waste management practices as presented in Figure 3.33. However, the introduction of a new landfill ordinance prohibiting the disposal of organic material changed the situation significantly.

3.2.4.2 Plastic Management from a Holistic View Point

In Figure 3.34, the advantage of an integrated MFA approach is visualized: If only MSW is considered ("MSW view"), 200 kt/year of plastic wastes are observed, with 80% being landfilled and 20% being incinerated. When public attention is drawn to packaging wastes, leading to legislation such as the Dual System in Germany or the Packaging Ordinance in Austria, a certain amount of plastic wastes (70 kt/year) is separately collected and thus not landfilled (–60 kt/year) or incinerated (–10 kt/year) anymore ("packaging view"). Due to inferior quality, not all separately collected plastic wastes can be recycled as polymers. Hence, a certain percentage is used as an alternative fuel, e.g., in cement kilns, leaving 50 kt/year for substance recycling.

If all plastic wastes are included in the assessment, a much larger amount of landfilled wastes is observed (590 kt/year) ("total waste management view"). It is important to note that without an investigation into the total national flows and stocks of plastic, it is not likely that the large amount of landfilled plastics can be identified. Only a balance of the process *consumption*, with estimates of the mean residence time of various plastic materials, allows a reliable assessment of wastes that are leaving consumption. It is a much more difficult, if not impossible, task to directly identify the amount

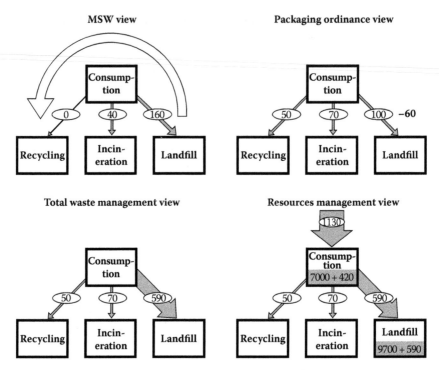

FIGURE 3.34
MFA as a decision-support tool enables different views of an issue such as management of plastic waste.

of plastics in the many wastes landfilled. Figure 3.34 shows clearly that rational decisions regarding plastic wastes have to be based on a complete set of flows and stocks of wastes in a national economy ("resource management view"). The sole focus on a single waste category such as packaging wastes results in solutions that are not optimized regarding resource and waste management.

The benefit of an MFA approach in resource management as discussed in this case study is as follows: a total plastic balance at a countrywide level shows the important flows and stocks of plastics and helps in setting the right priorities in resource management. First, the large and useful stock of plastics (and thus materials and energy) in consumption and landfills is recognized. Second, potential hazards due to toxic constituents of plastic materials are identified in both stock "consumption" and landfill; the toxics will have to be treated with care in the future. This knowledge is a precondition for controlling the flow of polymers and their hazardous additives to processes well suited for recovery and final disposal such as plastic recycling and waste to energy.

3.2.5 Case Study 9: Aluminum Management

Due to the buildup of substantial anthropogenic aluminum (Al) stocks, the sourcing of secondary raw materials from these Al stocks has become of major interest from an economic as well as from an environmental perspective (EC, 2014). Knowledge about past and present Al use patterns in society is essential to evaluate future anthropogenic resource potentials and provide recommendations on optimized Al management.

In this case study, stocks and flows of Al in Austria are analyzed using static as well as dynamic material flow analysis to create a basis for optimized Al resource management on a national level. By the static MFA, an in-depth understanding of Al use patterns in Austria in the year 2010 (Buchner, Laner, Rechberger, and Fellner, 2014a) is created. Next, the dynamic model is developed based on historical data about Al production and consumption in Austria for the time period between 1964 and 2012 (Buchner, Laner, Rechberger, and Fellner, 2015a). The dynamic MFA allows for determining the in-use stocks of Al in various sectors following a top-down approach (Laner and Rechberger, 2016) and estimating the end-of-life (EOL) Al flows from these sectors to waste management and exports. The dynamic model is calibrated by adjusting model parameters based on independent bottom-up estimates. The model results are cross-checked against independent estimates and data. In a following step, the future development of in-use stocks and old scrap generation is projected by combining the data from the historical dynamic material flow model with forecasts on future Al consumption (Buchner, Laner, Rechberger, and Fellner, 2015b). Future Al consumption is estimated using a stock-driven approach (i.e., stock development is the driver for consumption) for some sectors and an input-driven approach (i.e., based on projections of future consumption, for instance, by using annual growth rates) for others. The model projections are used to evaluate if the domestic Al scrap potentially can satisfy the demand for Al scrap in Austria given current trends in Al consumption and scrap generation. Finally, the quality of Al scrap is introduced in the modeling to account for constraints concerning the recycling of mixed scrap into specific alloy types, e.g., cast alloys cannot be recycled to wrought alloys (Buchner, 2015). The model thereby allows for investigating the potential of cast-alloy production to absorb mixed old scraps under different scenarios considering the application of advanced sorting technologies.

3.2.5.1 Static Al Balance

A static MFA of Al in Austria for the year 2010 is performed to investigate current Al use patterns on the national level. A particular focus is put on the waste management phase and on Al flows on the scrap market to provide a basis for evaluating the resource efficiency of national Al use (Buchner, Laner, Rechberger, and Fellner, 2014a,b). The static material flow model is developed using STAN. It comprises all main stages of the Al life cycle, from production and processing, to utilization and waste management. Foreign trade flows are considered for unwrought Al, semifinished products, and final products (indirect Al flows) to determine the total final Al demand in 2010. The total old scrap (i.e., postconsumer waste) generation is estimated by balancing national secondary production with the amounts of new scrap (i.e., preconsumer waste) and the net-import of foreign scrap and unwrought Al. Al scrap amounts are estimated for individual use sectors by combining top-down and bottom-up estimates (Buchner, 2015). Finally, data quality is assessed based on data quality indicators and then translated into uncertainty ranges for all input data, given by mean values and standard deviations.

The total output of domestic secondary Al production in 2010 is 572 Gg/year, which is either used domestically or exported (Buchner, Laner, Rechberger, and Fellner, 2014a). The Al input to the use phase in final products is around 218 Gg/year or 26 kg capita^{-1} year^{-1} in 2010. The major Al-consuming sectors, making up 86% of national Al consumption, are *buildings and infrastructure, transport,* and *packaging* (see Figure 3.35).

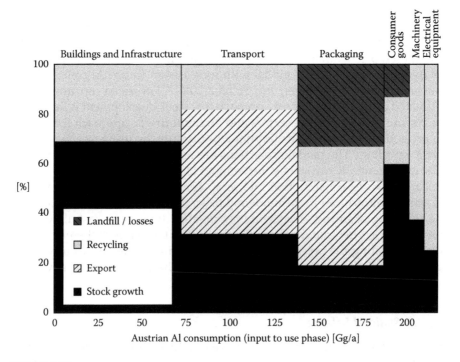

FIGURE 3.35
Total Al consumption of each use sector and partitioning into different pathways in the static model based on Buchner, Laner, Rechberger, and Fellner (2014a). Reading example: Al consumption of the sector *buildings and infrastructure* is around 70,000 metric tons/year, with nearly 70% of the consumed Al adding to the buildup of stock.

Forty-three percent of the input is accounted for as growth of Austrian in-use Al stock, with particularly strong Al stock growth in buildings. Old scrap generation is 7 kg capita^{-1} year^{-1}, of which 80% is recovered in waste management processes. The highest losses occur for Al in packaging waste, where roughly 30% of the Al in wastes is either landfilled or oxidized during thermal waste treatment processes. The largest share of EOL Al flows from the transport sector are not directed to waste management but exported in old vehicles for further use outside of Austria. From a production perspective, secondary Al production in Austria is highly dependent on net imports, which constitute around 40% of the production input. Due to this high share of foreign scrap, for which a distinction between new and old scrap is not possible, the qualitative resource demand of national Al is hard to evaluate, with a possible range of old scrap utilization in national production between 0% and 66%.

3.2.5.2 *Dynamic Al Flow Model*

The static MFA provides the basis for developing a dynamic material flow model, which enables a detailed investigation of the Al in-use stock developments and the trends in Al scrap generation over time. The focus on the national system provides the opportunity of increasing the confidence in model outcomes based on the comparison with other estimates, which is typically not possible for dynamic material flow studies on a large scale (cf. Buchner, Laner, Rechberger, and Fellner, 2014a). In-use Al stocks are calculated for six sectors following a top-down approach (see Equation 3.9). The growth of in-use stock in a specific year t is determined by subtracting the output $O(t)$ from the input $I(t)$. Summing up the change of stock over all previous time periods (from 1 to T) and accounting for the initial stock in year 0 $S(0)$ results in the total stock at the time T $S(T)$. This is a widely used approach in dynamic material flow modeling to derive estimates of in-use metal stocks based on historic production and consumption data, and to make projections on future secondary resource availability (e.g., Pauliuk, Wang, and Müller, 2013; Müller, Hilty, Widmer, Schluep, and Faulstich, 2014). However, as historic data on the outputs from use sectors are rarely available, the output of obsolete products is typically calculated using sector-specific lifetime functions. Such functions are defined for specific end-use sectors, with outputs being calculated by accumulation of the fraction of all former inputs becoming obsolete in a respective year. This is done by combining the input function $I(t)$ with the lifetime function $f_{lt}(t)$ in a convolution operation (Müller, Hilty, Widmer, Schluep, and Faulstich, 2014) as described in Equation 3.10, where T is the year for which the output is determined and d is the duration in years the material has been used. Because it is typically not possible to solve this convolution analytically, the calculations are performed for discrete time steps of single years. With respect to lifetime functions, several types of statistical distribution functions, such as normal, lognormal, beta, or

Weibull functions, are available for describing the residence time of material in the in-use stock (Melo, 1999). In the present study, Weibull functions with different sector-specific parameters (i.e., average lifetimes) have been chosen to model the obsolescence behavior of in-use Al products (cf. Buchner, Laner, Rechberger, and Fellner, 2015a).

$$S(T) = S(0) + \sum_{t=1}^{T} I(t) - O(t)$$ (3.9)

$$O(T) = (I \cdot f_{lt}) = \sum_{d=1}^{\infty} I(T-d) \cdot f_{lt}(d)$$ (3.10)

where S is stock, I is input, O is output, and T is the time for which the stock and the output is determined.

Apart from the choice of lifetime functions and corresponding parameters, many other model parameters (e.g., sector-split ratios, recycling rates) have to be defined. These model inputs are considered to be uncertain and potentially varying over time. In order to improve the initial parameter estimates, independent bottom-up estimates are used to calibrate the dynamic material flow model. Such estimates can be established for the input flows to the transport and packaging sectors, where it is then possible to adjust the respective sector-split ratios. Furthermore, the model outcomes can be cross-checked with results from other studies or statistical data to evaluate their plausibility. Implementing these calibration and validation steps creates a more reliable basis for assessing historical, current, and future Al resource use and scrap availability.

Projections on the future consumption of Al and the development of in-use stocks enable evaluations of the Al resource availability in Austria until the year 2050 (Buchner, Laner, Rechberger, and Fellner, 2015b). For three of the six in-use sectors (transport, buildings and infrastructure, and electrical equipment), a stock-driven approach is used to determine future Al consumption and calculate Al scrap flows. Although this approach is considered to be more robust in a long-term perspective than inflow projections, it cannot be applied to the other in-use sectors. In the case of the packaging sector, there is no substantial accumulation of Al in stocks, and for the sectors *machinery* and *consumer goods*, bottom-up stock estimates are not available. Therefore, the future development of Al consumption in these sectors is calculated by assuming a certain growth rate of annual consumption starting from current levels (in 2012).

The results of the dynamic Al flow model are shown in Figure 3.36 for the total inflow to and the outflow from the use phase and the development of in-use Al stocks over time. It is apparent that Al consumption has been on

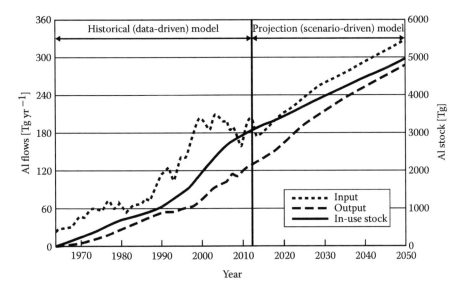

FIGURE 3.36
Total final Al demand (input to use phase) and EOL Al flows (output from use phase) as well as in-use stock development for the period from 1964 until 2050 in Austria based on Buchner, Laner, Rechberger, and Fellner (2015b).

the rise since the mid-1960s (beginning of the model period) until now (2012 in the model) and is expected to continue increasing also in the future. Al usage is expected to grow at a particularly high rate in the transport sector due to lightweight construction of cars. The generation of old scrap (output in Figure 3.36) will even grow at a slightly higher rate than consumption in the future and increase from a current level of around 130 Gg (14 kg/capita) per year to 210 Gg (24 kg/capita) per year in 2030 and 290 Gg (31 kg/capita) per year in 2050 (cf. Buchner, Laner, Rechberger, and Fellner, 2015b). The most substantial increases in old scrap generation are expected for the transport sector and the building and infrastructure sector. The total in-use stock of around 3.0 million metric tons (Tg) (360 kg/cap) in 2012 is projected to increase to 3.9 Tg (440 kg/cap) until 2030 and to 5 Tg (530 kg/cap) until 2050, which corresponds to an average annual growth rate of the in-use Al stock of 1% during the next 40 years.

3.2.5.3 *Potential of Anthropogenic Stock to Satisfy Demand*

The ongoing growth of Al resource flows and stocks highlights the significance of efficiently managing anthropogenic Al resources. From a national perspective, the question of whether domestic secondary resources can provide the basis for satisfying domestic Al demand is of particular interest. Therefore, the results of the dynamic model projections for the development of Al scrap generation (cf. Figure 3.36) are used to evaluate the future Austrian "self-supply potential" with respect to secondary Al.

The doubling of scrap generation from 2010 to 2050 is an opportunity for increasing the self-supply rate with respect to final Al consumption in Austria. Assuming no imports of unwrought Al and Al scraps and no exports of Al scrap as well as a ban on exports of Al-rich EOL products (i.e., end-of-life vehicles) and higher Al collection and recovery rates in waste management (sector-specific collection rates of 90–95%, processing losses of 2%), the final self-supply of Al for consumption is not expected to exceed 75% in 2050 [cf. scenario R_{high} in Buchner, Laner, Rechberger, and Fellner, (2015b)]. Hence, given the growth rates assumed for Al consumption, complete self-supply is not achievable in the foreseeable future, even with positive suppositions on recycling efficiency. If per capita consumption remained constant at the level of 2012 (approximately 23 kg capita^{-1} year^{-1}), still the available anthropogenic Al resources would not suffice to satisfy demand completely in 2050 [self-supply would rise to 83%; see Buchner, Laner, Rechberger, and Fellner, (2015b)]. Thus, satisfying final domestic Al demand based on domestically available secondary raw materials could only be reached through a decrease in consumption, which seems rather unlikely given historic developments.

A major issue with respect to a circular economy, apart from the quantity of recycled materials, is the quality of materials introduced to the recycling loop. In case of Al, the mix of old scraps from different applications contains various alloy elements in different concentrations, which may be a critical constraint to the use of old scrap in secondary production. Thus, increasing future Al self-supply may be jeopardized by qualitative limitations for Al recycling and secondary raw material utilization. In a first screening evaluation, wrought alloys are distinguished from cast alloys in the dynamic Al flow model, and the supply of Al scrap is projected with respect to these two major groups of alloys. Due to product specifications, cast alloys cannot be used to produce wrought alloys, but wrought alloys can be used to produce cast alloys. Consequently, the cast-alloy production represents a sink for Al scrap of different quality. Possible quality constraints regarding Al recycling in a closed national system (self-supply scenario) are investigated by comparing current and future cast and mixed scrap generation to national final cast Al demand (Buchner, 2015). The analysis of various scenarios using the dynamic Al flow model shows that a surplus of mixed scrap is expected to occur in the near future, if no separation between wrought and cast alloys is applied. This points out directions for future technological progress. The introduction of advanced separation technologies applied to old scrap could prevent a substantial surplus of mixed Al scrap compared to national final cast Al demand until 2040. Afterward, new technologies or foreign trade flows may be required to compensate for the misfit between the national final demand pattern and the scrap generation pattern in terms of alloy composition.

The MFA modeling results indicate clearly that the current recycling practice will lead to unsuitable Al scrap qualities on the domestic market, if high recycling rates and closed cycles are aspired to. Though these findings relate to aluminum and Austria, a single metal and a rather small country, they may essentially be transferable to other metals and other highly developed markets such as the European Union, because Al consumption patterns are rather similar. Consequently, moving toward a circular economy in terms of metal recovery requires intensified recycling and sorting of scrap. This implies technology development and appropriate commodity markets for secondary raw materials, because quality and composition of metal scraps are a determining factor for the properties of secondary metal products. Dynamic MFA represents a powerful tool for supporting policy development regarding Al, but also all other metals, in a circular economy.

PROBLEMS—SECTION 3.2

Problem 3.5:

Assume that the production of food by traditional agriculture can be replaced by "hydrocultures" that do not require soil for plant production. What will be the major change regarding total nutrient (N, P) requirements and losses from the activity *to nourish*? Use Figures 3.15 and 3.16 for your discussion.

Problem 3.6:

Use the following information to complete the four exercises listed afterward.

In 1996, about 8.1 million t/year of zinc (Zn) ores and 2.9 million t/year of Zn scrap are processed in order to produce 9.6 million t/year of Zn. Ore processing resulted in approximately 230 million t/year of tailings from milling with a content of about 0.3% Zn and 14 million t/year of slag from smelting with about 5% of Zn, each material flow representing a Zn flow of ca. 0.7 million t/year. Mining wastes are not considered. Zn is further manufactured into products that can be roughly grouped into five categories: galvanized products (3.3 million t/year), die castings (1.3 million t/year), brass (1.5 million t/year), Zn sheet and other semiproducts (0.6 million t/year), as well as chemicals and other uses (1.4 million t/year).

Galvanizing here stands for all kinds of technologies producing a coating of Zn on iron or steel in order to avoid corrosion. Die casting is a process to produce strong accurate parts in large quantities by forcing molten Zn alloy under pressure into a steel die (mainly used in the automotive industry). Brass is an alloy based on copper (Cu) and Zn. The Zn content ranges up to ca. 40%. Brass is used as sheets, wire, tubes, extrusions, and so on. Zn sheet is produced from Zn or Zn alloy rolled into thin sheets suitable for forming into roofing and cladding and other applications. The last category comprises mainly the dissipative uses, where Zn occurs as a trace metal, for example, in paints, automotive tires, brake linings, pesticides, animal feed and food additives, pharmaceuticals, cosmetics, etc. Manufacturing also results in production waste (Zn content, ca. 1.5 million t/year), mainly in the form of brass and galvanizing residues.

The total amount of Zn in products entering the use phase is 8.1 million t/year. The amount of Zn discarded is estimated at about 2.2 million t/year. Waste management separates 1.4 million t/year of Zn from the waste stream (Zn scrap). The remainder, which has a mean concentration of about 0.1% and comprises waste categories such as municipal solid waste, construction and demolition debris, wastes from electrical and electronic equipment, automotive shredder residues, hazardous wastes, industrial wastes, and sewage sludge, is landfilled (0.8 million t/year). This latter figure can only be regarded as a rough estimate. Mass flows of goods, their Zn content, and the resulting Zn flows are given in Table 3.30.

TABLE 3.30

Flows of Zn-Containing Materials, Their Zn Content, and Related Zn Flows for the World Economy

Goods	Mass Flow, Million t/Year	Zn Content, %	Zn Flow, Million t/Year
Zn ore	160	5	8.1
Tailings	230	0.3	0.7
Slag	14	5	0.7
Metal	9.6	100	9.6
Production waste	3.0	50	1.5
Zn scrap	17	11	1.4
Products	1500	0.54	8.1
Galvanized products	83	4	3.3
Die castings	1.3	99	1.3
Brass	4.3	35	1.5
Zn sheet and semiproducts	0.6	99	0.6
Chemicals and others	1400	0.1	1.4
Flow into stock	2200	0.27	5.9
(dissipative loss)	(1700)	(0.007)	(1.3)
Wastes	810	0.27	2.2
Landfilled wastes	800	0.1	0.8

Note: Values rounded to two significant digits.

a. Establish the flow diagram of the described Zn system.

b. Assign the material flows of the flowchart to stages and draw a diagram according to Figure 3.23 in Section 3.2.2.

c. Calculate the statistical entropy trend for the system. Is the trend sustainable?

d. Calculate what happens if 15% of consumed/used Zn neither is transformed to waste nor remains in the stock, but escapes to the environment (assume that the Zn flow is evenly dispersed in the soil.)

Problem 3.7:

Consider the following quantitative flowchart for the fluxes and stocks of construction materials within a fictitious region (see Figure 3.37).

a. Which stock will be most important for sand and gravel after 100 years (constant materials management assumed)?

b. Which conditions are required in order for recycling of construction materials to make a substantial contribution to the supply of construction materials (both buildings and underground)?

c. Which differences in material quality do you expect in the four stocks (which is the fourth stock)?

The solutions to the problems are given on the website http://www.MFA -handbook.info.

3.3 Waste Management

MFA is an excellent tool to support decisions regarding waste management for the following reasons:

1. In waste management, waste amounts and waste compositions are often not well known. MFA allows calculating the amount and composition of wastes by balancing the process of waste generation or the process of waste treatment. Thus, MFA is a well-suited tool for cost-efficient and comparatively accurate waste analysis.

2. As mentioned in the first paragraph of Chapter 1, inputs and outputs of waste treatment processes can be linked by MFA. Thus, if transfer coefficients are known, one can assess whether a given treatment plant achieves its objectives for a given input. Often, transfer

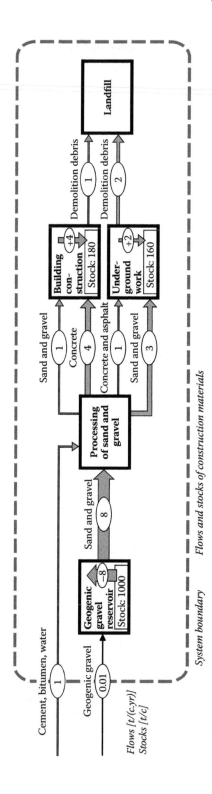

FIGURE 3.37

Fluxes and stocks of construction materials.

coefficients are not known in waste management, but they can be determined by MFA even if some inputs or outputs are not known.

3. Advanced waste management is a comparatively young branch of the economy. There is rapid development, driven by rising waste amounts, new technologies, and differing interests of stakeholders. There is a need for policy advice regarding future directions: What are the deficits of a given waste management system with regard to the goals set? What is the cost-effectiveness of a waste management system to reach the goals? And how can goal orientation as well as cost-effectiveness be measured and improved?

The following case studies are presented to exemplify how MFA can be used for waste analysis, optimization of waste treatment, waste policy analysis, and upporting policy decisions regarding waste management.

3.3.1 Use of MFA for Waste Analysis

Reliable information on waste composition and waste generation rate is crucial for the following objectives:

1. To identify potentials for recycling (biomass, paper, metals, plastics, etc.)
2. For the design and maintenance of waste treatment plants, including air and water pollution control technologies (recycling, incineration, landfilling)
3. To predict emissions from waste treatment and disposal facilities
4. To examine the effects of legislative, logistic, and technical measures on the waste stream

Because the composition and the generation rate of wastes are changing constantly, it is necessary to analyze them periodically. This is especially true when new consumer goods are being introduced to the market. Thus, routine, cost-effective determination of waste composition and of time trends is essential for waste management. In this chapter, selected methods of characterizing MSW are presented and discussed. These approaches were originally presented in a paper by Brunner and Ernst (1986).

The parameters that are used to characterize waste materials can be divided into three groups:

1. Materials or fractions of MSW (e.g., paper, glass, metals)
2. Physical, chemical, or biochemical parameters (e.g., density, heating value, biodegradability)
3. Substance concentrations (e.g., carbon, mercury, hexachlorobenzene)

To solve a particular problem of waste management, it is usually not necessary to analyze all parameters. For example, for recycling studies, information on the content of certain fractions in MSW such as paper or glass is required. To predict emissions, the elemental composition of MSW needs to be known.

Generally, there are three main methods for solid-waste analysis (see Figure 3.38). The first involves direct analysis of MSW, while the second and third are indirect methods based on MFA and the mass-balance principle.

1. Direct analysis, also known as the "sample and sort" method. A specified, statistically planned amount of MSW is collected. Samples are taken, screened, analyzed for waste goods, dried, pulverized, and finally analyzed for substances. The sample that is analyzed is usually small compared with the total MSW generated. This method has been used in many waste-characterization studies in the United States, Europe, and elsewhere (Barghoorn, Dobberstein, Eder, Fuchs, and Goessele, 1980; BUWAL, 1984; Maystre and Viret, 1995). Several manuals have been published describing how to conduct such analysis (Yu and Maclaren, 1995).

2. Indirect analysis of MSW composition by market-product analysis. This approach requires information about the production of goods and about the fate of these goods during use and consumption. Data collected from industrial sources such as key corporations, professional organizations, or government agencies are used to estimate flows of goods that are produced and consumed. The generation of MSW is calculated by measuring or assuming average life spans for these goods. Various adjustments are made for imports, exports, and

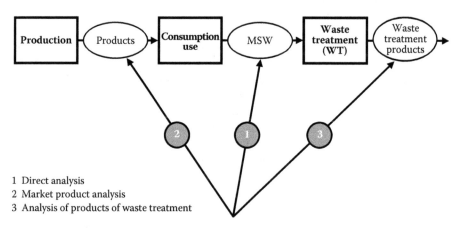

1 Direct analysis
2 Market product analysis
3 Analysis of products of waste treatment

FIGURE 3.38
Methods for MSW analysis. (From Brunner, P. H. and Ernst, W. R., *Waste Manage. Res.*, 4, 147, 1986. With permission.)

stocks in each product category. This method was developed in the early 1970s. Since then, data collection has improved, and databases have evolved. Results are compared with information about wastes that are landfilled, combusted, or recycled and with direct waste-analysis studies. The U.S. Environmental Protection Agency (EPA) applies this approach to estimate MSW generation (US EPA, 2002).

3. Indirect analysis using information about the products of waste treatment to calculate MSW composition. The advantage of this method is that the outputs of waste treatment are usually less heterogeneous than the input waste.

Especially for long-term monitoring, it is more cost-effective and accurate to determine waste composition by indirect methods (Morf and Brunner, 1998).

3.3.1.1 Direct Analysis

Direct waste analysis was the first approach used to determine waste composition. Waste samples are collected from different communities or regions based on statistical evaluations. The sample size usually varies between a minimum of 50 kg up to several tons. Samples are classified by hand into a selected number of fractions (paper, glass, etc.). Mechanical equipment is commonly used to separate magnetic metals and to sieve the remaining unidentified material into several additional fractions of different particle sizes. In order to determine the chemical and physical parameters of each fraction, representative samples are drawn from each material. These samples are further prepared (dried, pulverized, and sieved) for laboratory analysis.

The direct method is useful for

1. Measuring the concentration of most materials in MSW
2. Determining energy and water content of MSW and its fractions
3. Investigating the influence of geographic, demographic, and seasonal factors on the concentration of materials and some parameters in MSW
4. Assessing changes of waste composition with time
5. Evaluating the impact of separate collection measures on waste composition such as content of paper or glass or the impact of different collection systems (e.g., size of waste containers)

However, the direct method of waste analysis also has a number of limitations and disadvantages. First, it is labor-intensive and requires expensive equipment. Provided that adequate technical equipment and sufficient personnel are available, the analysis of one truckload takes at least half a day. A monitoring study on annual changes of MSW is estimated to consume

15 person-months of unpleasant and unhealthy labor. Second, the residue of separation that is not assigned to defined fractions such as glass, paper, etc., is usually quite large, often making up as much as 40–50% of the total MSW analyzed. As long as the composition of this fraction remains unknown, the value of the other results can be questioned, for example, the assessment of recycling potentials. Third, the determination of trace-element concentrations is problematic. If, for example, mercury batteries and their contribution to heavy metals in MSW are analyzed, one may find a few small batteries in 1 ton of MSW. This results in an average sample concentration of one to a few milligrams of mercury per kilogram of MSW. However, if only one or two MSW samples of 2 to 20 kg are collected, there is either a great chance of finding no mercury from batteries at all or of finding a high concentration if a single battery turns up in one of the randomly selected samples (Hg content up to 30% for Zn/Hg batteries).

This challenge is illustrated in Figure 3.39. If the chosen sample size is too small, the result of the analysis will probably be too low. The possible range of results increases with smaller sample sizes. Sufficiently large samples are needed to achieve results that reflect the actual content of unknown substances. Fourth, for technical and economic reasons, the metal fraction is often excluded from the chemical–physical analysis. However, this fraction

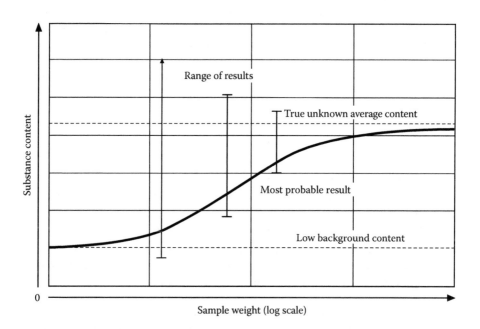

FIGURE 3.39
Drift of the most probable result as the sample size becomes very small. (From Pitard, F. F., *Pierre Gy's Sampling Theory and Sampling Practice*, Vol. II, *Sampling Correctness and Sampling Practice*, CRC Press, Boca Raton, FL, 1989, p. 159. With permission.)

may contain considerable amounts of heavy metals. Therefore, results of direct waste analysis may represent minimal values. A fifth problem is erosion and contamination from grinding and pulverization equipment.

These problems highlight some of the distinct limitations of the direct analysis of MSW with regard to the determination of chemical parameters, particularly of trace substances. They indicate that direct chemical analysis of waste materials represents the actual field concentrations only when large sampling campaigns are undertaken, resulting in extremely high costs. The direct approach is well suited for the determination of materials in MSW, but it seems of limited value in analyzing the elemental composition of MSW.

3.3.1.2 Indirect Analysis: Case Studies 11 and 12

The aforementioned problems and limitations of direct waste analysis led to the development of complementary methods, which yield more accurate results with less effort in terms of manpower and costs. Two case studies are presented to illustrate the use of MFA in indirect analysis.

3.3.1.2.1 Case Study 10: Waste Analysis by Market Analysis

Goods are produced and consumed. After use, they are either recycled or discarded as wastes. Since most industrial branches have accurate figures about their production, and since the pathways of many goods are well known, it is often possible to calculate the composition of MSW without field analysis and with high accuracy. This procedure, which can be used to analyze both the contents of materials and the elemental composition, is illustrated by the following examples for paper, glass, and chlorine content in MSW.

Paper: The most abundant single substance in MSW is cellulose, the main constituent of paper.

For paper recycling as well as waste treatment, it is of considerable interest to know the amount of paper in MSW. Figure 3.40 shows the flux of paper through the Austrian economy. Data are collected from pulp and paper manufacturers and checked against other available information. The amount of paper in MSW (48 kg/capita/year) is calculated as the difference between total paper consumption (179 kg/capita/year) and separately collected and recycled wastepaper (131 kg/capita/year). The Austrian population in 1996 was around 8.1 million inhabitants, and MSW generation amounted to 1.3 million t/year, which translates to 160 kg/capita/year of MSW for each resident. Based on these figures, a paper content of 30% (48 kg of wastepaper in 160 kg of MSW) can be calculated for average Austrian MSW. This figure has been confirmed by direct analysis.

Glass: A simple balance for the per capita glass flux in Switzerland in 2000 is given in Figure 3.41 (Kampel, 2002). Only packaging glass (bottles, beverage containers, etc.) is considered. The amount of glass in Swiss MSW (2.8 kg/capita/year) is calculated as the difference between glass consumed (46.6 kg/capita/year) and glass recycled (43.8 kg/capita/year). Glass with residence

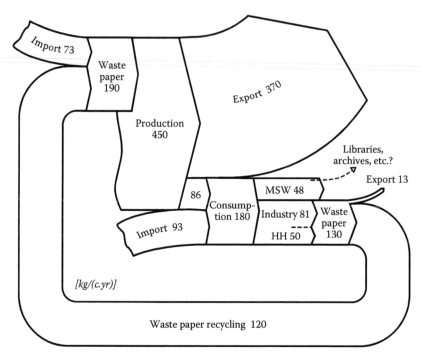

FIGURE 3.40
Paper flows in Austria (1996), kg/capita/year. (From Austrian Paper Industry, Personal communication, 1996.)

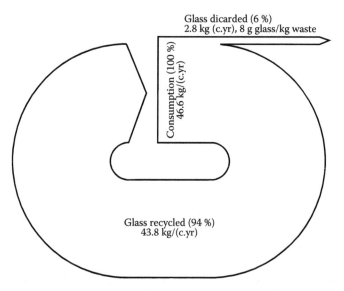

FIGURE 3.41
Recycling of packaging glass in Switzerland in 2000, kg/capita/year.

time greater than 1 year is not considered. Accumulation in the stock *household* is assumed to be less than 1% of consumption. With a Swiss population of 7.2 million inhabitants and 2.54 million t of MSW generated annually, the per capita allocation of MSW is 350 kg/capita/year. Based on these data, an average concentration of packaging glass in MSW of 8 g glass/kg MSW is calculated.

Paper and glass products have a short lifetime of less than 1 year. Therefore, it is reasonable to assume that inputs equal outputs over the balancing period. For other products with longer or even unknown residence times (e.g., wood in building materials), attempts to balance are more difficult. Yet, the US EPA studies on MSW generation show that this method is successful. Kampel used this approach to determine differences in waste-glass management among Australia, Austria, and Switzerland (Kampel, 2002).

Chlorine: The main sources of chlorine in MSW are assumed to be PVC and sodium chloride (NaCl). Minor amounts of Cl are contained in plant materials, other plastic materials, and other products. Thus, the content of Cl in MSW can be roughly estimated by the figures on consumption of PVC and table salt and by assumptions on the fate of these products during and after consumption and use. Data about goods such as NaCl and PVC are usually published in annual reports of the specific industrial branch, e.g., salt mine operators and plastics manufacturers. Sodium chloride in private households is utilized for dietary purposes mainly. It is assumed that not more than 10% of the NaCl purchased is discarded with MSW. Most salt is either eaten or discarded with wastewater while preparing food; in both cases, chloride leaves the household via sewage. Residence times of goods containing PVC are difficult to assess. It is estimated that $50 \pm 20\%$ of PVC is used in long-life products, and the other part is used for short-residence-time packaging material and consumer goods. Note that there is not yet a steady state for PVC flows. There is a large yearly growth rate on the input side. Because of the long residence time of some products, PVC is accumulated in the anthroposphere. Therefore, the amount of PVC-derived chlorine in MSW is calculated according to varying percentages of PVC. Despite the fact that chlorine estimates are based on several assumptions, the order of magnitude (5 to 10 g Cl/kg MSW) in Table 3.31 compares well with values resulting from product analysis of 7 to 12 g Cl per kg MSW.

Advantages and drawbacks: The main advantage of the analysis of MSW by a material balance of market products is the fact that no measurements are needed. MSW composition can be assessed quickly with little effort. In most cases, such rough estimates can give good results on a nationwide level. However, the method is not well suited to identify regional differences. It is usually more important to have reliable figures on the production/consumption side of a product than to have exact estimates of the proportion that enters the waste cycle. Another advantage of this method is the potential to predict trends in waste composition. Because today's products determine

TABLE 3.31

Determination of Chlorine in Swiss Municipal Solid Waste by Market Analysis

	NaCl	PVC Min.	PVC	PVC Max.
Consumption/use, kg/capita/year	5	8	8	8
Fraction discarded, %	10	30	50	70
Mass in MSW, kg/capita/year	0.5	2.4	4	5.6
Cl content, g/kg	610	580	580	580
Mass of Cl, kg Cl/capita/year	0.31	1.4	2.3	3.2
Contribution to Cl in MSW, g Cl/kg MSW	0.9	3.8	6.3	8.8
Total Cl in MSW (market analysis), g Cl/kg MSW			5–10	
Direct analysis, g Cl/kg MSW		3.4–4.2		
Product analysis, g Cl/kg MSW			7–12	

tomorrow's waste composition, this method is the only one that can be used to predict future waste composition.

Drawbacks of the method are (1) the dependency on production/consumption data, which are usually known on a national level only, and (2) that data are available only for a limited amount of materials and elements. It is not yet possible to characterize MSW from a physical point of view by this method (e.g., density and particle size).

3.3.1.2.2 Case Study 11: Analysis of Products of Waste Treatment

The analysis of the products of different waste-treatment processes is a powerful tool to characterize MSW (Brunner and Ernst, 1986). The main advantage is the homogenizing effect of treatment processes. This is particularly true if incineration is chosen for analysis. The incinerator acts as a large "thermal digester," separating substances from each other and releasing products that are of more uniform composition than the initial MSW. If all residues of the incinerator are analyzed and the total input and output mass flows are determined over a given period of time, the composition of the input into the plant can be calculated. This makes it possible to determine the flows of selected elements through an MSW incinerator and calculate the chemical composition of the waste input. The method has been successfully applied to several incinerators (Brunner and Mönch, 1986; Reimann, 1989; Vehlow, 1993; Belevi, 1995; Schachermayer, Bauer, Ritter, and Brunner, 1995; Morf, Ritter, and Brunner, 1997; Belevi and Mönch, 2000; Morf, Brunner, and Spaun, 2000).

Procedure: The procedure employed in a full-scale incinerator is as follows. The total mass flows of all input and output goods are determined during a given measuring period. Typical measurement campaigns last from several hours to several days. A (crane) balance measures the weight of the waste material fed to the incinerator. Consumption of water and chemicals is continuously recorded by incinerator control devices. The volume of air used for combustion is calculated based on a final mass balance and data about the energy consumption of the air blower. Solid incineration products are collected separately and weighed as received. Wastewater and off-gas are measured routinely by online flowmeters and translated into mass flows. To determine the chemical compositions, samples of bottom ash, filter cake, purified wastewater, and fly ash from the electrostatic precipitator (ESP ash) are taken and prepared for analysis. The bottom ash is the most heterogeneous product and requires extensive processing before analysis. First, it is separated from large pieces of iron, crushed, and sieved. The oversize material is weighed but usually not analyzed. It is assumed to consist mainly of iron (an assumption that is not justified for every incinerator). From the pretreated bottom ash, several composite samples are dried (at 105°C for ca. 24 h until a constant weight is achieved) and pulverized in a mill. Again, the oversize material is assumed to be of iron. For balance calculations, all fractions of the bottom ash are taken into account. Composite samples of fly ash are taken as close as possible to the filter device (to avoid time lag) and pulverized to laboratory sample size. Wastewater and filter cake, two rather homogeneous products, are sampled, too. In coordination with the sampling of solid and liquid incineration products, off-gas samples are taken to determine the flows of substances that are not measured continuously (mainly heavy metals). Detailed descriptions of effective sampling plans, procedures, the preparation of samples, and methods of analysis are presented by Morf, Brunner, and Spaun (2000) and Morf, Ritter, and Brunner (1997).

Figure 3.42 gives an example of appropriate locations for sampling and measurements in an MSW incineration plant.

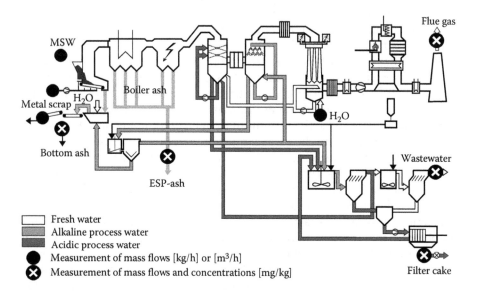

FIGURE 3.42
Locations of sampling and mass flow metering for indirect waste analysis in an MSW incinerator.

The element balances are calculated by multiplying the mass flow of goods by the respective concentrations of the elements for every period of the campaign. As mentioned before, input composition is not measured but is calculated indirectly by summarizing the mass flows of each element in the incineration products (minus substance inputs in other input goods such as water, air, or chemicals) and dividing by the mass flow of the waste input (see Equation 3.11).

$$c_{MSW,\,j} = \frac{\sum_{i=1}^{k} c_{ij} \cdot \dot{m}_i}{\dot{m}_{MSW}} \tag{3.11}$$

where
   k = number of incineration products
   j = substance

Concentrations of C, Cl, F, S, and several heavy metals in MSW have been determined by this method. Table 3.32 lists the results from six studies of five incinerators in Austria and Switzerland.

When analyzing wastes, it is highly important to consider uncertainty and to assess aspects of quality control. Bauer (1995) developed a method for quantifying the statistical uncertainty of such indirect waste analysis. Thus, it is possible to determine the effort that is necessary to obtain a given confidence interval for the waste composition. Higher efforts (more samples per time, larger sampling sizes) yield more reliable results (smaller uncertainties). A relationship between cost and accuracy can be established. Results with sufficiently small intervals (below ±20%) with a confidence of 95% can be obtained at reasonable costs. Morf and Brunner (1998) extended this approach. Based on MFA and transfer coefficients, they developed a method that allows routine measurement of MSW composition by analyzing only a single product of incineration per substance. They present procedures and examples of how to select the appropriate incineration residue to be analyzed, how to determine the minimum frequency for analyzing the residue, and how to measure the chemical composition of MSW routinely.

Results: Results of such investigations into MSW concentration are shown in Figures 3.43 and 3.44 (Brunner, Morf, and Rechberger, 2004). The monthly mean values of Cl and Hg vary by up to a factor two. The daily flows of the two selected elements Cl and Hg also vary. For Hg, these variations are quite substantial and up to a factor of four within a period of a few days. This emphasizes that random moment investigations are not a sufficient means of determining MSW composition.

The proposed MFA-based method for routine monitoring of waste composition by analyzing single incineration residues has significant advantages regarding data quality compared with the normally applied direct waste analysis. If waste composition were measured in the same way on several

TABLE 3.32

Results of Different Indirect Waste-Analysis Campaigns, g/kg

	Biel (CH) 1981	Müllheim (CH) 1984	St. Gallen (CH) 1991	Vienna (A) 1993	Wels (A) 1996	Wels (A) 1996
C	275 ± 55	n.d.	370 ± 40	190 ± 10	252 ± 25	265 ± 28
Cl	6.9 ± 1.7	n.d.	6.9 ± 1.0	6.4	12.2 ± 1.8	10.3 ± 1.2
S	2.7 ± 0.5	n.d.	1.3 ± 0.2	2.9 ± 0.2	4.2 ± 0.14	4.1 ± 0.17
F	0.14 ± 0.06	n.d.	0.19 ± 0.03	1.2 ± 0.1	0.054 ± 0.007	0.060 ± 0.002
Fe	67 ± 35	n.d.	29 ± 5	42 ± 1	37 ± 0.25	43 ± 0.2
Pb	0.43 ± 0.13	0.57 ± 0.43	0.70 ± 0.10	0.60 ± 0.10	0.40 ± 0.079	0.49 ± 0.088
Zn	2.01 ± 1.51	1.1 ± 0.5	1.4 ± 0.2	0.83 ± 0.07	1.2 ± 0.069	1.3 ± 0.14
Cu	0.27 ± 0.07	0.46 ± 0.19	0.70 ± 0.20	0.36 ± 0.03	0.59 ± 0.13	0.52 ± 0.076
Cd	0.0087 ± 0.0019	0.012 ± 0.0056	0.011 ± 0.002	0.008 ± 0.001	0.0107 ± 0.0028	0.0084 ± 0.0026
Hg	0.00083 ± 0.00081	0.002	0.003 ± 0.001	0.0013 ± 0.0002	0.0019 ± 0.00039	n.d.

Source: Morf, L. S. et al., *Güter- und Stoffbilanz der MVA Wels: Institut für Wassergüte und Abfallwirtschaft*, TU Wien, 1997.
Note: n.d. = not determined.

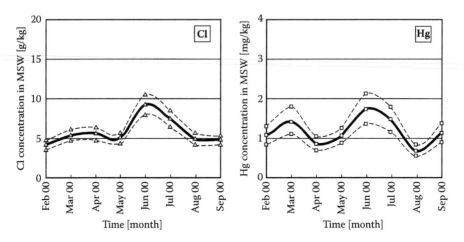

FIGURE 3.43
Time trends for monthly mean MSW concentrations of chlorine and mercury as determined for an incinerator (Spittelau) in Vienna, Austria, between February 1 and September 30, 2000. The figure shows means as well as the lower and upper limits for an approximately 95% confidence interval. (Reprinted from *Solid Waste: Assessment, Monitoring, and Remediation*, Twardowsky, I., Allen, H. E., Kettrup, A. A. F., and Lacy, W. J., Eds., Brunner, P. H., Morf, L., and Rechberger, H., Copyright 2003, with permission from Elsevier.)

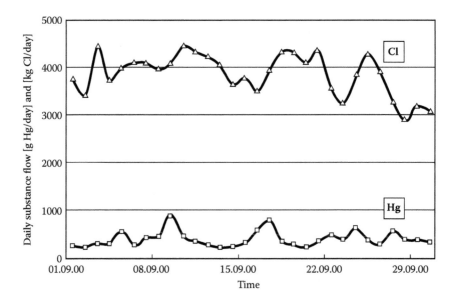

FIGURE 3.44
Time trends for daily flows of Cl (kg/day) and Hg (g/day) through the MSW incinerator (Spittelau) in Vienna, Austria, between September 1 and September 30, 2000. (Reprinted from *Solid Waste: Assessment, Monitoring, and Remediation*, Twardowsky, L., Allen, H. E., Kettrup, A. A. F., and Lacy, W. J., Eds., Brunner, P. H., Morf, L., and Rechberger, H., Copyright 2003, with permission from Elsevier.)

MSW incinerators, this would allow comparison of waste compositions in a more cost-effective and objective way than the present practice of direct waste analysis. Future MSW incinerators should be designed for and supplied with hardware and software to apply routine MFA for waste analysis. The additional costs would be small and the return on investment large when compared with the costs and accuracy of traditional approaches.

The main disadvantage of the waste product analysis is that waste fractions cannot be determined, e.g., it is not possible to calculate the contents of paper, plastic, or any other single fraction. This means that, in most cases, the product method is limited to the analysis of elemental composition and parameters like energy content, water content, and the content of total inorganic and organic matter.

Conclusion: It is highly important to choose the method of analysis that is most appropriate to solve a particular problem of waste management. In general, direct waste analysis yields good results on some fractions in MSW, but it is expensive and labor-intensive to determine reliably elemental concentrations by this method. Market-product analysis combined with MFA is an inexpensive and quick method to determine with sufficient accuracy the fraction-based and elemental composition of MSW. In many cases, this method of analysis can be applied in favor of direct waste analysis. However, the method is limited to those materials where information from the producing industries is available and where residence times in stocks are more or less known. MFA-based waste product analysis is well suited for determining element concentrations in MSW, but it does not allow analysis of material composition. It is the superior, cost-efficient method for determining time trends in elemental analysis of wastes.

3.3.2 MFA to Support Decisions in Waste Management

3.3.2.1 Case Study 12: ASTRA

In the case study ASTRA (a German acronym for "evaluation of different scenarios for waste treatment in Austria"), selected scenarios for the treatment of combustible wastes are compared in view of reaching the waste-management goals of "environmental protection," "resource conservation," and "aftercare-free landfills" (Fehringer, Rechberger, Pesonen, and Brunner, 1997). The incentive for this Austrian study is a new federal ordinance on landfilling that became effective in 1996 (Austrian Landfill Ordinance, 1996). The ordinance mandates that beginning in 2004, only wastes with a TOC <2–5% may be landfilled. The exact percentage depends on the type of landfill (e.g., monofill, landfill for construction waste, etc.). The reason for banning organic carbon in landfills is that organic carbon is transformed by microorganisms. The metabolic products are organic compounds that may be transferred to landfill leachates, and carbon dioxide and methane in landfill gas that contribute to global warming if not collected and treated properly. In addition, organic acids are produced that may mobilize heavy metals. Therefore, leachates of such reactor-type landfills are contaminated with a broad array of organic and inorganic pollutants. This requires treatment of the leachate over long periods (>100 years) and contradicts one of the Austrian objectives of waste management, which is to avoid shifting waste-related problems to future generations (aftercare-free landfills).

Because of the limit for organic carbon, treatment before landfilling is mandatory for most wastes such as MSW, sewage sludge, and construction debris. Combustion is an efficient means of transforming organic carbon to carbon dioxide. In order to ensure free choice of waste-treatment technologies, the landfill legislation allows an exemption from the TOC limit for the output material of mechanical–biological treatment facilities. These plants produce two fractions: (1) a combustible fraction that is mechanically separated and appropriate for further energy recovery and (2) a product derived from biological digestion. The biological degradation process cannot provide a residue with a TOC <5%, because persistent organic compounds such as plastics and lignin cannot be decomposed within months by microorganisms. Thus, an exception is stipulated for this fraction: it may be landfilled if the heating value is below 6000 kJ/kg. In contrast to the limit for TOC, which minimizes reactions in the landfill body and thus supports the objectives of waste management, the limitation of the heating value does not improve landfilling practice or reduce the need for aftercare. Rather, the exemption is based on political decisions. Both limits prevent direct landfilling of untreated MSW after 2004.

Some industry branches are eager to use combustible wastes as a substitute for fossil fuels. This helps to reduce costs of production, since wastes are usually cheaper than fuel. If the wastes are contaminated (e.g., with PCBs) or otherwise difficult to dispose of, they may even create revenue. Also, wastes made up of biogenic carbon are attractive fuels because they do not contribute to global warming.

The following treatment options are available for combustible wastes: incineration, cocombustion in industrial furnaces (using both conventional fuels and wastes), mechanical sorting, and biological digestion. All of these options have different environmental impacts and different contributions to the goals of waste management as stated in the Austrian Waste Management Act (AWG, 1990). In the case study ASTRA, various scenarios for the management of combustible wastes are developed and compared in view of the requirements of the new Landfill Ordinance and of the goals of the Waste Management Act.

3.3.2.1.1 Procedures

The ASTRA project consists of the following steps:

1. Selection of waste treatment processes and defining waste management systems and scenarios
2. Selection of substances

3. Selection of wastes

4. Establishment of mass balances (see Figure 3.45) and substance balances for the actual system

5. Development and selection of criteria to evaluate the scenarios

6. Development of an optimized scenario for improved management of combustible wastes (optimum assignment of wastes to treatment processes)

7. Establishment of total mass balance as well as substance balances for the optimized scenario

8. Comparison between actual system and optimized scenario

For brevity, not all steps of the comprehensive ASTRA study are presented here in detail. The only steps that are discussed are those relevant to the understanding of the results and implications of the case study.

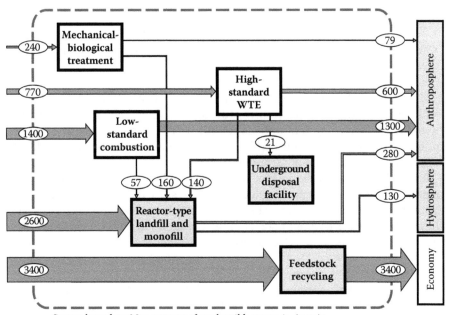

System boundary Management of combustible wastes in Austria

FIGURE 3.45
Mass flows of combustible wastes through the system waste management in Austria (1995), 1000 t/year. (From Fehringer, R. et al., *Auswirkungen unterschiedlicher Szenarien der thermischen Verwertung von Abfällen in Österreich (Project ASTRA)*. Vienna, Austria: Institute for Water Quality, Resource and Waste Management, Technische Universität Wien, 1997.)

TABLE 3.33

Goals of Waste Management and Assignment of Assessment Methods

Goals of Waste Management as Stated in the Austrian Waste Management Act	Assessment Methods and Criteria
1. Protection of human health and the environment	1. Critical air volume
2. Conservation of energy and resources	2. Efficiency of utilization of the energy content of wastes
3. Conservation of landfill space	3. Volume reduction through treatment
4. Aftercare-free landfill	4a. Total organic carbon in landfilled wastes
	4b. Fate of substances on their way to "final sinks"

3.3.2.1.2 Selection and Development of Criteria to Evaluate Balances

The starting point is the goals of waste management as listed in the Waste Management Act:

1. To protect human health and the environment
2. To conserve energy, resources, and landfill space
3. To treat landfilled wastes so that they do not pose a risk to future generations

The latter goal is part of the precautionary principle. Since the long-term behavior of landfills is not known, future emissions have to be prevented by today's waste treatment and immobilization. In general, these goals are quite abstract and therefore require focus: what are the indicators to decide if human health and the environment are protected?

Table 3.33 lists the criteria that have been applied in ASTRA. The chosen metrics or indicators are not absolute measures, since they cannot quantify the extent to which the goals of waste management have been reached. However, they do allow relative comparison of the actual situation with various scenarios, yielding statements such as "scenario X is Y% better than the actual situation."

3.3.2.1.3 Assessment Methods and Criteria

1. The critical air volume as used in ASTRA is adapted from the Swiss Eco-points approach. It is defined by the following equations:

$$V_{i,crit} = \frac{E_i}{L_i} \tag{3.12}$$

$$V_{crit} = \sum_{i=1}^{n} V_{i,crit} \tag{3.13}$$

where
E_i = emission of substance i into the air
L_i = concentration of substance i in ambient air
$V_{i,crit}$ = hypothetical air volume that is needed to dilute E_i to ambient air
concentration for substance i

The substance-specific critical volumes are added up to a final assessment indicator (V_{crit}), which has its optimum at small volumes.

2. The efficiency of utilization of waste energy content is calculated as follows:

$$\text{Efficiency} = \frac{\text{substituted fossil fuel } [\text{J} \cdot \text{yr}^{-1}]}{\text{energy content in waste } [\text{J} \cdot \text{yr}^{-1}]} \cdot 100 \qquad (3.14)$$

Fossil fuels can be substituted directly and indirectly by waste combustion. Direct substitution takes place when wastes replace fossil fuels, e.g., when coal is replaced by plastic waste to fire a cement kiln. It is assumed that wastes replace energy-equivalent units of fossil fuels. Strictly speaking, this is only true when the heating values of wastes and fossil fuels are similar (difference smaller than 20%). Indirect substitution is given when wastes are used in an incinerator to produce electricity and/or heat to feed into a network, conserving fossil fuel that would have been required without the MSW incinerator. An efficiency of 100% means that one energy unit of wastes replaces the equivalent energy amount of fossil fuels.

3. Volume reduction by waste treatment is expressed as the difference in landfill space required for the various scenarios.

4a. TOC of final wastes is assessed based on mass and substance balances.

4b. "Fate of substances" means that each substance will finally be transferred to intermediate A and final B sinks. These sinks are

1. (A) Recycling products or other new secondary products (e.g., cement, bricks)
2. (A) The atmosphere
3. (A) The hydrosphere
4. (B) The lithosphere as an underground disposal facility
5. (A + B) The lithosphere as a landfill (see gray boxes in Figure 3.45)

Sinks 1, 2, and 3 are intermediate sinks for most substances; sink 4 is designed as a final sink; and sink 5 is a sink that is leaching over very long periods of time. For each substance, a suitable sink must be defined. For example, only very minor amounts of cadmium should reach the atmosphere and the hydrosphere. Also, the transfer into recycled goods or cement is not desirable, since cadmium is not required in these products and has to be disposed of at the end of all life cycles. Cadmium in landfill poses a small long-term risk. The disposal in specially designed underground storage facilities that have not been in contact with the hydrosphere for millions of years (e.g., salt mines) represents a long-term solution with an extremely low risk of environmental pollution. Hence, from the point of view of finding an appropriate final sink for cadmium, the underground storage is the most preferred solution.

For nitrogen, recycling as a nutrient and emission into the air as molecular nitrogen (not nitrogen oxide) are positive pathways and sinks. Most other fates such as nitrate in groundwater or NO_x in air are considered negative paths. For chloride, transport in river systems to large water bodies such as oceans are acceptable solutions as long as the ratio of anthropogenic to geogenic concentrations and flows is small (e.g., <1%). The criterion for fate of substances is expressed as the percentage of a substance that is transferred into appropriate compartments. The reference (100%) is the total flow of the substance in combustible wastes.

3.3.2.1.4 Total Mass Balance for the Actual Situation (1995)

The generation and actual flows of combustible wastes in Austria are assessed by analyzing statistics and studies that were commissioned by the authorities responsible for waste-management issues in Austria. As a working hypothesis, combustible wastes are defined as wastes having a heating value >5000 kJ/kg dry substance. This is the range where autarkic combustion is possible. The result is given in Table 3.34. The total amount of wastes is 39 million t/year. It is dominated by construction and demolition debris, including soil excavation. But only a small fraction of this category is combustible (2%). Altogether, about 22% or 8.5 million t/year are combustible wastes. The most relevant fractions are waste wood (41%) and wastes from private households and similar institutions (26%). Waste wood comprises bark, sawdust, chips of wood, and other minor fractions. Wastes from water purification and wastewater treatment mainly consist of municipal and industrial sludge and screenings from the sewer and wastewater treatment plants. Other nonhazardous wastes comprise all sorts of industrial wastes. The composition of this fraction is comparatively unknown. Better statistics are available for hazardous wastes. Generally, one can say that for every average Austrian, about 1 metric ton of combustible wastes is produced per year. Three-quarters of this amount accrues elsewhere (industry, infrastructure) and is not directly visible for the consumer.

Figure 3.45 shows the flows of combustible wastes in Austria in 1995. Approximately 40% (3400 kt/year) is used for feedstock recycling. This can

TABLE 3.34

Total Waste Generation in Austria[a] and Combustible Fractions

Waste Category	Total Wastes, t/Year	Combustible Portion of Wastes		
		%	t/Year	% of Total Combustible Fractions
Wastes from private households and similar sources	2,500,000	87	2,170,000	26
Construction and demolition debris including soil excavation	22,000,000	2	500,000	6
Residues from wastewater treatment	2,300,000	41	940,000	11
Waste wood	3,500,000	100	3,500,000	41
Other nonhazardous wastes	7,800,000	14	1,130,000	13
Hazardous wastes	1,000,000	22	220,000	3
Total	39,100,000	22	8,500,000	100

[a] 8.1 million inhabitants.

be sawdust for chipboard production or wastepaper recycling. About 30% (2600 kt/year) is directly landfilled. After 2004, the disposal of this latter amount did not comply with the landfill ordinance of 1996. New methods of treatment and disposal were needed. Simple incinerators and boilers without advanced air-pollution standards utilize about 17% (1400 kt/year) of the combustible wastes for energy recovery. These plants are equipped with settling and baffle chambers, multicyclones, electrostatic precipitators (ESPs), or baghouse filters. The standard fuel is oil, coal, or biomass, and emission limits are not as stringent as for MSW incinerators. About 9% (770 kt/year) is incinerated in high-standard facilities. These plants are equipped with advanced air pollution control (APC) systems and easily surpass the most stringent emission regulations. Finally, some 3% (240 kt/year) is treated in mechanical–biological facilities.

3.3.2.1.5 Selection of Substances and Characterization of Wastes

The following substances are selected as indicators: carbon, nitrogen, chlorine, sulfur, cadmium, mercury, lead, and zinc. Carbon is selected because of the TOC limit that will apply beginning in 2004. Nitrogen is relevant as a potential nutrient and for the cement industry. Cement kilns are single sources (2.5%) of national NO_x emissions, along with traffic (62%), other industries (17%), and home heating (10%) (Hackl and Mauschitz, 1997; Gangl, Gugele, Lichtblau, and Ritter, 2002). Wastes rich in nitrogen may increase this emission load. Compounds of chlorine, sulfur, and heavy metals are major air pollutants. The heavy metals are also of interest for their potential as resources. An overview of the content of the selected substances in combustible wastes is given in Table 3.35. The ranges are broad and show that there are both extremes of

TABLE 3.35

Substance Concentrations of Combustible Wastes and Comparison with Other Fuels, mg/kg Dry Matter

	C	N	S	Cl	Cd	Hg	Pb	Zn
Mean	450,000	9100	2300	4300	5.7	0.8	230	520
Minimum	100,000	200	60	10	0.01	0.001	<1	1
Maximum	900,000	670,000	17,000	480,000	500	10	4000	16,000
MSW	240,000	7000	4000	8700	11	2	810	1100
Coal	850,000	12,000	10,000	1500	1	0.5	80	85
Fuel oil	850,000	3000	15,000	10	<1	0.01	10	20

wastes: "clean" wastes that have lower contamination than fuel oil and wastes that show a significantly higher level of pollution than MSW.

3.3.2.1.6 Criteria for Optimized Assignment of Wastes to Treatment Processes

The variance in chemical composition requires tailor-made assignment of combustible wastes to treatment processes. Not every facility is qualified to treat any waste if the goals of waste management are to be reached. The following criteria were developed to decide upon waste treatment.

First, wastes having lower contaminant concentrations than average coal are appropriate for production processes such as cement kilns or brickworks. Concentrations are not determined per mass but per energy content of the fuel (e.g., mg/kJ). The reason is that 1 ton of waste does not necessarily replace 1 ton of coal. Rather, equivalent energy amounts are substituted by waste utilization. The rationale for this criterion is that it prevents products from becoming a sink, for example, for heavy metals. It is not clear whether elevated concentrations in concrete, bricks, asphalt, etc. have an impact on the environment. Thus, the precautionary principle is applied by this criterion, and contamination of products is banned. A second argument is that once substances are transferred into such products, they cannot be recovered again. The second criterion considers the impact on air quality by waste combustion. Emissions of state-of-the-art MSW incinerators are smaller, for some substances orders of magnitude smaller, than modern air pollution regulation demands. Thus, MSW incineration has proved to be environmentally compatible and serves as a reference. The criterion says that emissions from any waste-combustion facility must not exceed typical emissions from state-of-the-art MSW incineration. This can be expressed by the following equation:

$$c_{max} = \frac{TC_I}{TC_{CP}} \cdot c_{MSW} \tag{3.15}$$

Transfer coefficients of state-of-the-art incineration (TC_I) for relevant substances into the air are known from several investigations. Also, average concentrations in MSW (c_{MSW}) are common. Typical values are given in Table 3.35. Transfer coefficients of the specific combustion process (TC_{CP}) have to be determined by an MFA. The maximum allowable concentration of a substance in a waste to be burned in a specific combustion process is then c_{max}.

3.3.2.1.7 Results of the Optimized Scenario and Comparison with the Actual Situation

Applying these criteria to the actual situation yields a new optimized scenario. In Table 3.36, the optimized assignment of wastes to combustion processes is listed. The capacity for combustion has to be more than doubled from 2.1 to 5.0 million t/year. Some of the required plants already exist (cement, pulp and paper, etc.). Most of them could manage the assigned wastes with little or no process adaptation. These changes can be carried out comparatively quickly. On the other hand, new incineration plants with advanced APC technology and a total capacity of 2.8 million t/year have to be erected. This may take up to 5 years, including the permitting process, financing, planning, and engineering.

The improvement between actual and optimized situations can be seen when the aforementioned assessment criteria are applied to the materials balances (Figure 3.46).

1. The critical air volume calculated for NO_x, SO_2, HCl, Cd, Hg, Pb, and Zn is reduced by 43%. This is surprising because the quantity of combusted wastes is increased by 140%. The reason for this paradox is that in the actual situation, a comparatively small quantity of wastes is combusted in simple furnaces that lack adequate APC. The new scenario assigns all wastes to appropriate plants. Noncontaminated wastes are utilized in furnaces that have a lower (but sufficient) standard in flue-gas cleaning. "Dirty" wastes are treated in well-equipped combustion plants.

TABLE 3.36

Assignment of Combustible Wastes in an Optimized Scenario and Changes Compared with the Actual Situation

	Optimized Scenario	Compared with Actual Situation
MSW incineration	1,500,000	+1,000,000
High-standard industrial combustion	2,000,000	+1,800,000
Hazardous waste combustion	70,000	±0
Wood industry	585,000	−30,000
Biomass cogeneration power station	110,000	+10,000
Pulp and paper industry	550,000	+31,000
Cement industry	170,000	+77,000
Total	5,000,000	+2,900,000

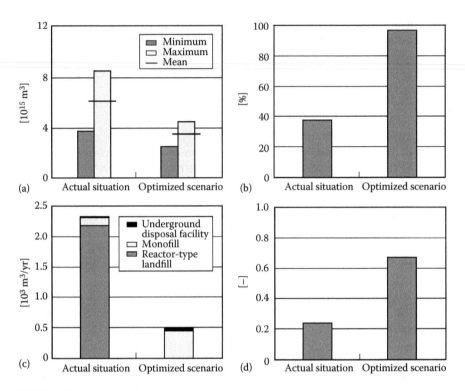

FIGURE 3.46
Comparison between status quo waste management and the optimized scenario based on selected criteria. (a) Critical air volume, (b) energy efficiency, (c) volume reduction, and (d) substance management and "final sink."

2. The efficiency of utilization of the energy content of wastes is improved by 150%. In the optimized scenario, one energy unit of waste replaces the energy-equivalent amount of fossil fuel almost completely. The main reason for this progress is that wastes are no longer landfilled without energy recovery; no waste is processed in a mechanical–biological treatment plant anymore.

3. Consumption of landfill space is reduced by 80%. Again, the main reason for this improvement is the ban of direct landfilling of wastes and the abandonment of mechanical–biological waste treatment. Combustion reduces the volume of wastes by 80–98%, depending on the ash content of the specific waste.

4a. The TOC of all residues landfilled is below 3%. This is an important step away from reactor-type landfills to "final storage" landfills that require no aftercare.

4b. The percentage of substances that are transferred into appropriate final sinks is increased by 180%. This indicates a relevant improvement in substance management.

3.3.2.1.8 Conclusions

This case study shows that MFA facilitates goal-oriented waste management. General goals on a high hierarchic level, as stated in various waste-management acts (e.g., European Waste Framework Directive, Swiss Guidelines for Waste Management, and German "Kreislaufwirtschaftsgesetz") can be translated into well-defined, concrete assessment procedures with appropriate criteria. ASTRA outlines a way for this to be achieved. Note that a single goal of waste management may require two or more assessment methods for comprehensive evaluation. MFA is used at several levels in the study:

1. To describe the actual situation of the system *management of combustible wastes*
2. To reveal deficits and develop criteria for waste assignment
3. To compile the optimized scenario
4. To demonstrate the differences between the actual and the optimized situation

The findings of the study reveal the capacities for new plants and may serve as a basis for planning and engineering. A next step is to assess costs (including uncertainties) for the scenarios. In fact, this is also done in ASTRA. The result is that the optimized scenario can be realized without significantly raising total costs for disposal (collection, separation, treatment, and landfilling). The main drawback of the optimized scenario, and also the main reason why this scenario will take a long time to be accomplished, is the following: A large proportion of the wastes that are landfilled at present will be incinerated in the future. For landfill owners and operators, this may cause a severe economic situation because the landfills will lose business. Considering that landfills are long-term investments with filling times between 25 and 50 years, it is clear that such a stern change cannot be pushed through in a short time. It is also clear that landfill operators will use every possible legal and economic means to postpone strategic changes endangering landfilling.

3.3.2.2 Case Study 13: PRIZMA

In the case study PRIZMA (German acronym for "positive list for utilizing of residues in the cement industry: methods and approaches"), the utilization of combustible wastes for energy recovery in cement kilns is investigated for Austria (Fehringer, Rechberger, and Brunner, 1999). In this country, production of cement requires about 10 million GJ/year to produce 3 million t/year of clinker that is further processed into cement. This amount of energy corresponds to some 400,000 tons of wastes with an average heating value of 25 MJ/kg. Today, wastes cover about 27% of the energy demand of the cement industry. Energy consumption for clinker production is quite high and represents a significant share of total production costs. For example, the European Cement Association estimates that energy accounts for 30–40% of the production cost of cement (CEMBUREAU, 1999). Hence, the cement industry strives to minimize energy-specific costs. One possibility is to reduce costs for fuel consumption. Wastes are alternative fuels. Compared with standard fuels such as natural gas, oil, or coal, they are less expensive. Other ways to reduce costs are to improve energy efficiency or to use different raw materials and technologies.

The Austrian cement industry has a long history of experience with energy recovery from wastes. Traditional waste fuels are used tires, waste oil, and solvents. Test runs have been carried out with sewage sludge, mixed plastics, and waste wood as well as meat and bone meal [as a result of the disposal crisis caused by bovine spongiform encephalopathy (BSE)], and others. Sorted fractions of MSW are also considered as a fuel alternative. The Austrian cement industry aims to cover 75% of its energy demand by wastes within a few years. Besides the expected cost relief, this goal contributes to the reduction of carbon dioxide emissions by the branch. Since direct landfilling of organic wastes was forbidden after 2004 in Austria, the cement industry supported the provision of the required capacities for treatment. On the other hand, not all wastes are appropriate for the cement process. Wastes with high contamination of heavy metals may lead to environmentally incompatible emissions and a polluted product (cement, finally concrete). Hence, operators as well as authorities want to have clear regulations specifying which kinds of wastes are appropriate for energy recovery. One possibility for establishing such an instrument is to generate a so-called positive list. The positive list specifies and characterizes waste types that are appropriate for recovery in cement kilns. The objective of PRIZMA is to develop criteria to establish such a list.

3.3.2.2.1 The Cement Manufacturing Process

The prevailing technology for cement manufacturing in Austria is the cyclone preheater type. A scheme of such a facility is displayed in Figure 3.47. The heart of each cement factory is a massive steel tube up to 100 m in length and up to 8 m in diameter. It is slightly inclined to the horizontal (3 to 4°) and slowly rotates at about one to four turns per minute. The main raw materials for the feed of the kiln are limestone, chalk, marl, and corrective materials (e.g., ferrous materials). The chemical properties of these materials and the desired properties of the clinker govern the correct mixture. Mixing is an important step in the process to ensure an even distribution of the properly proportioned components of the raw material so that the clinker will be of a uniform quality. The raw material is ground in a mill, from where it is pneumatically boosted into a mechanical predeposition (baffle) and a subsequent ESP. There, the raw material is collected and conveyed into a storage silo. Afterwards, the raw material runs through four stages: evaporation and preheating, calcining, clinkering, and cooling.

Evaporation and preheating remove moisture and raise the temperature of the raw material. This process takes place in the cyclones, where raw

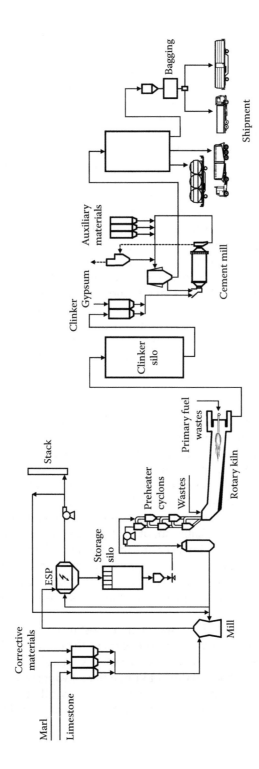

FIGURE 3.47
Scheme of a cement kiln (cyclone preheater type).

material counterflows the hot off-gas from the kiln. The raw material enters the kiln at the back (the upper end of the kiln), and gravity and the rotation of the kiln allow the mix to flow down the kiln at a uniform rate through the burning zone. The tube is lined with refractory bricks to avoid heat damage of the kiln. Calcining takes place at 600 to 900°C and breaks the calcium carbonate down into calcium oxide and carbon dioxide. Approximately 40% by weight of the raw material is lost by this process. Clinkering completes the calcination stage and fuses the calcined raw material into hard nodules resembling small gray pebbles. The clinker leaves the kiln at the front (the lower end of the kiln) and falls onto a reciprocating grate, where it is cooled. The primary fuel is introduced and burnt at the same end of the kiln. The flame is drawn up the kiln to the burning zone, where the heat intensity is highest and fusion of chemicals in the raw material takes place. Hot combustion gases continue to flow up the kiln and exit from the back end. Secondary firing at the upper end of the kiln maintains the energy-intensive calcination process. Product temperatures in the burning zone are around 1450°C. The primary flame has a temperature of 2000°C. The cooled clinker is stored in a silo. Cement is produced by grinding of the clinker and blending with gypsum and other materials (e.g., fly ashes from coal combustion) to produce a fine gray powder. The last stage is bagging of cement and preparing the product for transportation. The cooled flue gas (heat transfer to raw material in the cyclones) is cleaned by the ESP and emitted via a stack.

Generally, for any substance, there are only two ways to enter and to leave the process: enter via raw materials or fuels and exit via off-gas or the product. The partitioning for any substance A between raw materials and fuels can be determined by measurement. The result will be that X% of A enters the process via fuels and Y% via raw materials, with $X + Y = 100$%. For the off-gas and the product, only total amounts of substance A can be determined. It is not possible to say that $\tilde{X}$ of A stems from fuels and $\tilde{Y}$ stems from raw material in the clinker (again, $\tilde{X} + \tilde{Y} = 100$%). The same applies to the off-gas. Only some qualitative information is available about the behavior of substances in a clinker manufacturing process. However, for the given problem, which is waste combustion in cement kilns, it is mandatory to know how fuel-borne substances (i.e., substances that enter the process via fuel) behave in the process.

A special characteristic of the process is that two kinds of cycles evolve:

1. The so-called inner cycle arises when a substance i vaporizes in the kiln. It is then transferred to cooler parts in the cyclone, where substance i may condense at the surface of raw material particles. So substance i travels back to the hot kiln, where it vaporizes again. The cycle is closed and built up until some kind of equilibrium is established (theoretically). Operators try to break such cycles by bypassing the cyclones.

2. The so-called outer cycle develops because cyclones cannot intercept fine particles (<5 µm). But interception in the cyclones is a prerequisite for raw material particles to enter the kiln and leave the process as clinker. Fine particles are carried by the flue gas to the ESP. There they are removed from the flue gas with high efficiency (>99%) and reenter the cyclones, where they are, again, not intercepted. The loop is closed, and substance built up.

These cycles make it difficult to establish closed substance balances for the process and to predict the behavior of substances in the process. However, the following working hypothesis can be put forward: Fuel-borne substances (e.g., heavy metals) are predominantly embedded in an organic matrix. Organic substances are destroyed in the flame, which is an area with temperatures around 2000°C. This means that inorganic, volatile fuel-borne substances will vaporize to a high extent. Contrary to fuels, heavy metals in the raw material are fixed in a mineral matrix. The raw material is heated up to 1450°C, and it can be assumed that not all metals will vaporize. A portion will remain in the solid phase and contribute in the clinkering process (sintering). In other words, the possibility for a metal to reach the gaseous phase is higher for fuel-borne substances than for substances descending from raw material. When the flue gas is cooled down (in the cyclones), condensation of metals on particle surfaces (raw material, ashes) takes place. This process happens at the same rate for all metals, regardless of their origin (fuel or raw material). Let us assume that the partitioning of a substance χ between fuel and raw material is $X{:}Y$. Then the mentioned vaporization and condensation processes imply that there is a different ratio $\tilde{X}/\tilde{Y}$ of χ in particles in the cyclone with $X/Y < \tilde{X}/\tilde{Y}$. Fine particles will pass the cyclones, and a small fraction will also pass the ESP. Evidence of this process is that particulates emitted from cement plants are enriched with heavy metals compared with raw materials (see Table 3.37). The conclusion is that fuel-borne and raw-material-borne substances show different behavior in the process and therefore have different transfer coefficients. However, for the problem of waste combustion, it is crucial to know the transfer coefficients for fuel-borne substances.

TABLE 3.37

Mean Substance Concentrations in Raw Material and Emitted Particulates

	Cl	Cd	Hg	Pb	Zn
Raw material, mg/kg	150	0.15	0.15	15	37
Emission, mg/kg	46,000[a]	8	2000[a]	400	150
Enrichment	300	50	13,000	27	4

[a] Gaseous and solid emissions are related to the total mass of emitted particulates.

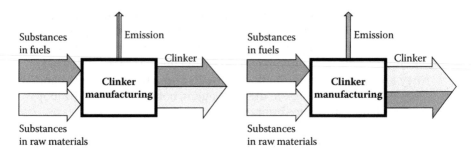

FIGURE 3.48
Assumptions for the fate of fuel-borne heavy metals in the clinker manufacturing process. The left/right assumption yields the lower/upper limit for the transfer coefficient of fuel-borne substances into the atmosphere.

3.3.2.2.2 Assessment of Transfer Coefficients

The uncertainty concerning transfer coefficients for fuels leads to the following approach: A range for transfer coefficients is established by determining hypothetical extreme values (see Figure 3.48). For the lower limit, it is assumed that the partitioning between fuel-borne and raw-material-borne substances is the same in the off-gas and the product. Based on the aforementioned considerations, this will underestimate the transfer of fuel-borne substances into the off-gas. The upper limit is given by the assumption that the emission only contains fuel-borne substances. This assertion certainly overestimates the influence of fuels for emissions. On the other hand, this is a reliable upper limit, since higher transfer coefficients are not possible. The real transfer coefficient has to be somewhere between these extremes. In cases where the range is large (e.g., one order of magnitude), it is safe to apply the upper limit, i.e., the higher value.

3.3.2.2.3 Criteria for Waste Fuels

Which substances should be incorporated into a "positive list" that defines wastes suited for cement kilns? In Section 3.3.2.1, criteria are presented for the selection of substances with regard to waste combustion. The list can be shortened for cement manufacturing because carbon compounds are most efficiently destroyed in the cement kiln. With the exception of carbon dioxide, emissions of carbon compounds are very small. The emission of nitrogen oxide is a problem of cement manufacturing, but this has little to do with waste recovery. The high temperatures and long residence times of gases in the process, which are essential for efficient mineralization of carbon compounds, are responsible for nitrogen oxide formation. The nitrogen content of wastes plays only a minor role in the formation of nitrogen oxide. Sulfur from wastes is efficiently contained into clinker. Recorded emissions of sulfur dioxide stem from certain kinds of raw material. As in the case of nitrogen, these emissions are not a result of waste combustion. Chlorine poses a limit for the quality of the product and may cause blockage in the cyclones. Therefore, it has to be part of the positive list. The heavy metals cadmium and mercury are chosen because of their toxicity and volatility. Lead and zinc are, on one hand, also potentially toxic but, on the other hand, are important resources, too.

Waste as a fuel for clinker manufacturing influences input as well as off-gas and clinker. Therefore, three criteria—A, B, and C—are developed for the positive list:

Criterion A deals with the off-gas and has been described in Section 3.3.2.1. The criterion says that emissions from clinker manufacturing must not exceed typical emissions from state-of-the-art incineration. This can be expressed by Equation 3.16:

$$c_{max} = \frac{TC_I}{TC_{CM}} \cdot c_{MSW} \tag{3.16}$$

TC_I is the transfer coefficient of state-of-the-art incineration (known). Also, the average concentrations in MSW (c_{MSW}) are commonly known (see Table 3.38). Transfer coefficients for clinker manufacturing (TC_{CM}) have to be determined according to aforementioned considerations about a reliable range for transfer coefficients. The result of criterion A, c_{max}, is the maximal allowable concentration of a substance in a waste.

Criterion B controls the quality of the product clinker. It is based on the principle that anthropogenic material flows must not exceed the natural fluctuations of geogenic flows (see Chapter 2, Section 2.5.8). The criterion now considers raw material chemically as a geogenic flow. As any natural material, raw materials show a certain variance in chemical composition. To apply the criterion, the clinker composition is assessed for (1) average and (2) maximal raw material concentrations. Criterion B says that the changes in clinker concentration caused by waste recovery must not exceed the calculated geogenic variance. For both scenarios 1 and 2, clinker is produced with an average fuel mix consisting of coal (52%), oil (21%), natural gas (3%), used tires (6%), plastics (5%), waste oil (9%), and others (percentages are based on energy equivalents). The calculation of criterion B requires the following assumptions: The mass ratio between raw materials (RM) and fuels (F) is ca. 10:1. Transfer coefficients for substances stemming from raw materials and

TABLE 3.38

Data for Criterion A: Transfer Coefficients of the Reference Technology Incineration, Typical Concentrations in the Reference Waste (MSW), and Extreme Transfer Coefficients for Off-Gas of a Cement-Manufacturing Process

	Cl	Cd	Hg	Pb	Zn
TC_I	0.0005	0.0005	0.02	0.0001	0.0002
c_{MSW}	10,000	10	2	500	1000
$TC_{CM,min}$	0.01	0.0002	0.4	0.0002	0.0001
$TC_{CM,max}$	0.02	0.0004	0.8	0.0008	0.0001

fuels are identical. The loss of carbon dioxide is 40%, and the ash content of the fuel is considered to be negligible (usually ca. 1% of raw material mass). Then the concentration of clinker (Cl) is

$$C_{Cl} = \frac{(c_F \cdot 1 + c_{RM} \cdot 10) \cdot TC_{Cl}}{10 \cdot (1 - 0.4)}$$

(3.17)

The data needed to calculate criterion B are summarized in Table 3.39. The result of criterion B yields the maximum load of a substance that can possibly be added to the clinker by waste recovery. The load (e.g., mg of a substance per ton of clinker) gives the dependence between the total mass of recovered wastes and the substance concentration in the waste. The result is a curve (see Figure 3.49).

Criterion C considers the input into cement manufacturing. If the total national consumption is assigned a value of 100%, then combustible wastes contain roughly 40% of cadmium and mercury (see Table 3.40). Hence, combustible wastes are important carriers of some heavy metals. This is one reason why waste management plays such an important role for the total turnover of several heavy metals. The consequential question for the cement industry is, Which share of these substances shall enter the processes and finally end up in the product cement? There is no uniform answer to this question. Some cement manufacturers do not want their products associated with hazardous materials and thus are cautious in using contaminated wastes as a fuel. Other manufacturers see a chance for economic advantage over their competitors by using inexpensive waste-derived fuel and use large amounts of wastes. In Table 3.40, results are presented assuming that cement manufacturers take over 15% of the metals contained in combustible wastes. As for criterion B, the result is a curve showing the dependency between total mass of wastes and substance concentration in the wastes (see Figure 3.49).

TABLE 3.39

Calculation of Maximum Allowable Load on Clinker through Waste Recovery

	Cl	Cd	Hg	Pb	Zn
Mean concentration in fuel mix, mg/kg	1100	0.9	0.4	60	65
Mean concentration in raw material, mg/kg	150	0.15	0.15	15	37
Maximum concentration in raw material, mg/kg	400	0.5	0.5	42	110
TC into clinker	0.99	0.99	0.6	0.99	0.99
Mean concentration in clinker, mg/kg	430	0.4	0.19	35	72
Maximum concentration in clinker, mg/kg	840	1.0	0.54	79	190
Maximum load through waste recovery, mg/kg	410	0.6	0.35	45	120
Maximum load through waste recovery, t/year	1200	1.7	1.1	130	360

Note: Based on a clinker production of 3 million t/year.

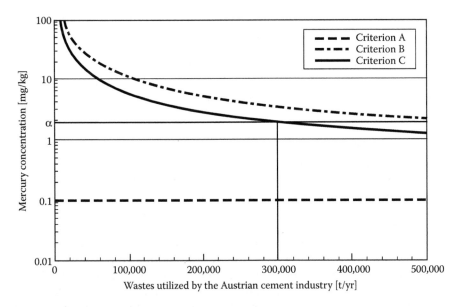

FIGURE 3.49
Results of criteria A, B, and C for Hg serve to support decisions regarding the utilization of wastes in cement kilns (Fehringer, Rechberger, and Brunner, 1999). Criterion A, emissions; criterion B, product quality; criterion C, dilution of metals.

TABLE 3.40

Estimated National Consumption of Selected Substances, Content in Combustible Wastes, and Maximum Flow into Cement Manufacturing

	Cl	Cd	Hg	Pb	Zn
National consumption, t/year	450,000	80	10	32,000	43,000
Combustible wastes, t/year	30,000	36	3.9	1600	3900
In combustible wastes, %	6.6	45	39	5	9
Input into cement, %	15	15	15	15	15
Input into cement, t/year	4500	5.4	0.6	240	590

3.3.2.2.4 Results

Criteria A to C are calculated for the selected substances and provide either maximum concentrations for wastes or maximum amounts of substances that can be transferred into clinker. Required parameters for calculation are transfer coefficients of waste-borne substances into the off-gas and into the clinker. They should be determined for each cement plant separately, as results may vary considerably among different technologies. In Figure 3.49, the result is given for mercury. Consider a waste (α) having a mercury concentration of 2 mg/kg. Criterion B allows waste recovery of 500,000 t/year. This means that, in practice, mercury does not pose a limit for the quality of clinker. Self-restriction of the cement industry (criterion C) still allows

300,000 t/year of waste α. A really severe limit poses criterion A: only wastes having Hg concentrations lower than 0.1 mg/kg are qualified as a waste fuel. This reduces the amount of potentially available wastes considerably. With respect to mercury, it should be evaluated whether the investment costs for improving APC technology are paid back within an acceptable time frame by the savings due to a cheaper waste fuel.

3.3.2.2.5 Conclusion

Case study PRIZMA exemplifies how MFA can be used to establish environmental regulations. The proposed criteria consider system-specific as well as plant-specific constraints. Criteria A and B guarantee that waste utilization in cement kilns does not pollute the atmosphere (and subsequently the soil) or lower the quality of clinker. The extended systems approach assures that only a limited amount of resources is transferred into cement and concrete (criterion C). Note that the selected substances are not required in cement and are lost for recovery and recycling. A limit for this kind of sink is therefore reasonable. Waste recovery in state-of-the-art cement kilns that fulfill criteria A, B, and C can be considered as environmentally compatible. Thus, the utilization of wastes in cement kilns can be a valuable contribution to goal-oriented waste and resource management.

3.3.2.3 Case Study 14: Recycling of Cadmium by WTE

In mineral ores of commercial value, cadmium is usually associated with other metals. Greenockite (CdS), the only cadmium mineral of importance, contains zinc, sometimes lead, and complex copper–lead–zinc mixtures. Hence, zinc and lead producers have no choice, and they usually produce cadmium, too. Most cadmium (>80%) is produced as a by-product of beneficiating and refining zinc metal from sulfide ore concentrates. An estimated 90–98% of the cadmium present in zinc ores is recovered in the zinc extraction process. About 3 kg of cadmium is produced for every ton of refined zinc. Small amounts of cadmium, about 10% of consumption, are produced from secondary sources such as baghouse dust from electric arc furnaces (EAFs) used in the steelmaking industry and the recycling of cadmium products. Total world production of cadmium in 2000 was about 19,300 tons (US Geological Survey, 2001a).

The International Cadmium Association has made the following estimates of cadmium consumption for various end uses in 2001: batteries, 75%; pigments, 12%; coatings and plating, 8%; stabilizers for plastics and similar synthetic products, 4%; and nonferrous alloys and other uses, 1% (U.S. Geological Survey, 2001b). Utilization of cadmium in developed countries is estimated at between 5 and 16 g/capita/year (Llewellyn, 1994; Bergbäck, Johansson, and Mohlander, 2001; US Geological Survey, 2001b). Currently, annual consumption of cadmium amounts to about 20,000 tons. In contrast to most other metals, production does not show an increasing trend, most likely due to increasing regulatory pressure to reduce or even eliminate the use of cadmium. This is a result of the growing awareness of cadmium being potentially toxic to humans and of the risks presented by its accumulation in the environment. Unlike other heavy metals (e.g., zinc, selenium), cadmium is not essential to the biosphere and has no known useful biological functions. It accumulates in the kidneys and liver and affects protein metabolism, causing severe disorder, pain, and even death. In addition, it may cause a variety of severe damages such as lung cancer (Heinrich, 1988; Waalkes, 2000) and bone diseases (osteomalacia and osteoporosis) (Verougstraete, Lison, and Hotz, 2002).

In the input of waste to energy (WTE) plants, the typical concentration of cadmium is between 8 and 12 mg/kg (Brunner and Mönch, 1986; Schachermayer, Bauer, Ritter, and Brunner, 1995; Morf, Brunner, and Spaun, 2000; Verougstraete, Lison, and Hotz, 2002). This is about 50 times the average concentration found in the Earth's crust (0.2 mg/kg). When MSW is landfilled, cadmium is quite immobile during the anaerobic phase but can be mobilized during aerobic periods by organic acids and reach the groundwater. Cadmium and some of its compounds such as chlorides have low boiling temperatures (Cd, 765°C; $CdCl_2$, 970°C). Therefore, cadmium belongs to the group of atmophilic elements like Hg, Tl, Zn, and Se. In combustion processes, these substances tend to volatilize and escape via the flue gas from the combustion chamber.

The average generation of MSW in Europe is between 150 and 400 kg/capita/year. In the United States, currently about 700 kg/capita/year is collected. There are several reasons to explain the differences in quantities. First, the term *MSW* is defined operationally. MSW usually designates all mixed wastes that are collected at the curbside on a daily, weekly, or biweekly basis. For example, in some areas, bulky wastes are collected separately and therefore not included in MSW data. In other areas, waste containers are large and allow collection of MSW together with bulky wastes. MSW includes mixed wastes from private households and may also include wastes from the service sector and small shops and companies. Decisive factors for waste generation include the kind of wastes the collector accepts, the collection frequency, the size of the bins or containers for collection, and how statistical data are compiled. A second reason for differences is the extent of recycling. Separate collection of paper, biowaste, glass, metals, etc. may reduce the weight of waste collected as MSW up to 50%. Third and fourth reasons are differences in lifestyle (e.g., small or large family and household size, packed food versus open food, etc.) and purchasing power of the consumer. At an average MSW generation rate of 250 kg/capita/year, about 2.5 g/capita/year of cadmium is collected via MSW. This is about 25% of the average national per capita consumption of cadmium. The remaining 75% is mostly incorporated in goods with long residence times that will enter waste management in the future. A smaller part of cadmium is expected in other wastes (estimated 20%, mainly contained in industrial wastes and the combustible fraction of construction and demolition waste), and a small quantity is lost to the environment via emissions and fugitive losses.

Modern WTE plants are equipped with advanced APC devices. As mentioned before, cadmium is transferred to the flue gas during incineration and removed by the APC devices. During off-gas cooling in the heat exchanger, volatile cadmium is condensed on very small particles that offer a large surface area. Consequently, more than 99.9% of cadmium can be removed by particle filters such as ESPs or fabric filters if these filters have been designed to capture small particles. Remaining quantities are removed upstream in wet scrubbers with high-pressure drops (venturi scrubbers) or adsorption filters. Very small amounts are emitted via the stack (<0.01%). Figure 3.50 shows transfer coefficients for cadmium in a state-of-the-art MSW incinerator.

Typical Cd concentrations in incinerator fly ash range from 200 to 600 mg/kg. In comparison, filter dust from EAFs that is recycled contains 500 to 1000 mg/kg Cd (Donald and Pickles, 1996; Stegemann, Caldwell, and Shi, 1997; Xia and Pickles, 2000; Youcai and Stanforth, 2000; Jarupisitthorn, Pimtong, and Lothongkum, 2002). This shows that incinerator fly ash could be used for cadmium recovery.

Thermal treatment of bottom ash and/or fly ash can improve the potential for recovery. For example, adequate thermal treatment of incinerator bottom ash results in three products:

1. A silicate product, with an average cadmium concentration similar to the Earth's crust that can be utilized for construction purposes
2. A metal melt containing mainly iron, copper, and other lithophilic metals (metals of high boiling points)
3. A concentrate of atmophilic metals

Applying such technologies to ashes makes cadmium and other metals accessible for efficient recovery. For cadmium, recycling efficiencies from

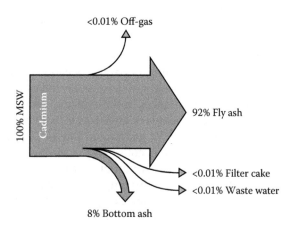

FIGURE 3.50
Transfer coefficients for cadmium in a state-of-the-art WTE plant. (From Schachermayer, E. et al., Messung der Güter- und Stoffbilanz einer Müllverbrennungsanlage, Monographien Bd. 56, Bundesministerium für Umwelt, Vienna, Austria, 1995. With permission.)

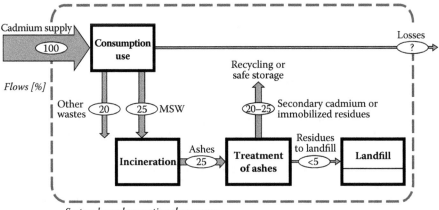

FIGURE 3.51
Recycling of cadmium by WTE of MSW. Percentage of flows of Cd in MSW, other wastes, and the rest may vary according to technological and economic situation.

MSW up to 90% are realistic (Figure 3.51). Another possibility is to use thermal processes, not to produce a concentrate for recovery but to immobilize metals in a ceramic or vitreous matrix (vitrification). Such residues may come close to final storage quality (Baccini, 1989).

Today, most of the obsolete cadmium enters landfills, where it remains a potential hazard for generations. The advantage of recycling is that the consumption of primary cadmium is reduced. Thus, the quantity of cadmium that enters the anthroposphere is reduced, facilitating the management and control of this resource. Technologies to immobilize cadmium are required to safely dispose of the large amounts already in the anthroposphere. The new task is to continually collect and transform cadmium that is stored in infrastructure and long-living goods into a form where it can be safely stored within the anthroposphere. Note that for both options, a concentration step is indispensable; thermal processes are proven technologies that can achieve such concentrations.

3.3.2.4 Case Study 15: Cycles and Sinks—The Case of PBDEs

3.3.2.4.1 Introduction

The purpose of case study 15 is twofold: First, the results demonstrate how MFA on the level of substances can be used to support waste management decisions on a city level. And second, it is used to explain a waste management strategy directed toward "clean cycles" and "safe final sinks." The object of investigation of this case study is PBDEs, a group of chemical compounds called polybrominated diphenyl ethers (Figure 3.52). They pose a challenge in plastic waste management because of their hazardous properties that impede polymer recycling.

PBDEs consist of two benzene rings linked by an oxygen bridge (*diphenyl ether*):

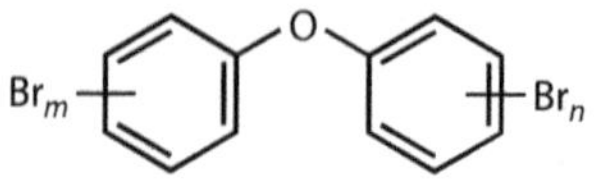

FIGURE 3.52
Polybrominated diphenyl ether, PBDE $[C_{12}H_{(10-x)}Br_xO_{(x=1,2,...,10=m+n)}]$.

Each benzene ring can be substituted with zero to five bromine atoms in five possible positions, yielding altogether 209 congeners (i.e., structurally related compounds). PBDEs are classified with respect to the number of bromine atoms, such as penta-, hexa-, hepta-, or octa-PBDE. On the one hand, they are useful chemicals and thus are widely applied as flame retardants for plastic materials and present in a wide array of products such as transportation vehicles (cars, airplanes), construction materials, home appliances, furnishings, electronic devices, thermal insulation foams, textiles, and much else. On the other hand, some of the congeners have been proven to pose a serious health hazard for humans and the environment. This is particularly the case for PBDEs averaging one to five bromine atoms. These lower-brominated PBDEs are regarded as more hazardous because they bioaccumulate, affect hormone levels in the thyroid gland, and have been linked to reproductive and neurological risks. Thus, the so-called Stockholm Convention (Stockholm Convention, 2004), which has the objective to protect human health and the environment from hazardous persistent organic pollutants (POPs), has restricted the production of some PBDEs (Sindiku et al., 2014) as well as produced recommendations about the management of wastes containing PBDEs (UNEP, 2015b). PBDEs are commonly used as commercial mixtures of several congeners and not in a pure form. Hence, often, the short form cPentaBDE is used, meaning commercially available pentabrominated diphenyl ether.

At the end of the lifetime of products containing PBDEs, these are either recycled or disposed of in incinerators and landfills. In order to fulfil the goals of waste management, namely (1) protection of human health and the environment and (2) resource conservation, information about the pathways of PBDEs from sources (industrial synthesis) to sinks (thermal destruction, landfilling) is instrumental: it is necessary to know which stocks of PBDEs have been accumulated in the past and are which still present, and which hazardous congeners will reach waste management by which collection system (separate collection of electronic wastes, plastic materials, construction wastes, and end-of-life vehicles). Also, the fate of PBDE-containing wastes during recycling and waste treatment must be known. MFA can supply and link such information about stocks and flows of individual congeners from the anthropogenic metabolism to the environment.

In case study 15, flows and stocks of commercial pentabrominated and octabrominated diphenyl ether (cPentaBDE, cOctaBDE) have been studied on a city level (Vienna, Austria) (Vyzinkarova and Brunner, 2013). The city level was chosen for analysis because of the following:

1. Cities are major hotspots of PBDE emissions. Concentrations on sites close to emission sources, such as the soil adjacent to urban roads (Luo et al., 2009; Gevao et al., 2011) and soils and sediments in the vicinities of landfills and wastewater treatment plants, are higher than at remote sites (Oliaei, 2010).

2. Flows of PBDEs to the hinterland are uncovered, and thus, the dependency of a city on its hinterland for disposing of hazardous substances becomes apparent.

3. Strategies for waste management are determined by the municipality and other urban stakeholders responsible for waste collection, recycling, and disposal. Often, the municipality controls by law the means to manage all hazardous as well as some beneficial waste materials within the city boundaries.

4. Because of the statistical data collected on the level of a city, such as MSW generation, collection rate, treatment capacities and recycling rates, and emissions from treatment processes, municipalities are often able to supply information about flows and stocks of goods and substances within their regime.

3.3.2.4.2 *Objectives*

The specific objectives of the case study are (1) to identify sources, pathways, stocks, and sinks of cPentaBDE and cOctaBDE in the city of Vienna, Austria, (2) to determine the fractions that either are recycled or reach appropriate final sinks, and (3) to develop recommendations for waste management ensuring a "clean cycles" and "safe final sink" strategy (Kral et al., 2013). Such a strategy demands that products from waste recycling are of high quality containing very low amounts of hazardous substances (clean cycles) and that the hazardous substances removed from the cycle are completely destroyed (e.g., thermal destruction by incineration) or are disposed of in a long-term safe storage without aftercare. The latter two processes, incineration and safe storage without aftercare, are called final sinks because there are no further flows of these materials into either the anthroposphere or the environment.

The motivation for these objectives stems from two key directives of the European regulation on cPentaBDE and cOctaBDE (European Council, 2003a, 2015). These regulations demand removal of pollutants as a fundamental rule for treatment. They prohibit recycling of waste electrical and electronic equipment (WEEE) containing cPentaBDE and cOctaBDE with more than 1 g/kg. On the other hand, the European waste hierarchy defines recycling for waste plastics as the preferred option. Thus, the art of recycling WEEE and other wastes containing PBDE is to remove PBDEs from such plastic wastes down to a concentration of 1 g/kg. Industrial efforts to achieve this goal with commercial mixtures are underway; it remains to be seen how successful they will be.

3.3.2.4.3 Methods and Data

To reach the objectives, substance flow analysis was applied to previously published data about flows and stocks of cPentaBDE and cOctaBDE (Morf et al., 2003; Tasaki et al., 2004). No additional laboratory analyses were performed. For modeling flows and stocks, the software STAN was used taking uncertainties into account. Scenario analysis was applied for addressing the issues of missing data and large uncertainties.

For MFA of cPentaBDE and cOctaBDE, the following system boundaries have been selected: the border in space was the administrative area of the municipality of Vienna, and the boundary in time was the year 2010. The

TABLE 3.41

Former Uses of Commercial Mixtures of PentaBDE and cOctaBDE in Various Sectors

Compound and Sector	Polymer	Application	Mass Flow Estimates
cPentaBDE			
Vehicles	PUR	Upholstery of seats, ceiling, headrest, textile back-coating	PUR foam in vehicles and construction accounts for 90–95% of total cPentaBDE use
Construction	PUR	Insulation foam	Major use
	PVC	Duroplastic sheeting	Minor use
Other	Various	Textiles, printed circuit boards, cable sheets, conveyor belts, etc.	Other applications less than 5% of total use
cOctaBDE			
Vehicles	HIPS	Dashboard and steering wheel	Estimates range from minor to major use
	PBT		
	PA		
Construction	PE	Thermoplastic sheeting	Minor use
EEE	ABS	WEEE categories 3 and 4, with focus on CRT computer monitors and TVs	Major use, estimated up to 95% of total cOctaBDE use in the EU

Source: Based on UNEP. (2015a). Draft guidance for the inventory of polybrominated diphenyl ethers (PBDEs) listed under the Stockholm Convention on Persistent Organic Pollutants. Retrieved from http://chm.pops.int/Implementation/NIPs/Guidance/Guidancefor theinventoryofPBDEs/tabid/3171/Default.aspx; Morf, L. S. et al., *Selected polybrominated flame retardants, PBDEs and TBBPA, substance flow analysis.* Bern, Switzerland: Swiss Federal Office for the Environment, 2003; flame retardants, PBDEs and TBBPA; Leisewitz, A., and Schwarz, W., *Erarbeitung von Bewertungsgrundlagen zur Substitution umweltrelevanter Flammschutzmittel Band II: Flammhemmende Ausrüstung ausgewählter Produkte—anwendungsbezogene Betrachtung: Stand der Technik, Trend, Alternativen.* Dessau, Germany: Umweltbundesamt, 2000; Lassen, C., and Løkke, S., *Brominated Flame Retardants—Substance Flow Analysis and Assessment of Alternatives.* Copenhagen, Denmark: The Danish Environmental Protection Agency, 1999.

Note: ABS, acrylonitrile butadiene-styrene; HIPS, high-impact polystyrene; PA, polyamide; PBT, polybutylene terephthalate; PE, polyethylene; PUR, polyurethane; PVC, polyvinyl chloride. Mass flow estimates vary in different sources, especially in case of cOctaBDE and are therefore only indicative.

flows and stocks were assessed for both levels of goods and of substances. Goods containing PBDEs are summarized in Table 3.41.

The following three main processes have been included in the STAN model: *consumption, waste management*, and *environment* (see Figure 3.53).

Each of these main processes acts as a subsystem and is subdivided into further processes. The process *consumption* serves to quantify anthropogenic stocks and flows of POP-PBDEs in goods that are in use, such as construction materials, vehicles, and electrical and electronic equipment (EEE), and it includes consumer emissions, too. Because of out-phasing of some of the brominated flame retardants, the processes *construction* and *vehicles* have no imports. With only exports, stocks are decreasing today, with transfers either to *waste management*, which is the biggest flow, or to the *environment*, a much smaller flow. Some of the PBDEs are recycled by waste management. It is not well known into which recycling products these flows are directed. In Figure 3.53b, all flows of recycled PBDEs are inputs to the process *use of EEE*. In reality, this may be different, with probably significant fractions being directed toward *constructions* and *vehicles*. Nevertheless, for an overall picture of the three main processes, the allocation of recycled PBDEs to the subsystem *consumption* is sufficient and correct. The flow of recycled PBDEs is of particular concern because it prolongs their lifetime. Thus, despite the intention of the authorities to out-phase POP-PBDEs, and regardless if the recycled PBDEs are contained in vehicles, construction materials, or EEEs, the consumer is exposed to these hazardous substances long after they have been abandoned.

PBDEs are contained in a variety of waste materials that are collected and in part treated in Vienna (see also Table 3.41): WEEE [CRT-PCs (computer monitors), CRT-TVs (televisions)]; plastic construction wastes (polyurethane, polyvinylchloride, and polyethylene), and end-of-life (EOL) vehicles. With regard to WEEE, 10 categories have been defined, with category 3 (information technology and telecommunication equipment) and category 4 (consumer equipment and photovoltaic panels) being relevant with respect to POP-PBDEs (Wäger, Schluep, Müller, and Gloor, 2011; UNEP, 2015a). For this case study, categories 3 and 4 have been taken into account. Because EOL vehicles are not treated within Vienna but are exported beyond the systems boundaries, they are not included in the subsystem *waste management*.

Imports into the "environment" consist mainly of consumer emissions into the air, from where they are subsequently transferred by dry and wet deposition to the soil and hydrosphere, including municipal wastewater. Literature values for consumer emissions are available in the literature and range from 0.054% of the stock for cOctaBDE to 0.39% for cPentaBDE (Morf et al., 2003; UNEP, 2015a). In this study, the existing PBDE stocks were multiplied by these emission factors. Because of the small contribution of the hydrosphere, sedimentation, and other minor flows, only the two major processes *atmosphere* and *soil* have been taken into account quantitatively within the subsystem *environment*. For more information about the calculations of emissions in Vienna, see Vyzinkarova and Brunner, 2013.

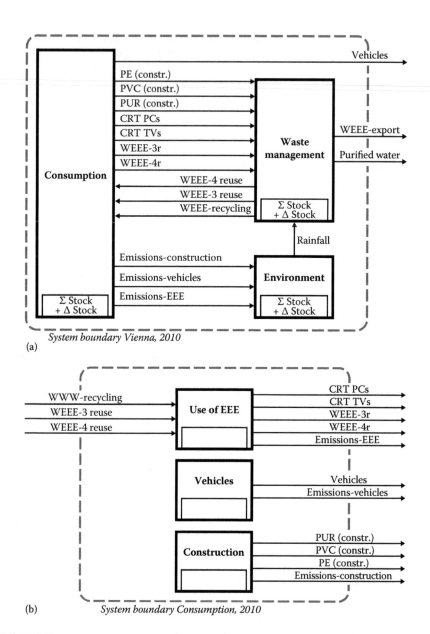

FIGURE 3.53

Model of flows and stocks of PBDEs in the city of Vienna, 2010. (a) Total system, and (b) subsystem consumption. CRT-PCs: cathode ray tube computer monitors; CRT-TVs: cathode ray tube televisions; EEE: electrical and electronic equipment; PE: polyethylene; PUR: polyurethane; PVC: polyvinyl chloride; WEEE: waste electrical and electronic equipment; WEEE-3: WEEE category 3; WEEE-3r: WEEE category 3 excluding screen devices ("rest"); WEEE-4: WEEE category 4; WEEE-4r: WEEE category 4 excluding screen devices ("rest"); WWTP: wastewater treatment plant.   *(Continued)*

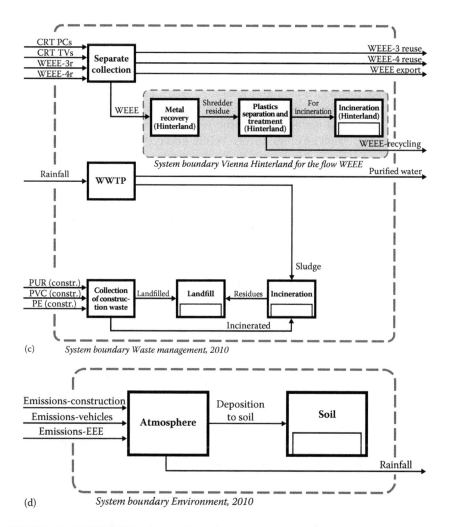

FIGURE 3.53 (CONTINUED)
Model of flows and stocks of PBDEs in the city of Vienna, 2010. (c) Subsystem waste management (including Vienna Hinterland) and (d) subsystem environment. CRT-PCs: cathode ray tube computer monitors; CRT-TVs: cathode ray tube televisions; EEE: electrical and electronic equipment; PE: polyethylene; PUR: polyurethane; PVC: polyvinyl chloride; WEEE: waste electrical and electronic equipment; WEEE-3: WEEE category 3; WEEE-3r: WEEE category 3 excluding screen devices ("rest"); WEEE-4: WEEE category 4; WEEE-4r: WEEE category 4 excluding screen devices ("rest"); WWTP: wastewater treatment plant.

The three key Austrian goals for waste management are protection of human health and the environment, resource conservation, and "after-care-free" waste management practice. The latter means no landfills requiring aftercare for decades to centuries and no recycling concepts that cycle hazardous substances so that the next generation has to take care of these risks. A large share of PBDEs and other plastic additives have entered waste management in the past; therefore, it is of high significance to focus on waste management as the major subsystem (UNEP, 2015b). In Vienna, the correct path for POP substances fulfilling the aforementioned objectives is incineration in one of the state-of-the-art municipal WTE plants. These plants can act as final sinks for most POPs and also for PBDEs (Vehlow and Mark, 1997) because they are designed for complete mineralization and for very low emissions to air and water. In 2010, the two major imports into the subsystem *waste management* of Vienna were WEEE and construction wastes. WEEE is collected and divided between reuse, export for treatment abroad, incineration, and recycling. Construction wastes are divided between incineration and landfilling. PBDEs contained in atmospheric deposition are a third, minor import. They are collected by the sewer system and transferred to the municipal wastewater treatment plant, where PBDEs, as a result of their low solubility and lipophilic character, accumulate in sewage sludge.

The data on the level of goods were collected from Statistik Austria (vehicles), Elektroaltgeräte Koordinierungsstelle Austria GmbH (EAK-Austria) (annual collection of WEEE in Austria, distribution of "new" and "historical" devices), and previously published literature in Germany and Switzerland (construction). The mean concentrations of POP-PBDEs in some goods were taken from UNEP (2015a). The emission factors were taken from previously published literature.

3.3.2.4.4 Uncertainty Treatment

Data uncertainties are taken into account by the software STAN. Normal distribution is assumed for data with mean value μ and standard deviation σ. This approximation is often not appropriate, but it offers the possibility to use error propagation and data reconciliation. In reality, however, the higher the uncertainties, the less symmetric the error intervals become (Hedbrant and Sörme, 2000). To overcome the challenge of missing information and highly uncertain data, scenario analysis has been introduced. Three cases were investigated, and for each case, the impact of selected scenarios (starting with most realistic assumptions and continuing to vary one parameter at a time) on the MFA system as a whole was evaluated. The cases are as follows:

1. The split of cOctaBDE contained in WEEE categories 3 and 4 between (a) CRT-PC monitors and TVs, and (b) other products excluding screen devices. For (a), higher concentrations of PBDEs are expected than for (b).

2. cOctaBDE occurrence in vehicles, which is poorly documented in the literature, with large deviations from source to source.

3. Flows of PBDE containing waste plastics in construction wastes, where the uncertainty about the path to incineration or landfilling can vary between 4:1 and 1:4.

Table 3.42 summarizes the outcomes of the scenario analysis for scenarios 1a and 1b. For more details about the sources of these data and the scenario analysis, see Vyzinkarova and Brunner (2013).

TABLE 3.42

Overview of Measured cOctaBDE Concentrations in Polymers of CRT-PCs, CRT-TVs, and WEEE Categories 3 and 4 without Screens (Wäger et al., 2010); in Housing Shredder Residues from CRT-Glass Recycling (Schlummer et al., 2007); and in CRT-PCs and -TVs Polymers (Single Housing Samples) of European Origin Imported to Nigeria (Sindiku et al., 2012)

Data Set Good	Wäger et al. (2010)										Schlummer et al. (2007)		Sindiku et al. (2012)			
	CRT-TVs		CRT-PCs		WEEE-3r		WEEE-4r		Mixed-3r&4r		HSR (CRT)		CRT-TVs		CRT-PCs	
cOctaBDE concentration in sample, g/kg	P41a	1.03	P31a	0.51	C3a	0.4	C4a	0.15	M3a	0.19	HSR1	0.00	S1-32	0.00	S1-22	0.00
	P41b	0.05	P31b	0.14	C3b	0.05	C4b	0.15	M3b	1.56	HSR2	0.00	S33	6.60		
	P41c	0.67	P31c	0.66	C3c	0.1			M3c	0.38	HSR3	6.39	S34	59.30		
	P41d	0.05	P31d	10.6							HSR4	8.10	S35	64.10		
	P41e	3.54	P31e	0.79							HSR5	2.88	S36	290.00		
	P41f	0.66									HSR6	13.84				
	P41g	0.1									HSR7	6.35				
Mean μ	0.87		2.54		0.18		0.15		0.71		5.37		11.67		0.00	
Median	0.66		0.66		0.1		0.15		0.38		6.35		0.00		0.00	
Standard deviation σ	1.14		4.04		0.15		0.00		0.61		4.55		49.12		0.00	
Coefficient of variation	131%		159%		84%		0%		85%		85%		421%		0%	

3.3.2.4.5 Results

The following three main outcomes were obtained: (1) Waste management plays a crucial role in the life cycle of PBDEs because it controls the path to the consumer and to the final sink. (2) Uncertainty of the data is high, pointing to future research needs. (3) Recycling plants are crucial to reach the objectives of clean cycles, and thus monitoring of plastic recycling products and emissions is required for quality control, in the same manner as, for example, monitoring of WTE residues and emissions.

The MFA shows clearly the key role waste management plays with regard to the objectives *protection of human health and the environment* as well as *resource conservation* (Figure 3.54): The largest amount of OctaBDE and PentaBDE flows from subsystem *consumption* to subsystem *waste management*. Vehicles are not treated within Vienna and are in part treated in Austrian car shredders outside of Vienna, partly exported. Consumer emissions to the environment are small and no longer play a significant role, neither for cPentaBDE (<10 kg per year [kg/yr]) nor cOctaBDE (<20 kg/yr). They will continue to decline as the consumption stock is depleted. Figure 3.54 shows the amounts of cPentaBDE and cOctaBDE in the consumption stock, estimated at 80 +/− 20 t for cPentaBDE and 20 +/− 40 t for cOctaBDE. Both stocks decrease at approximately the same velocities: dStock is −3 +/− 0.4 t/yr for cPentaBDE and −3 +/− 5 t/yr for cOctaBDE. If this trend continues in a static and linear way into the future, the two stocks of cPentaBDE and cOctaBDE will be depleted within 24 and 7 years, respectively, with high uncertainties for cOctaBDE.

Taking into account the aforementioned objectives, waste management should direct POPs into the *final sink* thermal treatment, either state-of-the-art WTE plants or cement kilns. Landfills are also sinks for POPs. However, in contrast to the complete thermal destruction during WTE or cement production, landfilling may release small amounts of POPs over a very long time period of several centuries. Thus, a landfill is not a final sink for PBDEs. MFA allows determining which fraction of cOctaBDE and cPentaBDE is directed toward a final sink, thus fulfilling the objectives of waste management. The largest waste flow of cPentaBDE (2 +/− 0.4 t/yr) is contained in construction waste plastics (PUR insulation foam). At the time of the case study, it was unidentified where these construction waste plastics end up. Thus, due to the lack of data, it is basically unknown if the objective of final sink has been reached for cPentaBDE.

The main flows of cOctaBDE are contained in WEEE (1.3 +/− 3 t/yr) and, possibly, EOL vehicles (2 +/− 0.9 t/yr), which leave Vienna for export, pointing to a supranational challenge. According to STAN modeling, 73% of cOctaBDE entering waste management ends up in WTE plants, with a high uncertainty of 1.2 +/− 5 t/yr. Five percent is exported, and 5% is landfilled. By recycling, 17% of cOctaBDE entering waste management returns back to *consumption*. However, this flow is highly uncertain (+/− 6 t/yr). Scenario analysis of case 1 shows that varying input concentrations of cOctaBDE in

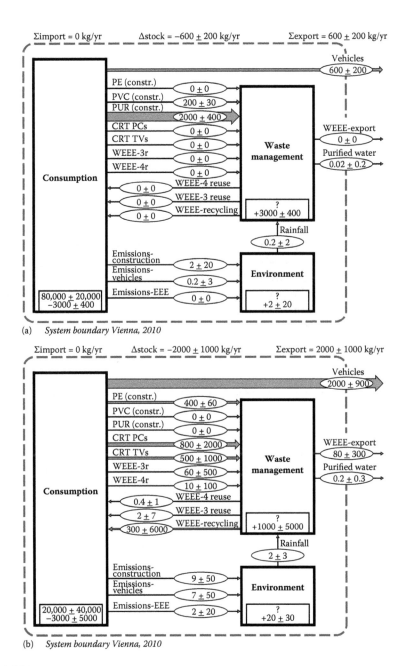

(a) *System boundary Vienna, 2010*

(b) *System boundary Vienna, 2010*

FIGURE 3.54
Stocks and flows of (a) cPentaBDE and (b) cOctaBDE in Vienna, 2010, as modeled by STAN in tonnes per year resp. tonnes, rounded to 1 significant digit. Numbers for both substances are given as commercial mixtures. "?" designates that the stocks of PBDEs in the environment (soil) and in waste management (landfill) are not known. EEE = electrical and electronic equipment; WEEE = waste electrical and electronic equipment.

polymers of CRT-PCs, TVs, and WEEE-3r and WEEE-4r largely affect the results (Table 3.42). The biggest impacts have variations in the concentration of CRT-PC monitors and WEEE-3r.

Both cases of cOctaBDE and cPentaBDE show high uncertainties. MFA clearly shows the need for better data. Part of the uncertainty of the process *use of EEE* is caused by the fact that there is no information about the stock of old EEE in Viennese households. Thus, statistical per capita data from other regions (Switzerland) had to be used to determine stocks of CRT-PCs and CRT-TVs in Vienna. The high uncertainty in stock changes influences estimates about recycling flows. Substance flows are calculated as (1) flow of goods multiplied by (2) polymer fractions and (3) substance concentrations in the polymer. Thus, the total uncertainty is additive and originates from three parameters.

To support the hypothesis that the current knowledge about the three afore-mentioned parameters is still small, available information about cOctaBDE concentration was reviewed. References include European flows of WEEE (Wäger, Schluep, Müller, and Gloor, 2011), housing, and mixed WEEE shredder residues (Schlummer, Gruber, Mäurer, Wolz, and Van Eldik, 2007), and CRT-PCs and TVs imported to Nigeria (Sindiku et al., 2012) (see Table 3.43). The data, which are further evaluated and discussed in Vyzinkarova and

TABLE 3.43

Results of the Scenario Analysis of Case 1, with Different cOctaBDE Input Concentrations in Polymer Fractions of (1a) CRT-PCs and -TVs, and of (1b) WEEE-3r and -4r

Scenario 1a and 1b		Average Flow of Polymer (t/yr), Average cOctaBDE Treated Fraction (%)	cOctaBDE Concentration in the Fraction (g/kg)	Impact on the System: cOctaBDE Recycling Flow Estimate (t/yr)
1a	CRT-PCs	994, 30	Min. $c = 0.14$	0.19 ± 6.36
			Max. $c = 10.6$	0.68 ± 6.63
			Mean $c = 2.54$	0.29 ± 6.06
			Median $c = 0.66$	0.20 ± 6.29
	CRT-TVs	1741, 30	Min. $c = 0.05$	0.23 ± 6.17
			Max. $c = 3.54$	0.52 ± 6.24
			Mean $c = 0.87$	0.29 ± 6.06
			Median $c = 0.66$	0.28 ± 6.08
1b	WEEE-3r	808, 42	Min. $c = 0.05$	0.24 ± 6.04
			Max. $c = 1.56$	0.96 ± 6.23
			Mean $c = 0.18$	0.29 ± 6.06
			Median $c = 0.38$	0.39 ± 6.07
	WEEE-4r	325, 24	Min. $c = 0.15$	0.29 ± 6.06
			Max. $c = 1.56$	0.46 ± 6.10
			Mean $c = 0.15$	0.29 ± 6.06
			Median $c = 0.38$	0.32 ± 6.06

Brunner (2013), show that the existing data sets are insufficient for advanced decision making and need to be amended by more reliable, large, and profound sampling and analysis. In particular, for goal-oriented waste management and recycling, PBDE concentrations in various wastes such as EOL vehicles, WEEE, and construction wastes must be assessed in a systematic and reproducible way.

Also, there is a need for information about recycling processes: transfer coefficients have to be determined for the various recycling techniques. Despite this discussion about the lack of sufficient data, the MFA displayed in Figure 3.54 indicates clearly that by WEEE management in Vienna, cOctaBDE is partly directed into consumer products. This has been observed by a similar study in Switzerland, too (Morf, Taverna, Daxbeck, and Smutny, 2003; Morf, Tremp, Gloor, Huber, Stengele, and Zennegg, 2005).

Hence, there is need for action. Austrian legislation requires federal states to control plants that treat hazardous waste at a minimum of every 5 years (AWG, 2002). This legislation could be expanded to include the flows of selected POP-PBDEs through recycling plants. If monitoring of products and emissions of PBDEs in recycling plants is introduced, recycling could reach the same high standards that WTE plants fulfill today. This would enable us to follow POP-PBDEs from sources to final sinks and to ensure that the goals of a *clean cycles* and *safe final sink* strategy can be reached by waste management.

3.3.2.4.6 Conclusions

The case study allows drawing conclusions with respect to (1) the application of MFA on one hand and (2) waste management decision making on the other hand.

1a. Regarding MFA, the case study focuses on the substance level and shows that defined mixtures of similar substances can be investigated by MFA and STAN, too. The substances investigated and balanced, e.g., cOctaBDE, are a commercial blends of several individual substances that are quite similar but not identical. As long as the commercial mixture contains similar congeners (in the case of cOctaBDE, from hexabromodiphenyl to decabromodiphenyl ether) with similar physical–chemical characteristics, it can be justified to treat the mixture as one substance. In case the mixture comprises also substances of different properties, it will be necessary to analyze and balance each substance individually.

The case study proves that even with little information about flows and stocks of substances and about transfer coefficients, it is still possible to establish an MFA on the substance level. Preconditions are a minimum data set about flows and stocks of goods containing POP-PBDEs and about concentrations of these chemicals in the corresponding goods. In order to reduce uncertainty, scenario analysis is useful. It allows us to identify the crucial parameters, and to focus on these. It is important to realize that even with very little data, an MFA/SFA can be established, although the uncertainty usually will be high. With increasing research, analysis, and expenditure, uncertainty can be reduced. There is a trade-off between costs and uncertainty: the higher the resource input, the lower the uncertainty. The art of designing an MFA system and collecting data in a cost-effective way is to reduce uncertainty only as much as is necessary to draw conclusions regarding the objectives of the project.

1b. For designing the MFA system and for data collection, the choice of system boundaries is crucial. In general, the system boundary in space should be selected with data collection in mind. *City* is an appropriate system boundary if the data are administered by a municipality or by a body that collects data on a city level. This is sometimes the case for goods but, unfortunately, rarely the case for substances. In hardly any city, data about the flows of PBDE are collected and managed. Thus, it is necessary to link various information sources: urban stocks and flows have to be reconstructed from national data, or from data from other urban regions where the missing information is available.

To choose a system boundary on an urban level makes sense for subjects that can be managed by the city. For instance, municipalities responsible for waste management want to know if their waste management practice fulfills federal regulations, and if not, what the most effective means would be to reach compliance. On a general level, the PBDE case study supports a *final sink* strategy of a municipality operating a WTE plant that is capable of completely destroying organic substances.

2a. Regarding waste management decision making, it is clear that without the MFA of PBDEs, it is not possible to identify those hotspots of PBDE flows and stocks that are offending legislation. MFA links sources and sinks, in this case PBDE containing consumer goods on one hand and recycling products, WTE plants, and emissions on the other hand. It is interesting to note that the main result of polluted cycles and incomplete flows to final sinks can be reliably estimated based on rather limited data. Thus, based on this SFA, measures to control the flows for compliance with regulations can be designed with confidence.

2b. A dilemma of modern waste management is the need for closing cycles, on one hand, and the fact that sometimes, hazardous substances are enclosed in wastes, rendering them unsuitable for recycling, on the other hand. MFA is instrumental for resolving this dilemma because it can show the level both of goods having a recycling potential and of substances comprising possible hazards for human health or the environment.

2c. In Vienna, the largest flows of POP-PBDEs are contained in three wastes: WEEE, construction wastes, and EOL vehicles. For cOctaBDE, WEEE and, possibly, vehicles are the main flows. Most EOL vehicles are exported from Vienna and pose a continental, rather than a local, challenge. According to the modeling, approximately 73% of cOctaBDE ends up in WTE plants with advanced APC, which represent safe final sinks. In view of the goals of waste management, namely, *protection of human health and environment*, cOctaBDE in WTE plants fulfills the objectives of complete destruction.

2d. A considerable fraction of POP-PBDEs containing waste is recycled. For cOctaBDE entering waste management, 17% is directed back to consumption, with little information about its fate during preparation and recovery. Secondary plastics, made from WEEE, may thus contain significant amounts of cOctaBDE; however, uncertainties are high. According to uncertainty analysis, the major cause is the lack of reliable values regarding cOctaBDE concentrations in European WEEE categories 3 and 4, including cathode ray tube monitors for computers and television sets. For adequate understanding and decision making in waste management, more information about the recycling processes is required.

In order to protect workers, human health, and the environment, a new, goal-oriented data set and mass balance of flows and stocks of polybrominated diphenyl ethers needs to be established. It must contain information about waste constituents, recycling plastic compositions, measured data about transfer coefficients, and emissions of POP-PBDEs in existing treatment plants, particularly recycling plants. Without the same set of information that, for example, WTE plants disclose, effective allocation of PBDE containing wastes to different waste treatment plants will not be possible. Dependable and sufficient information is particularly required because waste management is the key process for a region that has—due to successful regulations—no more inputs of POP-PBDEs but still has a large stock because of the legacies of the past.

2e. The main flows of cPentaBDE are contained in construction materials and are landfilled in construction waste landfills. They represent a long-term stock releasing minor amounts of PBDEs over long time periods. In view of the waste management goals *aftercare-free landfills*, this practice does not yet comply with legislation. Therefore, EOL construction materials made of plastic and containing POP-PBDEs, especially PUR foam insulation, PVC duroplastic sheeting, and PE roof sheeting, which may account for cOctaBDE flows into landfills, must be separated from construction wastes and properly treated, for example, in a state-of-the-art WTE plant. The example of PBDEs shows well the power of MFA to support a *clean cycle* and *safe final sink* strategy in waste management.

PROBLEMS—SECTION 3.3

Problem 3.8:

Plastic wastes have a high calorific value, which makes them a potential fuel for cement kilns, blast furnaces, and municipal incinerators. Packaging plastics have been successfully incinerated in cement kilns: production costs are reduced; the quality of cement does not change; and emissions are not altered significantly. Wastes from longer-lasting plastic materials (containing about 10% PVC) have a high chlorine content, rendering these wastes unsuitable as a fuel in cement kilns because they exceed the capacity of the process for chlorides. Using Table 3.29, evaluate whether nonpackaging plastics are better suited for blast furnaces or for MSW incinerators. Take into account environmental and resource considerations only, and do not consider economic (additional investments, fuel savings) or technological (pretreatment, adaptation of feed or furnace, etc.) aspects. Discuss *final sinks* for heavy metals. Use the transfer coefficients in this handbook for MSW incineration, Table 3.38, and look for data about blast furnaces in the library or the World Wide Web.

Problem 3.9:

Assess the paper content in MSW of a country of your choice. First, determine the appropriate system (processes, flows, system boundaries). Second, carry out an Internet search for the annual report of the pulp and paper industry of the selected country and determine the flows through and within your system. Third, find out the national MSW generation rate (e.g., contact the EPA website) and calculate your result.

Problem 3.10:

The combustion of biomass is described in Obernberger, Biedermann, Widmann, and Riedl (1997). Calculate the Cd concentration in cereals based on the information given in the paper. Using the approach described in Section 3.3.1.2, calculate the composition of the input (cereals) from the composition of the output (different ash fractions). Compare with Cd values for cereals you find in the literature.

Problem 3.11:

Figure 3.55 gives the Cd balance for the management of combustible wastes in Austria. Discuss the flowchart together with the total mass balance for combustible waste flows in Austria as given in Section 3.3.2.1, Figure 3.45. Consider resource potentials and potentially dangerous environmental loadings.

Problem 3.12:

Summarize the reasons why a cement manufacturer association might decide to limit the annual flow of heavy metals into cement kilns with 15% of the national consumption of heavy metals.

Problem 3.13:

Assume that incineration of 1 ton of MSW [copper (Cu) content, ca. 0.1%] yields the following solid residues: 250 kg of bottom ash, 25 kg of fly ash, 3 kg of iron scrap, and 3 kg of neutralization sludge from the treatment of scrubber water. About 90% of the Cu leaves incineration via bottom ash and 10% via fly ash. The Cu flow via other residues such as off-gas, iron scrap, etc. is <1% and can be neglected. Investigations show that by mechanical processing of bottom ash, approximately 60% of the Cu can be separated in the small fraction of metals concentrate. The Cu content of this fraction (ca. 50%) can be

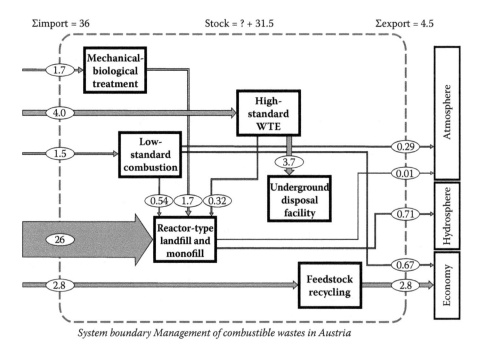

Σimport = 36   Stock = ? + 31.5   Σexport = 4.5

System boundary Management of combustible wastes in Austria

FIGURE 3.55
Flows of cadmium induced by the management of combustible wastes in Austria (1995), t/year.

recovered in a metal mill. (a) What is the recovery efficiency for the combined process *MSW and mechanical processing of bottom ash*? (b) Calculate the substance concentrating efficiency (SCE) for the process chain *incineration, mechanical processing, and metal mill*. (c) As a decision maker, would you support such a technology, and why?

The solutions to the problems are given on the website http://www.MFA-handbook.info.

3.4 Industrial Applications

MFA has a long-standing tradition in chemical engineering. Educts and resulting products and by-products have been balanced by stoichiometric methods for reasons of reaction design, optimization, and quality control of chemical processes. While this has been state of the art for many decades in the production of chemical substances, MFA has just recently been introduced to industrial processes in other fields, such as metal production and auto-motive or airplane engineering. Particularly in manufacturing, the advantage of applying MFA has been recognized when optimizing processes and process chains (Krolczyk et al., 2015). Krolczyk et al. use MFA as an analysis and optimization tool for reducing costs in an industrial company that manufactures composite elements for automotive, electric, and agricultural industries. They see the main advantage in the comprehensive picture that is produced by analyzing flows and stocks of materials through a production plant in a systematic way. Particularly, they point out how MFA can be used to create an internal transport program and to optimize the working stands arrangement. As a result of the reorganization of the manufacturing plant, the number of transport operations is reduced, material supply and transport are smoothened, and costs are reduced.

Another example of application of MFA in the processing industry has been conducted by Trinkel, Kienberger, Rechberger, and Fellner (2015). These authors attempt to balance different heavy metals in a blast furnace process in order to follow their path from source to products and emissions. However, they face various challenges, particularly because heavy metals are sometimes present at small concentrations in different input and output materials. The composition of these materials is often heterogeneous, making representative sampling and subsequent analysis difficult. In their case, the major challenge for performing an MFA of lead through a blast furnace is the analysis of the content of Pb in the metal produced. Different analysis methods result in different Pb concentrations. In addition, Pb proves to be unequally distributed in the metal product, calling into question current sampling and analysis procedures. This example shows well the power and limitation of MFA in supporting decisions on the production level: if adequate sampling and analysis methods are not available, balancing of substances in complex processing such as a blast furnace becomes a real challenge.

In the following chapter, an MFA of an industry manufacturing interior panels for airplanes is presented (Müller, 2013). The specific feature of this case study is the link between MFA and economic analysis. The flow of values is depicted in parallel to the flows of goods. Also, flows and stocks of materials are associated with working hours, thus enabling an economic optimization of the production lines with less idle time and more productivity. A similar approach has been attempted before by Kytzia (1989). This author linked flows of materials, energy, and financial resources into one model of the residential building stock. In her study, flows of financial resources consist of cost and revenue. The difficulty in such a model is how to allocate revenues to the various cost units.

The same problem is encountered by the MFA work of Eisingerich (2015) on open burning of rice straw. This author links material and substance flows on Thai farms to economic parameters in order to improve the economic situation of small rice farmers, and at the same time to decrease environmental loadings. The allocation of revenue to the individual farming processes could not be accomplished for all flows and processes. Case Study 16 by Müller avoids these difficulties by focusing on production costs only and neglecting the revenue (Müller, 2013). The rationale for this approach is that the goal of process optimization is to minimize cost of production, and not to increase revenue, because revenue is independent of production processes and cannot be increased by process optimization.

3.4.1 Case Study 16: MFA as a Tool to Optimize Manufacturing

To use resources efficiently and without environmental degradation is not only an economic objective of a company; it is also one of the goals of sustainable development. Hence, it is in the interest of both companies and society to minimize resource consumption, emissions, wastes, and cost per unit of good produced. The present case study demonstrates how MFA can be applied on the company level for minimizing resource use and optimizing economic benefits.

The main challenge for an entrepreneur is to identify those production processes that have the highest potential for resource conservation, environmental protection, and economic optimization. Key questions are which methodology to apply, how to get the data, and how to assess the effects of uncertainty on the results. To answer these universal questions, a case study on a state-of-the-art manufacturing system for an advanced product of the aircraft industry was performed (Müller, 2013). Because of the novelty of the linking of MFA with economic parameters, the case study also required new methodological development beyond traditional MFA and STAN.

The company involved in this case study is an internationally leading producer of insulating materials, laminates, and composites. The MFA covers a particular segment of production and focuses only on a small fraction of the entire company. For reasons of confidentiality, the name of the company as well as the names of goods and processes are undisclosed. All numbers of flows, stocks, and economic parameters are changed for reticence. Nevertheless, the results and conclusions serve well to demonstrate the power of MFA to support and optimize manufacturing processes.

3.4.1.1 Objectives

Case Study 16 aims at developing a method for mapping complex manufacturing systems in a transparent and comprehensible way in order to minimize production costs (primary objective) and wastes, and optimize resource use (secondary objective). The goal is to produce one or several models that take into account all relevant stocks and flows of materials, costs, and production time, including uncertainties (Müller, 2013). These models should test the feasibility of MFA for identification, analysis, quantification, and representation of production systems, and allow discerning of the production steps with the highest potential for improvement. They show flows and stocks of educts, products, wastes and emissions of each process, the associated costs, and the working hours required to produce a particular product or waste. Also, the possibility of STAN linking MFA and economic parameters such as costs and time represents an important new and original step.

This facilitates the understanding of the entire manufacturing process and represents the starting point for optimization in terms of resource efficiency, cost, and time.

The option of STAN to include the level of substances is not an objective of this study. However, if issues such as health protection or environmental pollution were to be addressed, hazardous substances could be added for investigation without methodological difficulties.

The following research questions are addressed in the case study:

1. Is it possible to jointly depict flows of material and money of manufacturing processes by STAN?
2. What is the main advantage of using STAN for this combination?
3. How can uncertainties be taken into account for risk assessment?
4. How can STAN diagrams be used by entrepreneurs for optimization of production systems?
5. How can STAN be improved for the specific purpose of mapping production processes physically and economically?

3.4.1.2 Procedures

Basically, three (MFA) models are created (Müller, 2013). In the first model, the whole production system for manufacturing one unit of output is described. In the second model, the effect of uncertainties of input flows is investigated. The third model allows the following of a semiproduct over a defined time period. These three detailed models facilitate comprehensive understanding of the manufacturing process and enable identification of the potential for optimization as well as the impact of uncertainties in the input values. The results of the three models have been compared with those of the enterprise resource planning (EPR) system that is installed in the company. This comparison serves as a plausibility check, too.

The entire manufacturing system is modeled in STAN following these steps:

1. Definition of system boundaries, units, balancing periods, and costs.
2. Structuring the manufacturing system into processes, material flows and stocks, and associated cost flows and working hours.
3. Implementation of steps 1 and 2 in STAN.
4. Collection of production and economic data about flows, stocks, and working hours, and input of this information into the STAN model. For this, all input goods are put in relation to one unit of output product.

5. Validation of the STAN model on the mass flow level by balancing the MFA system for all periods and correcting errors.

6. Validation of the STAN model on the money flow level by balancing the MFA system for all periods and correcting errors.

7. Checking for plausibility by comparing the results of STAN with another planning or quality control system, such as the ERP system implemented in this company.

8. Applying the results for optimizing the manufacturing process.

Simple examples for balancing mass flows, money flows, and working hours of a single process with the same STAN model are presented in Figures 3.56 through 3.58. They show that in principle, money and working hours can be treated the same way as mass flows. This offers the advantage that a

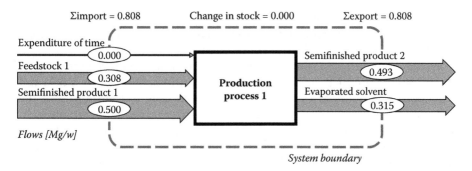

FIGURE 3.56
Example of flow of materials through production process 1. *Expenditure of time (working hours)* is expressed in STAN as a *virtual* material flow and thus is represented as 0 in the mass flow diagram. The numerical values for expenditure of time are given in Figures 3.58 and 3.62.

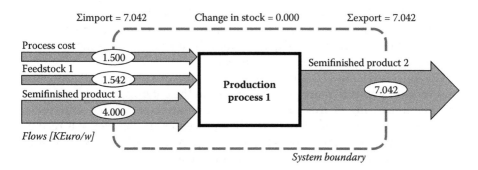

FIGURE 3.57
STAN representation of money flows associated with production process 1. The sum of the costs of individual educts plus process costs equals the cost of the product. Process costs are not associated with a material flow, and include all costs emerging from production except for educts such as feedstock and semifinished products. Labor cost is included in the process cost.

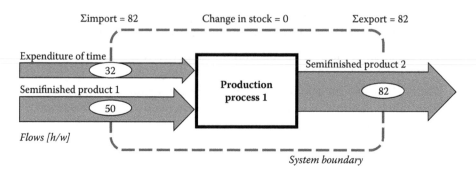

Σimport = 82   Change in stock = 0   Σexport = 82

Expenditure of time

32

Semifinished product 1

50

Production
process 1

Semifinished product 2

82

Flows [h/w]

System boundary

FIGURE 3.58
STAN representation of working hours associated with a production process. The sum of the
working hours used to produce the educts plus the time expenditure for producing the process
output yields the total working hours to produce a unit of output.

single STAN model allows combining of all three aspects of mass, monetary
values, and required working time. In this case study, the unit *energy*, implemented by default in STAN, was replaced by the unit *money*.

System boundaries in space comprise the area that is required for the production of the product, including machinery, equipment, space for stock, and
transport. The boundaries in time change according to the rhythm of the
production system, which is dependent of the external economic situation. In
this case study, 10 periods of 1-month duration each have been investigated
and balanced. Flows and stocks of materials are analyzed and balanced
according to MFA standards; for more information, see Müller (2013).

In order to include costs in STAN, the energy level offered by the software
was "abused" and exchanged for money flows, with the euro (€) replacing
the energy unit joule (J). The advantage of this procedure is that it is easily
possible to switch between the two STAN levels of mass and money flows
and that complete consistency is given between the two levels (cf. Figures
3.56 and 3.57). By a simple click, a manufacturing system depicted in STAN
can be viewed either as a physical material system of flows and stocks, or as
a money flow system. However, as of now, the energy level cannot be used
for simultaneously mapping energy flows.

The functional unit of this case study is one unit of a product. All processes
and flows of goods that are performed within the enterprise contributing to
the manufacturing of this unit are taken into account. Wastes, emissions,
and by-products are considered as well. If processes are complex, it is recommended to split them into subsystems. This prevents a black-box effect and
facilitates understanding of the underlying subsystems.

In order to evaluate the results, *operating numbers* (key figures) are defined.
They allow assessment of the effect of measures on the manufacturing system and are instrumental for comparing performance of different units
within and outside of the enterprise. The following two sets of operating
numbers are chosen for this case study:

Ecological operating numbers focus on resource efficiency, solvent use, and wastes because these three issues are the main environmental concern of the company: (1) total material efficiency: ratio of total material input per unit of product; (2) efficiency of solid auxiliary material utilization: ratio of solid auxiliary material input per unit of product; (3) solvent utilization: ratio of solvent used per unit of product; and (4) waste generation: ratio of waste produced per unit of product.

As economic operating numbers, the following five key figures are defined: (1) material costs: ratio of costs of raw materials versus total costs to produce a unit of product; (2) processing costs: ratio of process costs versus total costs; (3) solvent costs: ratio of solvent costs versus total costs; (4) solid auxiliary material costs: ratio of solid auxiliary material costs versus total costs; and (5) costs for waste management: ratio of costs for waste management versus total costs to produce a unit of product.

By taking into account uncertainties, production risks can be assessed. This allows us, for example, to set priorities for purchasing, to support the selection of cheaper or more environmentally sound substitutes, or to define tolerances for individual manufacturing processes. For this reason, values for uncertainty are implemented in STAN for all input flows on the level of mass flows as well as money flows. Various scenarios are calculated in order to analyze the effect of uncertainties on the results. For more information and practical application, see Müller (2013).

3.4.1.3 Results

The processes and flows of educts, auxiliary materials, solvents, and wastes for the production of one unit of product are presented in Figures 3.59 through 3.62. Figure 3.59 depicts all mass flows to produce one unit of final product, Figure 3.60 money flows per unit of final product, Figure 3.61 working time required to produce one unit of product, and Figure 3.62 the mass flows for producing one unit of semifinished product.

On the basis of Figures 3.59 and 3.60, the cost driving material flows can easily be detected, and the focus for economic improvement can be put on these flows, respective of the losses of the processes handling these flows. The calculation of the operating numbers yields the ratio of process costs versus material costs, thus allowing us to set priorities in optimization. In this case study, both costs are nearly equal [cf. Figure 3.60: sum of cost for total material import (235 €/P) minus cost of exported product (476 €/P); the difference of 241 €/P is the operating cost]. It is recommended to decrease the costs of auxiliary materials because they have the least impact on the final market product.

The scenario analysis allows checking of the effect of variations in manufacturing. The variations in Table 3.44 are chosen according to actual market and manufacturing conditions. Ten percent uncertainty of the import mass flow in scenario 1 yields only 3.7% uncertainty on the final product flow.

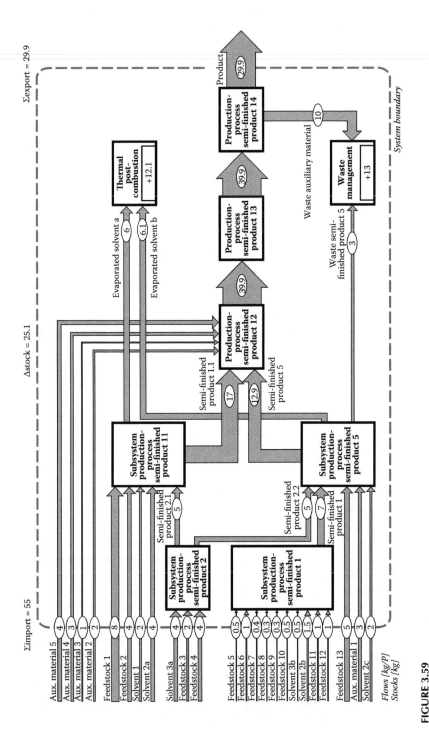

FIGURE 3.59

STAN mass flow diagram of the entire production from feedstock to product, including all imports, exports, semifinished products and products, and wastes. Exports such as off-gas from thermal post combustion and residues from waste treatment are not considered.

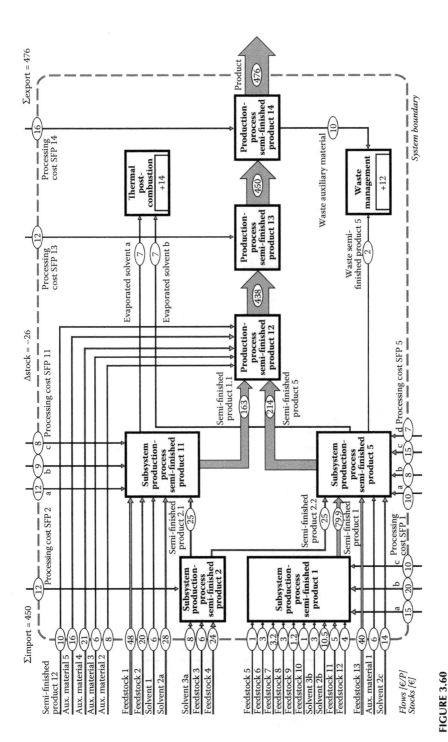

FIGURE 3.60

STAN money flow diagram of the entire production, including all material and operational costs, per unit of product.

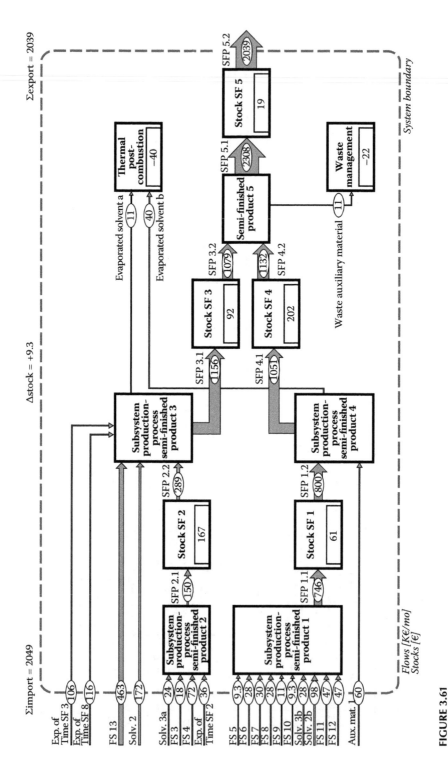

FIGURE 3.61
STAN cost-per-time-flow diagram (K€/month) to produce one unit of semifinished product during a period of 1 month.

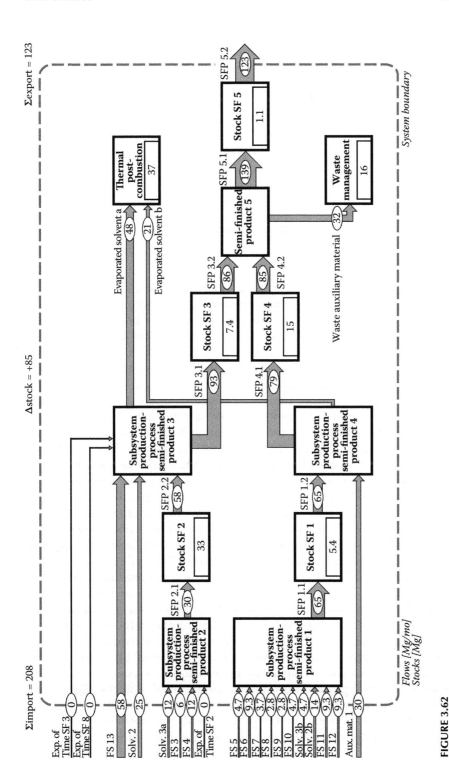

FIGURE 3.62
STAN mass-per-time-flow diagram for product manufacturing during a time period of 1 month.

TABLE 3.44

Uncertainty Assessment Based on Scenarios Analysis for Economic Optimization of Production

Scenario	Scenario Specification	Uncertainty of Final Product	
		Level of Goods, %	Level of Cost, %
1	10% uncertainty in all input mass flows	3.7	1
2	10% uncertainty in the purchase price of all input flows	0	1.8
3	Dependence from suppliers: 45% uncertainty in cost for raw materials	0	6.4
4	Variation in production: 20% uncertainty in processing cost	0	1.9

In order to maintain high-quality production, new specifications for import materials can be defined that allow a better performance on the output side.

Ten percent uncertainty in the purchasing price for imports (scenario 2) results in a rather small uncertainty of 1.8% for the cost of the final product. In order to assess the importance of unstable markets, uncertainties can be assumed to be much larger, and the corresponding effects on the total cost can be calculated. This enables a proactive business strategy anticipating future market volatility, e.g., in the resource or energy markets. If the price for the two most important raw materials fluctuates by 45%, the largest effect on the final product results (scenario 3). In scenario 4, the process costs are assumed to differ by 20%. The scenario analysis shows the largest effect on the final product for scenario 3. For the final revenue of the whole manufacturing process, plus or minus 6.4% cost of the final product is a significant number. Thus, to reduce the entrepreneurial risk, strategies must be developed to stabilize the cost of the two crucial raw materials that are, at the moment, purchased from a single supplier. Variations in operating costs are less relevant. To summarize, the scenarios displayed in Table 3.44 serve well for setting priorities for economic stabilization of the production.

The time required for the production of semifinished products is displayed in Figure 3.63. In this figure, the STAN results of 10 consecutive balancing periods of the manufacturing system are summarized for semifinished product 5. This allows, on one hand, identifying the individual workloads of the working places. On the other hand, the total time required to produce a semifinished product or a product can be calculated. Since some of the semifinished products are especially made within the company to supply the manufacturing process, information about available stocks (Figure 3.64) and time required for their production is instrumental for careful planning of the whole operation. Figure 3.64 presents a highly useful overview about mass flows and stocks and their changes over time, and shows that the manufacturing process is not a just-in-time operation yet but shows fluctuations.

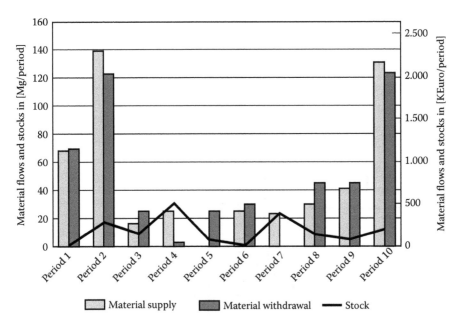

FIGURE 3.63
Flows and stocks of mass and money to produce a semifinished product over 10 balancing periods.

STAN diagrams and Figure 3.64 serve as means to raise awareness among the personnel for stock and flow issues in order to keep the stocks low and the wastes small. Also, they can be used to optimize labor force employment.

Tables 3.45 and 3.46 summarize the operating numbers determined in this case study. The ecological operating numbers show considerable promise for improvement of the manufacturing process. Overall material input is about 2.8 times higher than useful product output. Also, solid and liquid (solvents) auxiliary inputs are larger than the product output. Per 1 kg of product, 0.5 kg of wastes is generated. Table 3.45 shows that the production of semifinished product 4 yields the highest operating numbers and thus is of first priority when optimizing production as a whole.

Economic operating numbers in Table 3.46 show that efforts to minimize manufacturing costs should focus on the production of semifinished products 1, 4, and 5. The purchasing department is well advised to negotiate better purchasing conditions for semifinished products 2 and 3 because they have the largest potential for cost saving. The example shows that operating numbers are well suited to effectively support cost reduction in production processes.

3.4.1.4 Conclusions

The conclusions regarding application of MFA in manufacturing on the three levels mass flows, money flows, and time are summarized in Table 3.47.

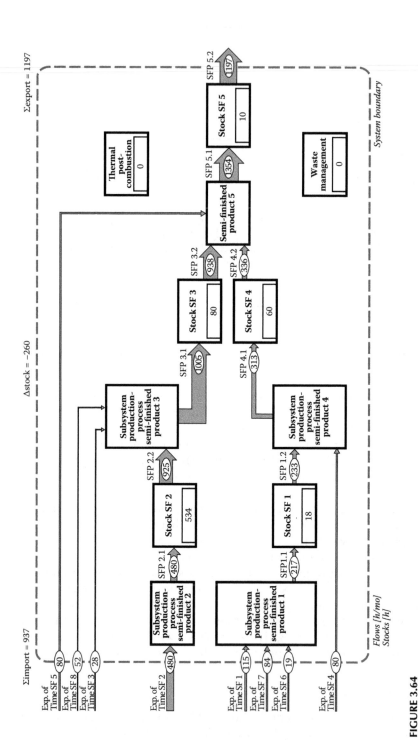

FIGURE 3.64

STAN diagram presenting the time required to produce semifinished product 5. The change in stock of minus 260 h signifies that during this period, more time was consumed for the production of the feed and semifinished products than was supplied and accomplished during that period. This can be due to a decrease in stock when material has been produced in a former period.

TABLE 3.45

Ecological Operating Numbers of the Semifinished Products (SFPs) and the Final Product (FP)

Ecological Operating Number	SFP1	SFP 2	SFP 3	SFP 4	SFP 5	FP
Total material input per product	140	167	150	200	125	276
Solid auxiliary material per product	40	67	50	100	25	125
Solvent utilization per product	40	67	50	40	0	60
Waste generation per product	0[a]	0[a]	0[a]	0[a]	0[a]	50

Note: The numbers stand for material flows per unit of product and are given in %.
[a] No waste generated.

TABLE 3.46

Economic Operating Numbers of the Semifinished Products (SFPs) and the Final Product (FP)

Economic Operating Number	SFP 1	SFP 2	SFP 3	SFP 4	SFP 5	FP
Material cost	27	60	55	20	36	33
Processing cost	56	24	24	57	43	34
Auxiliary material cost	17	16	18	19	18	28
Solvent cost	17	16	18	13	15	15
Solid auxiliary material cost	0	0	0	6	3	12
Waste disposal cost	0	0	3	4	4	5

Note: The numbers stand for, e.g., material cost per total costs to produce a unit of product, and are given in %.

TABLE 3.47

Application and Outcome of MFA in Manufacturing on the Three Levels of Goods (Mass Flows), Cost (Money Flows), and Time

Model	Level	Outcome
Entire production	Goods	Facilitates understanding of production system
		Supports the design of optimal material flows
	Cost	Points out processes and material flows of high costs
Uncertainty	Goods	Reveals the effect of inaccuracies of manufacturing on the final products
	Cost	Exposes the effect of fluctuations of cost of material and labor on cost of end product
		Exposes the effect of fluctuations of process cost on cost of end product
Semifinished product	Goods	Delivers actual material flows through the production system
		Shows the demand for material stock for each manufacturing step and process
	Cost	Depicts actual money flows through the manufacturing system
		Depicts capital required for each working place
	Time	Demonstrates workload of each working place
		Demonstrates minimum processing time of a semifinished product

The outcomes show the feasibility of this method for analysis and representation of production systems, and that STAN is well suited for decision support.

The answers for the five questions addressed in the beginning (cf. Section 3.4.1.1, "Objectives") are as follows:

1. Is it possible to jointly depict flows of material and money of manufacturing processes by STAN? Formally, STAN is not yet equipped with a feature that allows us to take economic parameters into account. However, it is possible to substitute *energy* (in J) for costs (in €). This exchange allows easy consideration of money flows. The drawback is that at the moment, it is not possible to work with both energy and costs. There is a need for a next version of STAN that will provide both possibilities.

2. What is the main advantage of using STAN for this combination? STAN delivers a total view of a production system including both mass flow and economic level. Switching from one to the other level is quick and easy. Full transparency and reproducibility are guaranteed. This facilitates fast comprehension of the entire production system.

3. How can uncertainties be taken into account for risk assessment? STAN is well suited to include and calculate data uncertainty (cf. Chapter 2, Section 2.4). Thus, based on the uncertainty of input data, output uncertainties can be assessed, and entrepreneurial decisions can be based on these uncertainties. Since values and uncertainties of data can be changed easily, STAN serves well for scenario analysis, too. However, at present, only data with standard distribution can be taken into account.

4. How can STAN diagrams be used by entrepreneurs for optimization of production systems? (a) Because both mass flow and economic levels are included, the STAN diagram yields an overview of production cost and allows detection of causes of high cost in a straightforward way. The graphs are very well suited for decision support in planning of future investments, for strategic priority setting, and for optimization of manufacturing systems. (b) The ratios of waste versus final product and resource use versus final product point out potential economic losses and allow the setting of priorities for improvement. (c) The working-hour diagram shows where the production line can be improved by decreasing idle time.

5. How can STAN be improved for the specific purpose of mapping production processes physically and economically? (a) Managing a production line requires appropriate information about mass flows, energy flows, money flows, and expenditure of time flows. For further development of STAN, it is recommended to incorporate these four levels into the software. This means amending the present version with two additional levels for money, and time and labor. (b) The possibilities to present results should be amended by

an evaluation showing the composition of the final product in terms of all mass inputs and money flow inputs. This allows exact assignment of the total materials and money flows used in manufacturing to one unit of product. (c) If operating numbers are implemented in STAN, they can be calculated automatically, saving time and effort. A graphical display of the development of operating numbers over time, or for different scenarios, would increase the value of STAN considerably for optimization of manufacturing. (d) At present, for uncertainty calculations, STAN assumes that data are normally distributed. STAN results could benefit by a feature that allows (i) choosing between various distributions when inserting data and (ii) determining a lower and upper limit for uncertainty. (e) If STAN can be linked to an ERP system, the application of STAN would be significantly facilitated, leading to a wider and regular use of STAN.

In summary, Case Study 16 shows that cost analysis and production time can be linked to material flows and stocks by STAN pursuing the following three steps: First is modeling the mass flows, money flows, and working hours of the entire production system for one unit of product. Second is modeling the flow of a semifinished product over a defined period of time from imports (educt, auxiliary materials, and solvents) to exports (semiproduct). In the third model, the effect of data uncertainties in input flows is investigated. These three highly detailed models allow the identification, quantification, and realization of optimization potentials for production systems in terms of resource efficiency, cost, and time.

For such a comprehensive study, the quality of input data is key because it determines the quality of the results and hence of the subsequent business decisions. In order to collect appropriate data of good quality, the expenditures to establish the three models are considerable. However, the graphic display of the results greatly increases the understanding of the entire manufacturing system. In addition, changes in production or input goods can be implemented quickly by STAN, and thus, the models serve as an excellent tool for improving and optimizing manufacturing. The addition of economic parameters such as cost and time in STAN represents an important step in decision support for efficient production systems. Due to the option of STAN to include the level of substances, too, this method can be further advanced to address environmental and health issues of a production line.

3.5 Regional Materials Management

The objective of regional materials management is to protect the environment, to conserve resources, and to minimize wastes in one combined effort. Regional materials management is an integrated approach that links all three issues and strives for an optimum solution. Instead of focusing on one topic alone, all three are taken into account at the same time. This comprehensive procedure requires less effort and results in more information than three separate studies. It also ensures that the results from the different fields are compatible and that conclusions regarding all three fields can be drawn. For regional materials management, it is essential to know the main anthropogenic as well as natural sources, conveyor belts (transport paths), stocks, and sinks of materials in a region. Without this information, regional materials management is not possible. In order to achieve the stated goals, a long-term view must be taken. Material flows and stocks have to be balanced over decades to centuries in order to examine whether harmful or beneficial accumulations and depletions of materials are taking place in the region. All materials used within a region must find a safe final sink. If a safe final sink is not available, use of the material should be phased out or controlled tightly and accumulated over long time periods with a clear purpose and economic plan for future reuse.

3.5.1 Case Study 17: Regional Lead Management

This example of regional lead management is drawn from Case Study 1 (Section 3.1.1), which described system definition and data collection. The following discussion covers only those results and conclusions that are important for regional materials management.

3.5.1.1 Overall Flows and Stocks

A total of 340 t/year of lead is imported into the region, and 280 t/year is exported (see Figure 3.1). The main import consists of used cars that are crushed in a car shredder. The main exports are filter residues from a steel mill that produces steel for construction from the shredded cars, lead contained in construction steel, and lead in MSW. The difference between imports and exports amounts to 60 t/year, which is accumulated mainly in landfills. The geogenic stock *soil* includes about 400 t of lead (the term *geogenic* is not actually precise here, since a certain fraction of lead in soils is of anthropogenic origin (Baccini, von Steiger, and Piepke, 1988). The anthropogenic stock *landfill* is much larger and amounts to >600 t (>10 years of landfilling 60 t/year). Like most materials in urban regions (Chapter 1, Section 1.4.5.4), lead is accumulated in this region. Imports and exports of lead by geogenic conveyor belts (air, water) are marginal and are <1%. From the point of view of resource management, shredder residues and filter dusts are of prime interest. From an environmental point of view, depositions on the soil as well as potential leaching of lead from landfills to surface water and groundwater are important.

The advantage of a regional material balance is that with one single balance, present and future hot spots for environmental, resource, *and* waste management can be detected. For example, the potentially large but, at present, unknown flow of lead from local landfills to the hydrosphere cannot be hypothesized without information about local landfills and their constituents. Quantitative information about landfills in general is not available, but it is known that most of the shredder residue is landfilled within the region. By a simple balance of the car shredder, assuming a certain lead input based on car manufacturers' information and the number of cars treated in the shredder, and the lead in the metal fraction used and analyzed by the smelter, it is possible to roughly assess the amount of landfilled lead.

3.5.1.2 Lead Stock and Implications

The existing stock of lead in landfills totals >600 t. A doubling time (t_{2x}) for the lead stock of ≈ 10 years can be calculated. In other words, if the regional anthroposphere remains the same for the next 100 years, the stock will have increased from 600 to 7000 t. According to Chapter 1, Section 1.4.5.1, there are no indications yet that waste lead flows will decrease. What makes this case study special is the huge extent of the accumulation. Nearly 20% of the lead imported does not leave the region and stays there probably for 10,000 years until erosion slowly removes the landfill. All lead landfilled and deposited on the soil is of no further use. The concentration is comparatively low, and the heterogeneity of the landfilled materials is much larger than that of lead ores. Hence, economic reuse of this stock is, at present, not feasible. Emissions from this stock are likely but not known. A conscientious approach to regional materials management would dictate that, in the future, this lead stock be managed in a different way, turning it from a hazard into a positive asset. Means for upgrading and reuse have to be explored (see item 2 in Section 3.5.1.4).

3.5.1.3 Lead Flows and Implications

Lead flows can be divided into flows in products, in wastes, and in emissions. The management goal is to maximize the use of lead in products, to reuse lead in wastes, and to reduce emissions to an acceptable level. MFA shows where the large lead flows are and thus points out key processes and goods for control and management. For each environmental compartment (water, soil, air), potential sources are identified and sometimes quantified. Thus, priorities can be set when measures for the protection of the environment are taken.

Figure 3.1 shows that lead increases by 1.4 t/year in the river between the point of entry in and exit out of the region. While 0.31 t/year is due to leaching from soils, and 0.14 t/year due to treated wastewater, 1 t/year has not been accounted for by MFA. This flow is so large that it is not likely due to an error in measuring soil leaching or effluent from WWTP. The most probable source of this large amount of lead is leachates of shredder residue from landfills. It is not efficient to reduce the comparatively small amount of lead emitted by WWTP effluents. The first step is to investigate the hypothesis that shredder residue landfills are really leaching such a large amount of lead to the surface waters. The second step is to reduce the loadings of the soil, e.g., by banning leaded gasoline (as was done in the late 1980s) or by incinerating sewage sludge and landfilling the immobilized ash.

MFA supports environmental impact assessment and serves as a design tool. Figure 3.1 shows that emissions from landfills (and any other point source) are not relevant if they are in the range between 0.002 and 0.02 t/year (0.1–1% of present aquatic export flow). Considering a total stock of ≈1000 t of lead landfilled, one can calculate that no more than about 2 to 20 ppm (mass) of lead may be mobilized in the landfill if there is to be no significant effect on the river water concentration. This figure can serve as a goal for the design of waste treatment such as immobilization or solidification. Note that this calculation does not consider groundwater pollution. If a local groundwater flow is small and the residence time is high, the lead flows from landfills calculated previously may be large enough to exceed drinking water standards. Hence, it is important to take groundwater into consideration, too.

3.5.1.4 Regional Lead Management

For the region, it is more efficient to manage lead in a comprehensive way than to segregate the lead issue into different problem areas. This is exemplified by the following three conclusions:

1. Lead not in use should be accumulated actively and purposefully in safe, intermediary stocks with residence times of several decades. The objective is to build up concentrated stocks of lead and other metals and to reuse these stocks once they have reached a size that makes them viable for economic reuse. In order to concentrate lead as much as possible, the shredder residue should be treated in an incinerator with advanced air pollution control. Mineralization will increase lead concentration by at least a factor of 10. Many materials are well suited for such accumulation. The region could offer to take back filter residues from MSW incineration and to accumulate these materials together with car shredder residues. Intermediary lead stocks are distinctly different from fluff or MSW in landfills. They are highly concentrated in metals, and the chemical form is such that economic metallurgical reuse is facilitated. Hence, solidification with cement is not recommended. The intermediary stocks are engineered sites that are designed and constructed to last for a predefined period during which they have to be maintained. The period

is calculated according to economic considerations. Due to the economy of scale, and depending on the technology applied, there is a minimum size required for economic reuse of materials. This minimum size divided by the waste generation rate yields the time span needed for accumulation of an amount of material for economic recycling.

2. Accumulation and mineralization for reuse ensure that most lead is controlled within the anthroposphere and no longer poses a threat to the environment. Information about anthropogenic lead flows allows identification of those processes that emit lead. Based on the regional flows and depositions and on dispersion models, the acceptable flows and depositions for lead in water and soil can be calculated. *Acceptable* can be defined from a toxicology point of view (limiting value for lead content in water or soil) as well as a precautionary principle point of view (lead input into soil or water equals lead output). In any case, concentrations *and* flows of lead have to be taken into account. Also, potential accumulation of lead in downstream regions needs to be considered. Information about input flows and acceptable output flows of processes is useful in designing transfer coefficients that ensure regional environmental protection over long periods of time.

3. Monitoring based on materials accounting allows one to track accumulation or depletion as well as harmful flows of lead. Efficient monitoring points are as follows:

   a. The products *construction steel* and *filter residue of the smelter.* These two goods are routinely analyzed for production and quality control purposes. The results allow the determination of lead in shredded cars and indicate whether a change in landfilled lead is to be expected.

   b. Concentration of lead in gasoline. This figure is supplied by gasoline producers.

   c. Filter residues from MSW incineration. This information combined with known transfer coefficients allows calculation of lead flows in MSW.

   d. Sewage sludge. Routinely sampled and analyzed sewage sludge yields information about the sewage network as a potential source. This analysis is instrumental for identifying new emissions or for confirming that loadings to the sewer have been successfully eliminated.

   e. Surface waters. For water quality assessment, sampling surface water at the outflow of the region yields adequate information about the total load of the hydrosphere, especially if the same information is available from upstream regions. Monitoring of soil samples may be adequate initially to get an overview of lead in soils. However, as mentioned in Section 3.1.1, routine monitoring by soil sampling is expensive and inefficient, and it does not allow early recognition of harmful accumulations or depletions in soils.

3.5.2 Case Study 18: Accounting of Phosphorus as a Tool for Decision Making

This case study exemplifies how materials accounting can be performed on a routine basis, thereby increasing the power of MFA to understand complex systems and to detect fields of action for the optimization of a region's metabolism. If the MFA of a region is periodically repeated (e.g., yearly), a resource accounting scheme is obtained. Zoboli, Laner, Zessner, and Rechberger (2016) established a retrospective accounting scheme for the region Austria and the resource phosphorus (P) by compiling yearly P budgets from 1990 to 2011 to demonstrate the feasibility of such a scheme. Their work delivered several important findings:

First, workload and number of budgets (years) are not linearly correlated. Most of the time had to be used to establish the basic system and identify the data sources. Once this is accomplished, the budgets for adjacent years were produced comparably fast.

Second, even in a relatively short and economically stable period of 22 years, the national P budget of Austria, consisting of 122 flows and 8 stock change rates (Figure 3.67), has undergone unexpected significant and partially

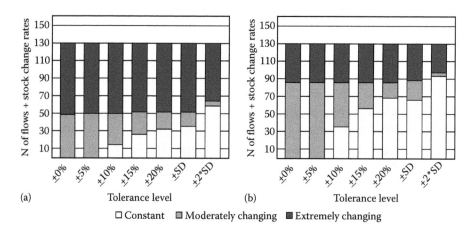

FIGURE 3.65

Degree of temporal change of 122 flows and 8 stock change rates: (a) categorization according to the change with respect to the reference year 1990; (b) categorization according to annual change. Results are shown for different tolerance levels (uncertainty thresholds used to determine whether temporal changes can actually be detected or not). The y-axis indicates the number of flows and stock change rates in each category. (From Zoboli, O. et al., Added values of time series in material flow analysis: The Austrian phosphorus budget from 1990 to 2011. *Journal of Industrial Ecology*, 2015. doi: 10.1111/jiec.1238.)

abrupt changes. This is illustrated in Figure 3.65, where the outcomes of the analysis of the degree of change of the budget with respect to the reference year 1990 are shown (Figure 3.65a). Here, flow changes are divided into three categories, namely, constant, moderate, and extreme. Constant means that a flow did not change since 1990 (rather unrealistic). Extreme means that a flow more than doubled or more than halved compared to 1990. Changes in between are considered as moderate. In order to compare uncertain flows of different years, different tolerance levels were applied (see Figure 3.65). For example, the P flow via import of mineral fertilizer to Austria in 1990 was 44.000 t/year ± 8%. In 2003, the same flow amounted to 32,000 t/year ±8%. Applying tolerance levels of ±0% to 15% and ±σ would classify the change as moderate, while ±20% and ±2σ would give no (significant) change as a result. Consequently, the results are partly sensitive to the applied tolerance levels. If tolerance levels between 0% and ±5% are applied, Figure 3.65 indicates that one-third of the flows and stock change rates changed moderately, and two-thirds were affected by an extreme variation, whereas with ranges from ±10% to ±20%, the fraction of moderately changing flows and stock rates gradually decreases until 15%. The specific standard deviation shows outcomes very similar to the ±20% range, whereas the level of twice the standard deviation decreases both the extreme and moderate fractions to 50% and 5%, respectively. In conclusion, the analysis reveals that half of the flows and stock change rates changed substantially, with certain flows that appeared or disappeared and others that at least doubled or halved their initial value.

The second component of this analysis (Figure 3.65b), instead explores to what extent the flows and stock change rates changed from a given year to the following one, to provide an overview of whether the changes took place gradually or rather abruptly. This analysis reasonably suggests that a large proportion of the flows were affected by gradual and moderate changes, but between 24% and 33% of the flows (depending on the considered tolerance level) recorded at least one extreme variation, indicating the noteworthy presence of substantial and sudden changes. The outcomes also highlight the difficulty of detecting smaller annual changes when uncertainty ranges are applied. However, the main conclusion from this analysis is that national anthropogenic material systems tend not to be stable over time; at least, for P this is the case. This means that classical 1-year MFA studies help to get a common understanding of a system's metabolism but have to be regularly updated for robust decision making. Zoboli's work shows that such updating is feasible. Additionally, the multiyear approach also improves the understanding of a system and helps making the model more comprehensive and more suitable to constitute the basis of materials accounting and monitoring.

The analysis of MFA time series directly leads to relevant actions in decision making. This is demonstrated in Figure 3.66 for phosphorus and Austria. In the upper-left diagram, one can see that the total P inputs into the Austrian waste management sector increased considerably since 1990.

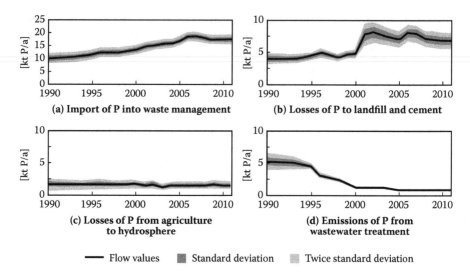

FIGURE 3.66
While the import of P into waste management increased constantly over the past years, (a) the losses of P to landfills and cement increased even more, (b) indicating a clear field for required action. The losses of P from agriculture to the hydrosphere remained rather constant, (c) indicating that efforts for optimized fertilizing and farming practice have been rather inefficient. Contrarily, (d) the emissions of P from wastewater treatment works could be reduced significantly, showing the effectiveness of technical solutions. (From Zoboli, H., Novel approaches to enhance regional nutrients management and monitoring applied to the Austrian phosphorous case study (PhD Thesis). Vienna: Technische Universität Wien, 2016.)

One of the major tasks of waste management is to collect materials; therefore, such a development can be regarded as positive, in any case showing the rising importance (and responsibility) of the sector. On the other hand, the upper-right time series of Figure 3.66 reveals that large amounts of the waste P are lost in landfills and in concrete. The latter is due to cocombustion of sewage sludge and meat and bonemeal (slaughter waste) in cement kilns. Comparing the two time series reveals that the ratio of losses versus input rather increased over the years, a clear negative trend that requires counteraction(s).

The other two time series of Figure 3.66 provide information on emissions of P to the hydrosphere. While emissions from point sources (here, wastewater treatment plants) could be substantially reduced, areal emissions, which stem from agricultural soils (diffuse, nonpoint sources), stayed rather constant and are now even becoming dominant. A general conclusion is that point sources are easier to control than areal emissions, a finding that has been made several times before, e.g., by Bergbäck (1992), for some heavy metals. The specific conclusion for P is that effective water protection has to put more emphasis on the agricultural sector. This is another hint for decision makers where action (adequate policy) is required.

Zoboli and colleagues (Zoboli, Zessner, and Rechberger, 2016) determined how and to what extent the management of P in Austria could be optimized. They used a detailed national model, obtained for the year 2013, as a reference system (Figure 3.67). Then they selected a range of measures of decision making aimed at reducing consumption, increasing recycling, and lowering emissions of P and discussed them with regard to applicability and

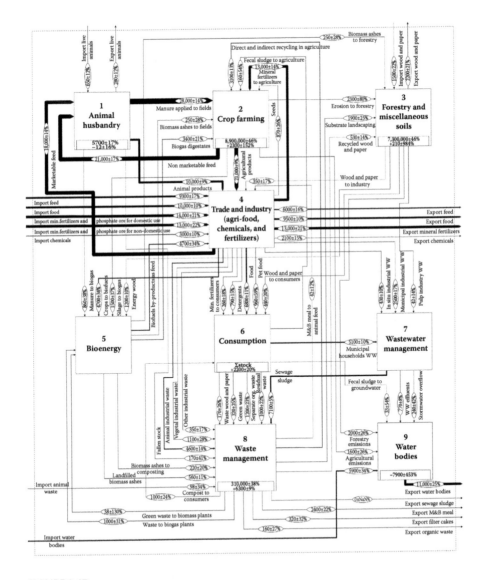

FIGURE 3.67
Austrian phosphorus balance for the reference year 2013 (unit: tP/year). (From Zoboli, O., Zessner, M., and Rechberger, H., *Science of the Total Environment.* 565, 313–323, 2016.)

TABLE 3.48

Relative Effect of the Fields of Action on the National P Management, Expressed through Three Indicators

Field of Action	Scope for Reduction of Import Dependency	Scope for Reduction of Mineral Fertilizer Consumption	Scope for Reduction of Emissions to Water Bodies	Main Data Gaps	Main Challenges
Increase of P recycling from meat and bonemeal	16%	23%	—	P concentration	Legal framework and market uncertainties for recovered fertilizers
Increase of P recycling from sewage sludge	23%	32%	—	Performance and product quality for new recovery technologies	Legal framework and market uncertainties for recovered fertilizers
Increase of P recycling from compost	11%	15%	—	Current use shares; P concentration	Regulation/coordination of sales in large number of composting plants
Increase of P recycling from digestates	—	—	—	Feedstock amounts and composition	Large number and heterogeneity of biogas plants
Increase of P recycling from biomass ashes	2%	3%	—	Current recycling rate; ash quality	Lack of economic incentives that offset logistical costs
Increase of P recycling from manure	—	—	—	Livestock excretion factors; use efficiency of manure as fertilizer	Enhancement of agricultural advice services
Improvement of municipal and industrial organic waste management	2%	3%	—	P concentration in MSW; current use of industry. by-products; food waste prevention potential	Resistance of households and similar establishments to further increase separate collection; increase of logistical effort and costs for the municipalities

(Continued)

TABLE 3.48 (CONTINUED)

Relative Effect of the Fields of Action on the National P Management, Expressed through Three Indicators

Field of Action	Scope for Reduction of Import Dependency	Scope for Reduction of Mineral Fertilizer Consumption	Scope for Reduction of Emissions to Water Bodies	Main Data Gaps	Main Challenges
Achievement of a balanced and healthy diet	20%	–	5–6%	Complexity of system feedbacks	Resistance to behavioral change; opposition of meat producers
Increase of the use efficiency in crop farming	8%	11%	–	Livestock excretion factors; P concentration in crops	Enhancement of agricultural advice services
Optimization of P content in feedstuff	20%	–	–	Current state of optimization; complexity of system feedbacks	Enhancement of agricultural advice services
Reduction of P use in detergents	4%	–	2%	–	–
Reduction of P use in other industrial processes	–	–	–	Materials flows in industrial applications	Substitutability of P
Reduction of surplus accumulation in private and public green areas	11%	15%	–	Home composting; sales of compost to privates	Resistance to behavioral change; coordination of large number of people
Reduction of point discharges	–	–	10%	Loads and perform. of in situ industrial treatment plants	Higher Fe levels in sewage sludge would pose a problem for several P recovery technologies
Reduction of erosion from agricultural soils	12%	17%	13%	Retention processes; long-term behavior of "legacy" P	Implementation at large scale; identification of hot spots
Indicator value in 2013	18,600 tP/y $2.2\ kgP\ cap^{-1}\ y^{-1}$	13,200 tP/y $1.6\ kgP\ cap^{-1}\ y^{-1}$	4600 tP/y $0.54\ kgP\ cap^{-1}\ y^{-1}$		

Source: Zoboli, O., Zessner, M., and Rechberger, H., *Science of the Total Environment*. 565, 313–323, 2016.

Note: Percentage values indicate the estimated improvement with respect to the reference year 2013.

limitations. The potential effect of each field of action on the reference system was quantified and compared using three indicators: import dependency, mineral fertilizer consumption, and emissions to water bodies. Table 3.48 presents the potential gain that can be achieved through each field of action, expressed as percentage of the indicators values in the reference year 2013.

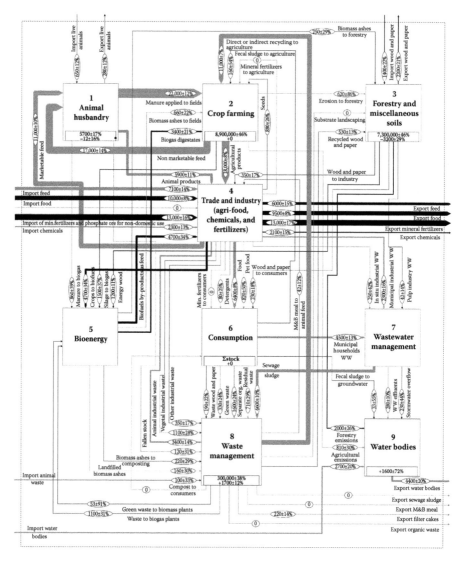

FIGURE 3.68
Optimized Austrian phosphorus balance based on the reference year 2013 (unit: tP/year). Objectives for the optimization are reduction of import dependency, consumption of mineral fertilizers, and emissions to water bodies. (From Zoboli, O., Zessner, M., and Rechberger, H., *Science of the Total Environment.* 565, 313–323, 2016.)

In a next step, all the gains that could be obtained through the measures (fields of action) were integrated in the reference system to generate an ideal target system (Figure 3.68). The fact that this is characterized by an extremely low import dependency of 0.23 kgP cap^{-1} y^{-1} (2.2 kgP cap^{-1} y^{-1} in 2013), zero consumption of mineral fertilizer for domestic use, and a 28% decline of emissions to water bodies indicates that governance in Austria offers a large scope for P stewardship.

The systemic approach of MFA in this study allowed quantification of the relative effect of each field of action on the national performance measured with different indicators, and thus performance of a proper comparative assessment. Further, it has made possible the generation and visualization of a target system, obtained through the integration of all potential gains in the reference model. The resulting concise though exhaustive overview can be very useful in supporting decision makers in designing national governance strategies and setting priorities, as well as in assisting domain experts in fitting their work into a broader context. As a next step, such studies need to be complemented with the analysis of the different costs involved in implementing each field of action—therefore another need and chance for interdisciplinary research.

PROBLEMS—SECTION 3.5

Problem 3.14:

Taking the lead example in Figure 3.1 as a starting point, establish an MFA for cadmium in the same region, assuming no major industrial application of cadmium. Use data given in this handbook, such as Table 3.29 and Figure 3.50, and data from the Internet on cadmium in soils, MSW, etc. Assume that MSW from 280,000 persons is incinerated in the region.

a. What are the major flows and stocks of cadmium in the region with "old" incineration and air pollution control technology (transfer coefficient to air = 0.10)?

b. How do these flows and stocks change if advanced air pollution control equipment is applied and the transfer coefficient is changed to 0.00001?

c. Evaluate environmental and resource implications arising from the two technologies in the region.

The solutions to the problems are given on the website http://www.MFA -handbook.info.

4

Outlook: Where to Go?

As indicated by the case studies presented in the previous chapters, the systematic investigation into material flows and stocks of anthropogenic systems allows a new view of the anthroposphere. Material flow analysis (MFA) and STAN are key tools for this new view: they allow linking of anthropogenic activities with resource consumption and environmental loadings. If the important flows and stocks of materials used by man are uniformly analyzed by MFA as described in this handbook, and if information of many individual MFAs is linked together to large databases on material sources, pathways, intermediate stocks, and final sinks, a more efficient use of resources will become possible. On one hand, MFA can be used to explore different cultural systems and economies with diverse metabolisms. On the other hand, MFA can be applied within one cultural system to analyze the various sectors of primary and secondary production, the service sector, consumers, and governments. By combining the two approaches, completely new schemes of utilizing resources become feasible. Also, the combination serves as a powerful tool for policy decision support in the fields of resource efficiency, urban planning, and environmental protection. This chapter points out some of the fascinating future potentials associated with MFA.

4.1 Vision of MFA

The development of MFA was driven by two ideas (Brunner, 2001): First, materials accounting becomes a standard analytical procedure for enterprises, regions, and nations. The goals in this case are to save resources by conserving materials and energy, and to produce less waste and minimize environmental loadings. The means to achieve these objectives are to analyze, evaluate, and control the flows and stocks of goods and substances within the anthroposphere. Like in financial accounting, stakeholders who account for materials will have a comparative advantage over those who do not use MFA and materials accounting. Hence, MFA is progressively applied by all actors of a modern society, and resource conservation and environmental protection are accomplished in an efficient way.

The second vision is that MFA will become a major design tool to optimize the metabolism of the anthroposphere (MoA). MFA will be instrumental to explain and understand the metabolic processes of complex cultural systems that have developed over time. This tool will support the development of new urban systems. It will assist decisions concerning the design of more efficient processes in view of energy and material consumption. It will be essential for the design of new goods and for the choice of new materials. It will be employed as a tool for early recognition of human impacts on the environment. Thus, MFA will be instrumental for the continuous improvement of anthropogenic systems, by supporting the choice and design of the set of processes and materials that are necessary to drive human activities.

How far are these visions from today's reality? Much methodological development has been undertaken, particularly in modeling flows and stocks and in the field of evaluation of MFA results. MFA is applied in diverse areas such as industrial ecology (IE), environmental management and protection, resource management, and waste management. It is a basis for life cycle assessments, ecobalancing, environmental impact statements, and waste management concepts. There are several groups around the world who apply MFA routinely for these tasks. Plenty of data are collected in studies ranging from polyvinyl chloride accounting to heavy metal inventories and nutrient balances of large watersheds.

By linking mining, production, consumption, and waste management, MFA has become instrumental for the transition from waste management to resource management. Recent MFA works point to waste management measures that are inefficient, and allow identifying those measures within the total economy that are more efficient and goal oriented. Thus, in the future, MFA will be even more important to reach a new, advanced level in waste management, and to define the limit of waste management beyond which measures in other economic sectors are more effective and powerful.

However, on the level of production, manufacturing, trade, and commerce, as well as the institutional level, MFA has not experienced a major breakthrough yet. It is not used as a standard analytical tool in everyday decisions on materials management. While the chemical industry has always practiced mass balancing of chemical reactants as a standard procedure, other branches so far do not see an advantage in using this tool. Few MFAs are being applied in other sectors. Case Study 15 proves that this tool is well suited for environmental and resource management on the corporate level, too. However, because the accounting of substances is a new, laborious, and costly task, so far, most companies do not see an economic advantage in applying MFA to their enterprise. Hence, there is still a long way to go until the various sectors of society come close to the visions presented previously.

At the beginning of the twenty-first century, it appears that the motivation to use technologies and systems that conserve resources is still comparatively small. Economic signals that indicate immediate resource scarcities do not exist. The warnings in *Limits to Growth* by the Club of Rome (Meadows, Meadows, Randers, and Behrens, 1972) seem to be falsified by reality: instead of running out of resources at the turn of the twentieth century, resources have never been as cheap. New studies even suggest "growth without limits" (Becker-Boost and Fiala, 2001). On the other hand, many MFA studies (cf. Chapter 3) point to severe deficiencies of the current industrial economy in terms of environmental impact and inefficient use of resources. MoA relies on fossil fuels that are causing climate change and will becoming scarce in the distant future; the specific energy demand is high. Many materials are dispersed after use as off-products; some of them are overstraining the carrying capacity of water, air, and soil. Detrimental impacts have been observed at the regional as well as the global level. The hypothesis, that the limits to growth appear first at the back end of the anthropogenic system, has not been falsified yet. On the contrary, problems like global accumulation and DDT and PCBs, ozone layer deterioration due to CFCs, and climate change (in part due to anthropogenic carbon emissions) all support this hypothesis. Natural reservoirs such as soils, lakes, and the atmosphere, which serve as sinks, have limited carrying capacities. They have to be regarded as finite natural resources that have to be conserved. Applying the material balance principle to the anthroposphere, it becomes clear that safe final sinks are needed for all materials used for industrial and consumption activities (Brunner and Kral, 2014). It is a major task of the coming decades to bring sources and sinks together; the materials exploited from the Earth's crust must match the capacity of the final sinks. A common concept is needed (1) to control material flows in a way that leads materials to appropriate and proven final sinks, (2) to monitor the grade of exploitation of the carrying capacity of final sinks, and (3) to gradually phase out those substances that have no such sink.

In the past as well as today, MoA is constantly improved and adapted to changing boundary conditions by many stakeholders of various disciplines. The community of research groups and organizations working for the increase in resource productivity, for environmental protection, for the decrease of ecological footprints, for sustainable development, and for the definition of progress is constantly growing. The result may be a significantly different use of resources, namely, energy, materials, and information. The latter is an important resource that has not been specifically addressed in this book yet. Information includes all knowledge of a society, be it in science, technology, culture and art, education, warfare, or administration. Without information, the resource *material* is a resource without value. Societies as we know them today depend highly on stocks and continuous expansion of information. The rise and fall of cultures such as the Roman Empire in Europe, the Egyptian dynasties in Africa, or the Mayas in the Americas was always accompanied by a tremendous development and loss in the resource *information*. Today, looked at from a global perspective, this resource stands on a very high level. The so-called global village embraces all cultures and thus is the greatest pool of information that ever existed on this globe. All our complex systems (supply of energy, water, food, construction materials, transport of goods, persons, information, etc.) are large assemblies of information and know-how and would not function without a comprehensive information base. MFA is an important means to identify, collect, and supply information about the "material world." MFA links the resource *information* with the resource *materials* (Schwab, Zoboli, and Rechberger, 2016), and is instrumental to better understand, design, and control materials within the anthroposphere. At present, it is often difficult to find good information and data, in particular about regional issues. It is necessary to develop a new set of tools for data acquisition about historical and present flows and stocks of materials, such as data reconstruction, back-channeling, and the use of redundancy to investigate regional systems (Eklund, 1995; Eklund, Bergbäck, and Lohm, 1995). In the future, MFA should become an important tool for the advancement of the resource *information*, too.

What can be done to foster a broad use of MFA and thus to come closer to the aforementioned objectives? A whole array of measures can promote future application of MFA. But first of all, a general awareness for the importance of materials as resources, as environmental loadings, and as constituents of final sinks must be created on all levels. Integration of MFA into the academic curriculum and research agenda is needed. This will promote broader application and further development of MFA. Another requirement is standardization of the method. An important step forward was the introduction of STAN. This user-friendly freeware supports the use of a common language of the MFA community. Also, linkage of MFA to other disciplines is necessary, e.g., the combined analysis and illustration of materials and financial flows known as economically extended MFA (EE-MFA) (Müller, 2015; Kytzia, 1998; Kytzia and Faist, 2002). This again will stimulate further methodological development. And finally, there is a need to incorporate MFA into selected parts of legislation related to environment, wastes, and resources.

4.2 Standardization

Standardization is seen as one requirement to establish MFA as an ordinary tool for materials management decisions. An ultimate goal could be the development of a common materials database, which is publicly available, that includes information about the most important anthropogenic processes and flows and stocks of goods and substances, and that is constantly updated with newly available information. If such a database is standardized, individual interpretative material flow accounting tools and their evaluation methods [ecological footprints, SPI, life cycle analysis (LCA), etc.] can be attached too.

Prerequisites for successful standardization are as follows:

1. National and international bodies define terms and definitions in a uniform way.

2. Students and professionals are taught and trained in standardized MFA methodology (terms and procedures). This will create a

new generation of professionals familiar with MFA. They comprise civil engineers, process engineers, architects, urban planners, landscape planers, environmental engineers, resource managers, waste management experts, product designers, and others. These experts will design the goods, buildings, and infrastructure of the future anthroposphere.

3. Tailor-made software (including statistical tools) such as STAN is needed to support the application of MFA and to facilitate modeling of metabolic systems. This will advance active application in academia; in consulting, engineering, and industry; and by authorities. The software should include interfaces for links to MFA databases and to evaluation systems (LCA, footprinting, and resource efficiency).

4. Professionally developed and maintained MFA databases are necessary. They include worldwide available information about specific generation rates and composition of materials, transfer coefficients of technologies, and other relevant data. It is indispensable that data are collected, processed, organized, and stored uniformly so that they can be easily linked to commercial MFA software and utilized to solve different challenges in environmental, resource, and waste management. These databases have to be made known and should be accessible for professionals as well as the public at affordable costs. Up to now, there do not yet exist concepts on how, where, and by whom such databases can be established.

A drawback of standardization might be that if a standard exists, further competition, development, and progress are less encouraged. Once a method is standardized, most experts will use the same methodology, the variety of approaches decreases, and users are less likely to have new and creative ideas. The history of science and technology as well as of fine arts and music has shown that standardization and rules are crucial for all disciplines but could never prevent talented people from further developing their field by extending or breaking rules and routines. It seems timely to standardize the application of MFA even if some undesirable effects cannot be ruled out.

4.3 MFA and Legislation

If legislation in the fields of environmental management, resource management, and waste management is based on MFA, decisions and measures are likely to be more cost-effective. Especially for goal-oriented legislation, MFA in combination with social science methods such as cost–benefit or cost-effectiveness analysis can serve as a powerful tool to analyze whether the objectives of regulations are met. Such combined approaches have been used in the past, e.g., to analyze quantitatively how far the objectives of waste management can be reached by several management scenarios (Brunner et al., 2001).

Banning hazardous substances by legislation requires MFA-based design and control. When the stocks and flows of a specific substance are known, the future effects of the ban can be predicted by scenario analysis. The modeled development for the substance can then be compared with measurements of the actual situation. In case of a relevant difference, decision makers have the possibility to interfere and correct again. For example, a stop in production of a substance may not involve instant relief for the environment because of stocks that serve as reservoirs for continuing emissions. If the stocks are not known, the effectiveness of actions may be overrated. One example for such a case are CFCs, where, for example, foams used in construction represent a stock that is responsible for ongoing CFC emissions despite the total ban on production (Obernosterer, 1994). Besides the prediction of the decline of substance flows and stocks, a complete MFA also helps to detect appropriate points for monitoring the trend of a banned substance. Monitoring points should allow for easy analysis of the relevant flows and guarantee that the information processed is significant. This will ensure a maximum in accuracy and cost-effectiveness of the monitoring process. This also applies to other legal regulations such as taxes on energy from fossil fuels, nonrenewable material resources (e.g., copper, gravel), or hazardous substances (e.g., cadmium).

Regional and national waste management plans can be improved when based on MFA. Today, these plans focus on the treatment and disposal of single waste streams that are defined by the way they are collected [e.g., separate collection of paper or industrial wastes, mixed municipal solid waste (MSW) collection] or by the functions of the goods that have become wastes (packaging function). In the future, the collection and management of wastes should focus more on the substances contained in individual wastes: it is more efficient to collect polyethylene wastes or cellulose wastes than to collect packaging plastics or packaging paper.

Also, investigations have shown that considerable amounts of both valuable and hazardous substances (e.g., zinc, copper, mercury, cadmium) are collected by waste management. As a result of rising waste streams (e.g., construction and demolition debris, wastes from electrical and electronic equipment), this potential will become even more important in the future. MFA supplies substance-based information for waste management decisions. This is important because the main objectives of waste management, namely, protection of human health and the environment as well as resource conservation, are substance-oriented and not good-oriented objectives. Waste management strategies should focus primarily on the level of substances, and only secondarily on goods. This facilitates the solving of problems such as finding appropriate recycling and treatment technologies and appropriate final sinks for residues. Information on substances is also needed to facilitate a "clean cycle" and "safe final sink" strategy (Kral, Kellner, and Brunner, 2013), preventing mixing of contaminated and clean wastes, and to decide which wastes and substances are to be collected separately.

Such MFA-based new developments are starting to take place at the political and administrative level. The following two examples from Germany and Austria are first encouraging signs on how MFA and materials accounting can become routine instruments for planning of effective waste management.

Since 1999, the German province North Rhine–Westphalia has required an MFA for permitting the operation of waste treatment facilities. According to an ordinance, the state grants licenses for the construction and operation of new waste treatment plants only if a complete MFA of the total plant is given. All input and output goods and concentrations of relevant substances in these goods have to be accounted for. Finally, a material balance with transfer coefficients for selected substances has to be established. In order to support the implementation of this ordinance, the authorities have prepared a database, containing waste compositions, emission data, and transfer coefficients for selected waste treatment processes (Alwast, Koepp, Thörner, and Marton, 2001). The purpose of this ordinance is to make sure that the same level of information is available for the decision maker about all processes that are taken into account in waste management. So far, information about state-of-the-art MSW incinerators has been much more abundant than on any other thermal, mechanical, and biochemical process, making it an uneven basis for decisions regarding the selection of technologies.

Another example of new advances on the administrative level is the 2001 edition of the Federal Waste Management Plan of Austria (Bundesministerium für Land- und Forstwirtschaft, Umwelt und Wasserwirtschaft, 2001). It contains several models of how waste management can be improved by the use of MFA and points out how to gradually implement MFA as a tool to optimize goal-oriented waste management. The reason for including MFA in the waste management plan is that MFA-based studies on packaging materials, on plastic waste management (Fehringer, 1997), and on scenario evaluation (Brunner et al., 2001) have had a considerable impact on the improvement of the Austrian packaging ordinance and on the design of new waste management concepts.

4.4 Industrial Ecology and Metabolism of the Anthroposphere

MFA protrudes in two fields of application: IE and MoA. While the emphasis of IE is on industrial systems and their linking, MoA takes an overall approach focusing on human activities and their satisfaction as a whole

(Baccini Brunner, 2012). For both fields, MFA has already become a widely applied tool. This will be even more pronounced in the future because of the many challenges that arise regarding resource management and environmental protection. Substances, goods, and processes necessary for sustaining human society are constantly changing due to innovation in technology and economy. At present and from a material point of view, this transformation happens more or less at random. Besides economic criteria, there are no clear objectives yet for the choice and use of materials and processes. If long-term efficient and ecological utilization of resources is set as a goal, MFA can serve as an important tool to support decisions regarding the reconstruction, design, and maintenance of the anthroposphere.

4.4.1 MFA and IE

IE is a rapidly growing field that systematically examines local, regional, and global materials and energy uses in industrial sectors and economies. It focuses on the potential role of industry in reducing environmental burdens throughout the product life cycle, from the extraction of raw materials to the production and use of goods, and to the management of the resulting wastes. Thus, many examples exist where IE experts have successfully applied MFA, e.g., for investigating global metal flows and stocks or for improving resource efficiency. The objectives and the area of application of MFA are quite congruent with IE. Both IE and MFA address various scales, ranging from local to global systems, from single products to entire economies. IE demands a systemic approach, the consideration of the total life cycle from cradle to the final sink, and farseeing thinking, which can be all realized by appropriate MFA system boundaries. Therefore, in many IE challenges, MFA will be part of the initial problem and system analysis. MFA is instrumental to assess the quality and completeness of available data and identify data requirements. It will strengthen the system understanding, and the MFA result will be the basis to determine spots of required action, which can be losses in industrial processes during beneficiation, manufacturing, and fabrication; accumulation or depletion of stocks; wanted, unwanted, or nonavoidable losses via diffuse or point emissions to the environment; and losses into landfills, dumps, and the like. A useful MFA variant in this context is materials accounting, which establishes MFA time series of the system of study. First of all, it deepens the understanding about the system. Second, it helps to identify data deficiencies such as gaps or systematic errors. The latter are likely to be observed if data reconciliation is applied to the time series of a system, showing that one or several streams are always reconciled in the same direction. Third, analyzing the trends of problematic flows provides information about the urgency of counteraction. In general, MFA helps to identify spots for successful intervention and also to rank different actions with respect to their effectiveness.

4.4.2 MFA as a Tool for the Design of Products and Processes

Anthropogenic systems for infrastructure, transportation, communication, etc., have lifetimes of 25 to 100 years. Hence, within this time range, new systems have to replace old ones. In this section, a three-step procedure is outlined on how MFA can be used in a comprehensive way to plan and design the anthroposphere of tomorrow. The result points out new ways of managing materials. The purpose of this section is not so much to give a preview of the future (which is not possible) but to stimulate readers to arrive at their own thoughts about the potential of MFA to create new ways of resource management.

First, MFA should become a standard tool for the design of products and processes. It can support industry, consumers, government, and public interest groups in their endeavor for better goods and services. MFA should be used for environmental impact assessments, life cycle assessments, and other tools for environmental management. Decisions about the choice of waste management scenarios, about nutrient management, and about management of metals and other resources should also be based on MFA. This will improve decisions by engineers, designers, communities, and authorities. The goal is to create a critical mass of users and applied case studies, and to show the advantages of using MFA for the design process. Also, an increasing number of users as well as communication between users will stimulate and promote further application of MFA in this and neighboring fields. Readily available, user-friendly, and well-supported software for MFA such as STAN will further advance the use of MFA.

4.4.3 MFA as a Tool to Design Systems

In a second step, knowledge and information gained by these individual applications of MFA is collected and evaluated by a systemic approach. Based on this new know-how, future design takes into consideration the whole system that is influenced by a good, process, or service. Material balances become a requirement for goal-oriented design and management of larger systems such as regional transport, supply, and disposal systems. Uniform census and processing of data facilitate the establishment of these larger and more complex systems. Materials accounting is introduced on a regional and national level. National material flow and stock accounting indicators are used for regular reporting, and the results are used in policy decision making. For selected substances, materials accounting is implemented also on the industrial level. Companies collect data and use them to optimize their performance. Public as well as private databases are established. All these developments assist the transition from a system that mainly relies on end-of-pipe measures to a resource-oriented materials management system. The control of material flows to the environment by end-of-pipe measures is then supplemented by a more comprehensive approach focusing on the optimization of the total anthropogenic resource flows.

4.4.4 MFA as a Tool to Design the Anthropogenic Metabolism

The third level is the design of the total anthropogenic metabolism based on MFA. Human activities such as to nourish, to transport, to reside, and to clean require materials and energy that are limited on one hand and are polluting the environment on the other hand. The objective is to supply appropriate material flows and stocks in view of the goals of resource conservation and long-term environmental protection. In a goal-oriented approach, MFA is instrumental for data collection, processing, and evaluation on a qualitatively high level and in a consistent way. The puzzle of the anthropogenic metabolism can be systematically put together step by step. Comparative metabolic studies are performed in an organized way, and comprehensive databases for complex systems and for many materials are set up.

On the third level, primary ore mining is replaced by secondary "urban mining" (Obernosterer and Brunner, 2001; Johansson, Krook, Eklund, and Berglund, 2013; Lederer, Laner, and Fellner, 2014). The amounts of some resources in the urban stocks are in the same order of magnitude as in natural deposits of reserves. Hence, material supply for urban systems has to be shifted from primary resources, which are extracted from natural deposits, to secondary resources, so-called urban ores, recovered from the urban stock itself. New methodologies need to be developed for how the urban stock can be used and maintained with low consumption of energy and primary materials. First approaches for restructuring urban systems have been developed by Baccini, Oswald, and colleagues (Baccini, Kytzia, and Oswald, 2002; Baccini and Oswald, 1999; Oswald and Baccini, 2003). In their studies, they link architectural and social structures and metabolic processes of urban regions and develop new strategies for resource-oriented cities. The result of their interdisciplinary combination of MFA with architecture, planning, and socioeconomic issues is a product called "network-city," a restructured urban region with a more sustainable metabolism.

The demand of urban systems for new virgin resources can be increasingly covered by renewable materials. Dissipation of materials while being processed, used, recycled, or disposed of is controlled and reduced to long-term acceptable levels. *Acceptable* refers to environmental and resources issues. Materials are used in a cyclic manner if resulting environmental loadings are smaller, energy requirements are low, no further substance needs arise, and economic benefits exist or can be created. For all losses from cycles as well as linear use, appropriate final sinks have to be found or designed. If they are not available, the losses have to be minimized further to meet the capacity of the available sinks, or the substances have to be phased out. Optimized materials management prevents long-term, harmful accumulation of hazardous substances in environmental compartments and depletion of finite resources within a few generations.

Many substances are not sold any longer; they are leased to manufacturers and end users, ensuring minimum dissipation due to the strong interest of the substance owner not to lose his/her stock. This will stimulate the development of new products with better resource performance (product life cycles, logistic systems). New objectives for the design process arise: minimize waste, emissions, and wear during use; enable easy dissembling for reuse of components or substances after consumption; facilitate refurbishment and reuse. This is achieved by standardization of materials, avoidance of complex composite materials, minimized employment of hazardous materials, reduction of the number of substances incorporated in a product, and use of mainly recyclable or abundant substances.

Actions are possible on all levels and scales: new products, new processes, and new systems and structures; regional, national, and international scale. The regional level is especially attractive because it is well suited to trying out new strategies and concepts to gain experience. Restructuring of the anthroposphere has an impact on lifestyle. Vice versa, lifestyle has a controlling effect on material flows and stocks. For future materials management, it will be crucial how consumer lifestyle develops and whether customers can be motivated to choose a resource-conserving lifestyle. According to Steindl-Rast and Lebell (2002), there are several ways to deal with goods: they can be taken in and given out in balance, leading to wealth and happiness, or they can be accumulated only, leading to hostile and deteriorating conditions. The old question of what is important—to be or to have—deserves new answers in view of resource management. New concepts are wanted that show how to improve the quality of life by consuming less energy and resources. Eventually, the focus is shifted from material growth to nonmaterial welfare and development. MFA is a necessary element for such a change in paradigm. Nevertheless, it is a tool and not a driving force. It is up to the user to apply MFA in a beneficial way.

References

Abrahamson, N. (2007). Aleatory variability and epistemic uncertainty. Retrieved from http://www.ce.memphis.edu/7137/PDFs/Abrahamson/C05.pdf

Ahbe, S., Braunschweig, A., and Müller-Wenk, R. (1990). Methodik für Ökobilanzen auf der Basis ökologischer Optimierung. *Schriftenreihe Umwelt No. 133*, Bern, Switzerland: BUWAL.

Allenby, B. (1999). Culture and industrial ecology. *Journal of Industrial Ecology, 3* (1), 2–4.

Allesch, A., and Brunner, P. H. (2015). Material flow analysis as a decision support tool for waste management: A literature review. *Journal of Industrial Ecology, 19* (5), 753–764.

Alonso, E., Sherman, A. M., Wallington, T. J., Everson, M. P., Field, F. R., Roth, R., and Kirchain, R. E. (2012). Evaluating rare Earth element availability: A case with revolutionary demand from clean technologies. *Environmental Science and Technology, 46* (6), 3406–3414.

Alwast, H., Koepp, M., Thörner, T., and Marton, C. (2001). Abfallverwertung in Industrieanlagen. *Berichte für die Umwelt, Bereich Abfall (7)*. Düsseldorf, Germany: Ministerium für Umwelt und Naturschutz, Landwirtschaft und Verbraucherschutz des Landes Nordrhein-Westfalen.

Amatayakul, W., and Ramnäs, O. (2001). Life cycle assessment of a catalytic converter for passenger cars. *Journal of Cleaner Production, 9* (5), 395–403.

Anderberg, S., Bergback, B., and Lohm, U. (1989). Flow and distribution of chromium in the Swedish environment: A new approach to studying environmental pollution. *Ambio, 18* (4), 216–220.

Andersen, J. K., Boldrin, A., Christensen, T. H., and Scheutz, V. (2010). Mass balances and life-cycle inventory for a garden waste windrow composting plant (Aarhus, Denmark). *Waste Management and Research, 28* (11), 1010–1020.

Andersson, K., Ohlsson, T., and Olsson, P. (1998). Screening life cycle assessment (LCA) of tomato ketchup: A case study. *Journal of Cleaner Production, 6* (3–4), 277–288.

Anonymous. (2000). Facts and figures for the chemical industry. *Chemical and Engineering News, 78* (26), 49.

Applegate, J. S. (2000). The precautionary preference: An American perspective on the precautionary principle. *Human and Ecological Risk Assessment (HERA), 6* (3), 413–443.

Atkins, P. W., and Beran, J. A. (1992). *General Chemistry*. New York: Scientific American Inc.

Auer, M. (2012). MonteCarlito 1.10. Retrieved from http://www.montecarlito.com/

Austrian Landfill Ordinance. (1996). DVO, BGBl, No. 164/96. Vienna: Austrian Ministry for Agriculture, Forestry, Water and Environment.

Austrian Paper Industry. (1996). Personal communication.

Austrian Standard. (2005). *Stoffflussanalyse—Teil 1: Anwendung in der Abfallwirtschaft— Begriffe*. Berlin, Germany: Beuth Verlag.

Ayres, R. U. (1994). Industrial metabolism: Theory and policy. In R. U. Ayres and U. E. Simonis (Eds.), *Industrial Metabolism: Restructuring for Sustainable Development*. Tokyo: United Nations University Press.

Ayres, R. U. (1995). Life cycle analysis: A critique. *Resources, Conservation and Recycling, 14* (3–4), 199–223.

Ayres, R. U. (1998). Eco-thermodynamics: Economics and the second law. *Ecological Economics, 26* (2), 189–209.

Ayres, R. U., and Ayres, L. W. (1999). Accounting for Resources. *Vol. 2. The Life Cycle of Materials*. Cheltenham, U.K.: Edward Elgar.

Ayres, R. U., and Martinás, K. (1994). *Waste Potential Entropy: The Ultimate Ecotoxic*. 94/05/EPS. Fontainbleau, France: Centre for the Management of Environmental Resources, INSEAD.

Ayres, R. U., and Nair, I. (1984). Thermodynamics and economics. *Physics Today, 37* (11), 62–71.

Ayres, R. U., and Simonis, U. E. (1994). *Industrial Metabolism: Restructuring for Sustainable Development*. Tokyo: United Nations University Press.

Ayres, R. U., Martinás, K., and Ayres, L. W. (1996). *Eco-thermodynamics: Exergy and Life Cycle Analysis*. Fontainebleau, France: Working Paper 96/04/EPS, INSEAD.

Ayres, R. U., Norberg-Bohm, V., Prince, J., Stigliani, W. M., and Yanowitz, J. (1989). *Industrial Metabolism, the Environment, and Application of Materials-Balance Principles for Selected Chemicals*. Laxenburg, Austria: International Institute for Applied System Analysis (IIASA).

Baccini, P. (1989). The landfill-reactor and final storage. *Lecture Notes in Earth Sciences* (Vol. 20). Berlin: Springer.

Baccini, P., and Bader, H. P. (1996). *Regionaler Stoffhaushalt—Erfassung, Bewertung, Steuerung*. Heidelberg, Germany+A77: Spektrum Akademischer Verlag GmbH.

Baccini, P., and Brunner, P. H. (1991). *Metabolism of the Anthroposphere (1st Edition)*. Heidelberg, Germany: Springer.

Baccini, P., and Brunner, P. H. (2012). *Metabolism of the Anthroposphere—Analysis, Evaluation, Design (2nd Edition)*. Cambridge, MA: The MIT Press.

Baccini, P., and Lichtensteiger, T. (1989). Conclusions and outlook. In P. Baccini (Ed.), *The Landfill, Reactor and Final Storage* (pp. 427–431). Berlin: Springer.

Baccini, P., and Oswald, F. (1998). *Netzstadt—Transdiziplinäre Methoden zum Umbau urbaner Systeme*. Zürich, Schweiz: Vdf Hochschulverlag AG an der ETH.

Baccini, P., Kytzia, S., and Oswald, F. (2002). Restructuring urban systems. In F. Moavenzadeh, K. Hanaki, and P. Baccini (Eds.), *Future Cities: Dynamics and Sustainability*. Dordrecht, The Netherlands: Kluwer Academic Publishers.

Baccini, P., von Steiger, B., and Piepke, G. (1988). Bodenbelastung durch Stoffflüsse aus der Anthroposphäre. In K. E. Brassel and M. C. Rotach (Eds.), *Die Nutzung des Bodens in der Schweiz*, Zürich, Schweiz: Zürcher Hochschulforum.

Baehr, H. D. (1989). *Thermodynamik (7th Edition)*. Berlin: Springer.

Barghoorn, M., Dobberstein, J., Eder, G., Fuchs, J., and Goessele, P. (1980). *Bundesweite Müllanalyse 1979–1980, Forschungsbericht 103 03 503*. Berlin, Germany: Technical University Berlin.

Barin, I. (1989). *Thermochemical Data of Pure Substances*. Weinheim, Germany: VCH.

BAS. (1987). *Statistisches Jahrbuch der Schweiz 1978/88*. Basel, Schweiz: Bundesamt für Statistik, Birkhäuser.

Bauer, G. (1995). *Die Stoffflußanalyse von Prozessen der Abfallwirtschaft unter Berücksichtigung der Unsicherheit*. (Dissertation), TU Wien, Wien.

Becker-Boost, E., and Fiala, E. (2001). *Wachstum ohne Grenzen—Globaler Wohlstand durch nachhaltiges Wirtschaften*. New York: Springer.

Beer, B. (1990). *Schlussbericht RESUB Luft*. Dübendorf, Switzerland: Eidgenössische Anstalt für Wasserversorgung, Abwasserreinigung und Gewässerschutz EAWAG, Abteilung Abfallwirtschaft und Stoffhaushalt.

Belevi, H. (1995). Dank Spurenstoffen ein besseres Prozessverständnis in der Kehrichtverbrennung. *EAWAG News, 40 D*, Dübendorf, Switzerland: EAWAG.

Belevi, H., and Moench, H. (2000). Factors determining the element behavior in municipal solid waste incinerators. 1. Field studies. *Environmental Science and Technology, 34* (12), 2501–2506.

Bergbäck, B. (1992). *Industrial Metabolism: The Emerging Landscape of Heavy Metal Emission in Sweden*. (Ph.D. thesis), Linköping University, Sweden.

Bergbäck, B., Johansson, K., and Mohlander, U. (2001). Urban metal flows—A case study of Stockholm. *Water, Air and Soil Pollution: Focus, 1* (3), 3–4.

Bertram, M., Graedel, T. E., Rechberger, H., and Spatari, S. (2002). The contemporary European copper cycle: Waste management subsystem. *Ecological Economics, 42* (1–2), 43–57.

Beven, K., and Young, P. (2013). A guide to good practice in modeling semantics for authors and referees. *Water Resources Research, 49* (8), 5092–5098. doi: 10.1002/wrcr.20393.

BMUJF. (1990). *Österreichisches Abfallwirtschaftsgesetz, BGBl, No. 325/1990*. Vienna, Austria: Federal Ministry for Environment, Youth, and Family.

BMUR. (1994). *Gesetz zur Förderung der Kreislaufwirtschaft und Sicherung der umweltverträglichen Beseitigung von Abfällen KrW-/AbfG—Kreislaufwirtschafts- und Abfallgesetz*. Bundesministerium für Umwelt und Reaktorsicherheit. Berlin, Germany: Juristisches Informationssystem für die Bundesrepublik Deutschland.

Boltzmann, L. (1923). *Vorlesungen über Gastheorie*. Leipzig: Barth.

Brand, G., Braunschweig, A., Scheidegger, A., and Schwank, O. (1998). Bewertung in Ökobilanzen mit der Methode der ökologischen Knappheit—Ökofaktoren 1997. *Schriftenreihe Umwelt No. 297*. Bern, Switzerland: BUWAL.

Braunschweig, A., and Müller-Wenk, R. (1993). *Ökobilanzen für Unternehmungen: Eine Wegleitung für die Praxis*. Bern, Switzerland: Haupt.

Brentrup, F., Küsters, J., Kuhlmann, H., and Lammel, J. (2001). Application of the life-cycle assessment methodology to agricultural production: An example of sugar beet production with different forms of nitrogen fertilisers. *European Journal of Agronomy, 14*, 221.

Bringezu, S. (1997). Material Flow Analyses for Sustainable Development of Regions. *Proceedings of the ConAccount Conference, 11–12 September 1997, Wuppertal Special 6*, 57–66.

Bringezu, S., Fischer-Kowalski, M., Kleijn, R., and Palm, V. (1997). Regional and national material flow accounting: From paradigm to practice of sustainability. *Proceedings of the ConAccount workshop 21–23 January, 1997, Wuppertal Special 4*, Leiden, The Netherlands: Wuppertal Institute for Climate, Environment, Energy.

Bringezu, S., Fischer-Kowalski, M., Klejin, R., and Palm, V. (1998). *The ConAccount Agenda: The concerted action on material flow analysis and its research and development agenda. Wuppertal Special, 8*. Wuppertal, Germany: Wuppertal Institute.

Bringezu, St., Schütz, H., and Moll, St. (2003). Rationale for and interpretation of economy-wide materials flow analysis and derived indicators. *Journal of Industrial Ecology, 7* (2), 43–64.

Brunner, P. H. (2001). Materials flow analysis: Vision and reality. *Journal of Industrial Ecology*, 5 (2), 3–5.

Brunner, P. H. (2010). Clean cycles and safe final sinks. *Waste Management and Research*, 28 (7), 575–576.

Brunner, P. H., and Baccini, P. (1981). Die Schwermetalle, Sorgenkinder der Entsorgung? *Neue Zürcher Zeitung (Beilage Forschung und Technik)*, 70, 65.

Brunner, P. H., and Baccini, P. (1992). Regional material management and environmental protection. *Waste Management and Research*, 10 (2), 203–212.

Brunner, P. H., and Bauer, G. (1994). Abfallwirtschaft und Stoffflußbewirtschaftung—Zieldefinition einer nachhaltigen Wirtschaft in Österreich *Forschungs-und Entwicklungsbedarf für den Übergang zu einer nachhaltigen Wirtschaftsweise in Österreich (Endbericht SUSTAIN)*. Graz: Institut für Verfahrenstechnik, TU-Graz.

Brunner, P. H., and Ernst, W. R. (1986). Alternative methods for the analysis of municipal solid waste. *Waste Management and Research*, 4 (2), 147–160.

Brunner, P. H., and Kral, U. (2014). Final sinks as key elements for building a sustainable recycling society. *Sustainable Environment Research*, 24 (6), 443–448.

Brunner, P. H., and Lampert, C. (1997). "Nährstoffe im Donauraum, Quellen und letzte Senken," ("The Flow of Nutrients in the Danube River Basin"). *EAWAG News*, 6 (43D+E+F), 15–17.

Brunner, P. H., and Mönch, H. (1986). The flux of metals through municipal solid waste incinerators. *Waste Management and Research*, 4 (1), 105–119.

Brunner, P. H., and Rechberger, H. (2001). Anthropogenic metabolism and environmental legacies. In T. Munn (Ed.), *Encyclopedia of Global Environmental Change* (Vol. 3, pp. 54–72). West Sussex, UK: John Wiley & Sons.

Brunner, P. H., and Rechberger, H. (2004). *Practical Handbook of Material Flow Analysis 1st Edition*. Boca Raton, FL: Lewis Publishers/CRC Press.

Brunner, P. H., and Stämpfli, D. M. (1993). Material balance of a construction waste sorting plant. *Waste Management and Research*, 11 (1), 27–48.

Brunner, P. H., Allesch, A., Färber, B., Getzner, M., Grüblinger, G., Huber-Humer, M. et al. (2016). *Benchmarking für die österreichische Abfallwirtschaft (Projekt Benchmarking)*. Bericht für das Bundesministerium für Land- und Forstwirtschaft, Umwelt und Wasserwirtschaft inkl. Ko-Finanzierungspartner. Retrieved from https://www .bmlfuw.gv.at/greentec/abfall-ressourcen/Benchmarking-Studie.html.A156

Brunner, P. H., Allesch, A., Getzner, M., Huber-Humer, M., Pomberger, R., Müller, W. et al. (2015). *Benchmarking für die österreichische Abfallwirtschaft - Benchmarking for the Austrian waste management*. Vienna, Austria: Technische Universität Wien, Institut für Wassergüte, Ressourcenmanagement und Abfallwirtschaft.

Brunner, P. H., Baccini, P., Deistler, M., Lahner, T., Lohm, U., Obernosterer, R., and van der Voet, E. (1998). *Materials Accounting as a Tool for Decision Making in Environmental Policy—Mac TEmPo Summary Report* (4th European Commission Programme for Environment and Climate, Research Area III, Economic and Social Aspects of the Environment, ENV-CT96-0230).

Brunner, P. H., Daxbeck, H., and Baccini, P. (1994). Industrial metabolism at the regional and local level. In R. U. Ayres and U. E. Simonis (Eds.), *Industrial Metabolism Restructuring for Sustainable Development*. Tokyo: UN University.

Brunner, P. H., Daxbeck, H., Hämmerli, H., Rist, A., Henseler, G., Gajcy, D. et al. (1990). *RESUB—Der regionale Stoffhaushalt im Unteren Bünztal, Die Entwicklung einer Methodik zur Erfassung des regionalen Stoffhaushaltes (Development of a methodology to assess regional material management*). Dübendorf, Switzerland: Eidgenössische Anstalt für Wasserversorgung, Abwasserreinigung und Gewässerschutz EAWAG, Abteilung Abfallwirtschaft und Stoffhaushalt, CH-8600.

Brunner, P. H., Döberl, G., Eder, M., Frühwirth, W., Huber, R., Hutterer, H., and Wöginger, H. (2001). Bewertung abfallwirtschaftlicher Maßnahmen mit dem Ziel der nachsorgefreien Deponie (Projekt BEWEND). *Monographien des UBA Band 149*, Vienna, Austria: Umweltbundesamt GmbH Wien.A130.

Brunner, P. H., Morf, L. S., and Rechberger, H. (2004). Thermal waste treatment— A necessary element for sustainable waste management. In I. Twardowska, H. E. Allen, A. A. F. Kettrup, and W. J. Lacy (Eds.), *Solid Waste: Assessment, Monitoring and Remediation 1st Edition* (pp. 783–806). Kidlington, Oxford: Elsevier Ltd.

Buchner, H. (2015). *Dynamic material flow modelling as a strategic resource management tool for Austrian aluminium flows.* (Doctoral thesis), Wien, Austria: Technische Universität Wien.

Buchner, H., Laner, D., Rechberger, H., and Fellner, J. (2014a). In-depth analysis of aluminum flows in Austria as a basis to increase resource efficiency. *Resources, Conservation and Recycling, 93* (0), 112–123.

Buchner, H., Laner, D., Rechberger, H., and Fellner, J. (2014b). Material flow analysis as basis for efficient resource management—The case of aluminium flows in Austria. *Metallurgical Research and Technology, 111* (6), 351–357.

Buchner, H., Laner, D., Rechberger, H., and Fellner, J. (2015a). Dynamic material flow modeling: An effort to calibrate and validate aluminum stocks and flows in Austria. *Environmental Science and Technology, 49* (9), 5546–5554.

Buchner, H., Laner, D., Rechberger, H., and Fellner, J. (2015b). Future raw material supply: Opportunities and limits of aluminium recycling in Austria. *Journal of Sustainable Metallurgy, 1* (4), 253–262.

Bundesamt für Umweltschutz. (1984). Ökobilanzen von Packstoffen, Zusammenfassender Übersichtsbericht. *Schriftenreihe Umweltschutz, No. 24.* Bern: Bundesamt für Umweltschutz.

Bundesministerium für Land- und Forstwirtschaft, Umwelt und Wasserwirtschaft. (2001). *Bundes-Abfallwirtschaftsplan—Bundesabfallbericht 2001*, Vienna, Austria: BMLFUW.A136.

Burgess, A. A., and Brennan, D. J. (2001). Application of life cycle assessment to chemical processes. *Chemical Engineering Science, 56* (8), 2589–2604.

BUWAL. (1984). *Abfallerhebung Schriftenreihe Umweltschutz 27.* Bern, Switzerland: Bundesamt für Umweltschutz.

Canter, L. W. (1996). *Environmental Impact Assessment.* New York: McGraw-Hill.

CEMBUREAU. (1999). *Alternative Fuels in Cement Manufacture: Technical and Environmental Review.* Brussels, Belgium: European Cement Association.

Cencic, O. (2004). Software for MFA. In P. H. Brunner and H. Rechberger (Eds.), *Practical Handbook of Material Flow Analysis 1st Edition* (pp. 80–132). Boca Raton, FL: Lewis Publishers/CRC Press.

Cencic, O. (2016). Nonlinear data reconciliation with software STAN. *Sustainable Environment Research.* [in press].

Cencic, O., and Frühwirth, R. (2015). A general framework for data reconciliation—Part I: Linear constraints. *Computers and Chemical Engineering, 75*, 196–208. doi: http://dx.doi.org/10.1016/j.compchemeng.2014.12.004

Cencic, O., and Rechberger, H. (2008). Material flow analysis with software STAN. *Environmental Engineering Management, 18* (1), 3–7.

Cencic, O., Rechberger, H., and Kovacs, A. (2006). STAN—Software for substance flow analysis (Version 2.5.1302) [Computer software]. Vienna: TU Wien. Retrieved from http://www.stan2web.net

Chen, P.-C., Crawford-Brown, D., Chang, C.-H., and Ma, H.-w. (2014). Identifying the drivers of environmental risk through a model integrating substance flow and input–output analysis. *Ecological Economics, 107,* 94–103.

Chen, W.-Q., and Graedel, T. E. (2012). Anthropogenic cycles of the elements: A critical review. *Environmental Science and Technology, 46* (16), 8574–8586.

Chertow, M. R. (2000). Industrial symbiosis: Literature and taxonomy. *Annual Review of Energy and the Environment, 25,* 313–337.

Clausius, R. (1856). Über verschiedene für die Anwendungen bequeme Formen der Hauptgleichungen der mechanischen Wärmetheorie. *Poggendorff's Annalenm,* (125), 353.

Cleveland, C. J., and Ruth, M. (1998). Indicators of dematerialization and the materials intensity of use. *Journal of Industrial Ecology, 2* (3), 15–50.

Commoner, B. (1997). The relation between industrial and ecological systems. *Journal of Cleaner Production, 5* (1–2), 125–129.

Connelly, L., and Koshland, C. P. (2001). Exergy and industrial ecology—Part 1: An exergy-based definition of consumption and a thermodynamic interpretation of ecosystem evolution. *Exergy, An International Journal, 1* (3), 146–165.

Consoli, F., Boustead, I., Fava, J., Franklin, W., Jensen, A., de Oude, N.,... Vignon, B. (1993). *Guidelines for life-cycle assessment: A "code of practice."* Brussels: SETAC.

Costa, M. M., Schaeffer, R., and Worrell, E. (2001). Exergy accounting of energy and materials flows in steel production systems. *Energy, 26* (4), 363–384.

Daly, H. E. (1993). Sustainable growth: An impossibility theorem. In H. E. Daly and K. N. Townsend (Eds.), *Valuing the Earth* (pp. 271). Cambridge, MA: MIT Press.

DeFur, P. L., and Kaszuba, M. (2002). Implementing the precautionary principle. *Science of the Total Environment, 288* (1–2), 155–165.

Desrochers, P. (2000). Market processes and the closing of 'industrial loops': A historical reappraisal. *Journal of Industrial Ecology, 4* (1), 29–43.

Dittrich, M., Bringezu, S., and Schütz, H. (2012). The physical dimension of international trade, Part 2: Indirect global resource flows between 1962 and 2005. *Ecological Economics, 79* (1), 32–43.

DKI Deutsches Kupferinstitut. (1997). *Kupfer, Vorkommen, Gewinnung, Eigenschaften, Verarbeitung, Verwendung Informationsdruck.* Duesseldorf: Deutsches Kupferinstitut.

Do, N. T., Trinh, D. A., and Nishida, K. (2014). Modification of uncertainty analysis in adapted material flow analysis: Case study of nitrogen flows in the Day-Nhue River Basin, Vietnam. *Resources, Conservation and Recycling, 88,* 67–75.

Döberl, G., Huber, R., Brunner, P. H., Eder, M., Pierrard, R., Schönbäck, W. et al. (2002). Long-term assessment of waste management options—A new, integrated and goal-oriented approach. *Waste Management and Research, 20* (4), 311–327.

Donald, J. R., and Pickles, C. A. (1996). A kinetic study of the reaction of zinc oxide with iron powder. *Metallurgical and Materials Transactions B, 27* (3), 363–374.

Du, X., and Graedel, T. E. (2011). Uncovering the global life cycles of the rare Earth elements. *Nature Scientific Reports, 1,* 145.

Dubois, D., Fargier, H., Ababou, M., and Guyonnet, D. (2014). A fuzzy constraint-based approach to data reconciliation in material flow analysis. *International Journal of General System, 43* (8), 1–23.

Dutta, P., and Ali, T. (2012). A hybrid method to deal with aleatory and epistemic uncertainty in risk assessment. *International Journal of Computer Applications, 42* (11), 37–43.

Duvigneaud, P., and Denayeyer-De Smet, S. (1975). L'Ecosystème Urbs, in L'Ecosystème Urbain Bruxellois. In P. Duvigneaud and P. Kestemont (Eds.), *Productivité Biologique en Belgique* (pp. 581–597). Bruxelles: Travaux de la Section Belge du Programme Biologique International.

EC. (1975). *Council Directive 75/442/EEC on waste (so-called EC-Waste Framework Directive).* Brussels, Belgium: Official Journal of the European Union.

EC. (2014). *Communication from the Commission to the European Parliament, the Council, the European Economic and Social Committee and the Committee of the Regions—Towards a circular economy: A zero waste programme for Europe.* Brussels: European Commission.

Eder, P., and Narodoslawsky, M. (1996). *Input–output based valuation of the compatibility of regional activities with environmental assimilation capacities.* Paper presented at the Inaugural Conference of the European Branch of the International Society for Ecological Economics, Paris.

Ehrenfeld, J. R. (1997). Industrial ecology: A framework for product and process design. *Journal of Cleaner Production, 5* (1–2), 87–95.

Eidgenössische Kommission für Abfallwirtschaft. (1986). Leitbild für die Schweizerische Abfallwirtschaft. *Schriftenreihe Umweltschutz No. 51.* Bern: Bundesamt für Umwelt, Wald und Landschaft.

Eisingerich, K. (2015). *MFA as a Decision Support Tool for Resource Management in Emerging Economies—The Case of Optimizing Straw Utilization on Small Farms* (PhD Dissertation). Vienna, Austria: Technische Universität Wien.

EKA. (1986). *Leitbild für die Schweizerische Abfallwirtschaft der Eidgenössischen Kommission für Abfallwirtschaft. Schriftenreihe Umweltschutz No. 51.* Bern, Switzerland: Bundesamt für Umwelt, Wald und Landschaft.

Eklund, M. (1995). *Reconstruction of historical metal emissions and their dispersion in the environment (Doctoral thesis).* Linköping, Sweden: Linköping Studies in Arts and Science.

Eklund, M., Bergbäck, B., and Lohm, U. (1995). Reconstruction of historical cadmium and lead emissions from a Swedish aluminium works, 1726–1840. *Science of the Total Environment, 170* (1–2), 21–30.

Erkman, S. (1997). Industrial ecology: A historical view. *Journal of Cleaner Production, 5* (1–2), 1–10.

Erkman, S. (2002). The recent history of industrial ecology. In R. U. Ayres and L. W. Ayres (Eds.), *A Handbook of Industrial Ecology.* Northampton, MA: Edward Elgar.

Ertesvåg, I. S., and Mielnik, M. (2000). Exergy analysis of the Norwegian society. *Energy, 25* (10), 957–973.

European Commission. (1985). *Directive on the Assessment of the Effects of Certain Public and Private Projects on the Environment, 85/337/EEC. Official Journal of the European Communities.* Brussels, Belgium.

European Commission. (2014). *Report on critical raw materials for the EU.* Brussels, Belgium: Ad hoc Working Group on defining critical raw materials, European Commission, DG Enterprise.

European Council. (2003). Directive 2002/96/EC of the European Parliament and of the Council of 27 January 2003 on waste electrical and electronic equipment (WEEE). *Official Journal of the European Union L, 37,* 24–38.

European Council. (2015). Commission Delegated Directive (EU) 2015/863 of 31 March 2015 amending Annex II to Directive 2011/65/EU of the European Parliament and of the Council as regards the list of restricted substances. *Official Journal of the European Union L, 137,* 10–12.

European Union. (1975). *Council Directive 75/442/EEC on Waste (so-called EC-Waste Framework Directive).*

European Union. (2000). Directive 2000/76/EC of the European Parliament and of the Council of 4 December 2000 on the Incineration of Waste. *Official Journal of the European Communities, Brussels.*

Eyerer, P. (Ed.) (1996). *Ganzheitliche Bilanzierung—Werkzeug zum Planen und Wirtschaften in Kreisläufen.* Berlin, Heidelberg: Springer Verlag.

Fehringer, R., and Brunner, P. H. (1996). *Kunststoffflüsse und die Möglichkeiten der Kunststoffverwertung in Österreich.* Vienna, Austria: Umweltbundesamt Wien GmbH.

Fehringer, R., and Brunner, P. H. (1997, Jan.). *Flows of Plastics and their Possible Reuse in Austria.* Paper presented at the ConAccount Workshop, Wuppertal Institute and Science Centre, Leiden, The Netherlands.

Fehringer, R., Rechberger, H., and Brunner, P. H. (1999). *Positivlisten in der Zementindustrie: Methoden und Ansätze.* Vienna, Austria: Vereinigung der österreichischen Zementwerke (VÖZ).

Fehringer, R., Rechberger, H., Pesonen, H.-L., and Brunner, P. H. (1997). *Auswirkungen unterschiedlicher Szenarien der thermischen Verwertung von Abfällen in Österreich (Project ASTRA).* Vienna, Austria: Institute for Water Quality, Resource and Waste Management, Technische Universität Wien.

Fellner, J., Lederer, J., Purgar, A., Winterstetter, A., Rechberger, H., Winter, F., and Laner, D. (2015). Evaluation of resource recovery from waste incineration residues— The case of zinc. *Waste Management, 37,* 95–103.

Ferson, S., and Ginzburg, L. R. (1996). Different methods are needed to propagate ignorance and variability. *Reliability Engineering and System Safety, 54* (2–3), 133–144. doi: http://dx.doi.org/10.1016/S0951-8320(96)00071-3

Fiedler, H., and Hutzinger, O. (1992). Sources and sinks of dioxins: Germany. *Chemosphere, 25* (7–10), 1487–1491.

Finnveden, G. (1999). Methodological aspects of life cycle assessment of integrated solid waste management systems. *Resources, Conservation and Recycling, 26* (3–4), 173–187.

Finnveden, G., and Ekvall, T. (1998). Life-cycle assessment as a decision-support tool—The case of recycling versus incineration of paper. *Resources, Conservation and Recycling, 24* (3–4), 235–256.

Finnveden, G., Björklund, A., Ekvall, T., and Mosberg, A. (2006). Models for Waste Management: Possibilities and Limitations. *Proceedings of the ISWA/DAKOFA Congress 1–5 October 2006.* Copenhagen, Denmark.

Fischer, T. (1999). *Zur Untersuchung verschiedener methodischer Ansätze zur Bestimmung entnommener mineralischer Baurohstoffmengen am Beispiel des Aufbaus von Wien (Diploma Thesis).* Technische Universität Wien.

Fischer-Kowalski, M., Haberl, H., Hüttler, W., Payer, H., Schandl, H., Winiwarter, V., and Zangerl-Weisz, H. (1997). *Gesellschaftlicher Stoffwechsel und Kolonisierung von Natur: Ein Versuch in Sozialer Ökologie.* Amsterdam: Gordon and Breach Fakultas.

Frosch, R. A., and Gallopoulos, N. E. (1989). Strategies for Manufacturing. *Scientific American, 261* (3), 144–152.

Furuholt, E. (1995). Life cycle assessment of gasoline and diesel. *Resources, Conservation and Recycling, 14* (3–4), 251–263.

Gangl, M., Gugele, B., Lichtblau, G., and Ritter, M. (2002). *Luftschadstoff-Trends in Österreich 1980–2000*. Vienna: Federal Environmental Agency Ltd.

Georgescu-Roegen, N. (1971). *The Entropy Law and the Economic Process*. Cambridge, MA: Harvard University Press.

Gevao, B., Ghadban, A. N., Uddin, S., Jaward, F. M., Bahloul, M., and Zafar, J. (2011). Polybrominated diphenyl ethers (PBDEs) in soils along a rural-urban-rural transect: Sources, concentration gradients, and profiles, *Environmental Pollution, 159* (12), 3666–3672.

Glöser, S., Soulier, M., and Tercero Espinoza, L. A. (2013). Dynamic analysis of global copper flows. Global stocks, postconsumer material flows, recycling indicators, and uncertainty evaluation. *Environmental Science and Technology, 47* (12), 6564–6572.

Goedkoop, M. (1995). *The Eco-Indicator 95*. Amersfoort, The Netherlands: Pré Consultants.

Gordon, R. B. (2002). Production residues in copper technological cycles. *Resources, Conservation and Recycling, 36* (2), 87–106.

Gorter, J. (2000). Zinc balance for the Netherlands. *Material Flow Accounting: Experience of Statistical Institutes in Europe* (p. 205). Brussels: Statistical Office of the European Communities.

Gottschalk, F., Sonderer, T., Scholz, R. W., and Nowack, B. (2010). Possibilities and limitations of modeling environmental exposure to engineered nanomaterials by probabilistic material flow analysis. *Environmental Toxicology and Chemistry, 29* (5), 1036–1048.

Gove, P. B. (1972). *Webster's Seventh New Collegiate Dictionary*. Springfield, MS: G&C Merriam.

Graedel, T. E. (1998). *Streamlined Life-Cycle Assessment*. Englewood Cliffs, NJ: Prentice Hall.

Graedel, T. E. (2002). The contemporary European copper cycle: Introduction. *Ecological Economics, 42* (1–2), 5–7.

Graedel, T. E. (2004). Characterizing the cycles of metals. *Minerals and Energy, 19*, 271, A315.

Graedel, T. E., and Allenby, B. R. (2002). *Industrial Ecology 2nd Edition*. Upper Saddle River, NJ: Prentice Hall.

Graedel, T. E., Barr, R., Chandler, C., Chase, T., Choi, J., Christoffersen, L., ... and Zhu, C. (2011). Methodology of metal criticality determination. *Environmental Science and Technology, 46* (2), 1063–1070.

Graedel, T. E., Bertram, M., Fuse, K., Gordon, R. B., Lifset, R., Rechberger, H., and Spatari, S. (2002). The contemporary European copper cycle: The characterization of technological copper cycles. *Ecological Economics, 42* (1–2), 9–26.

Graedel, T. E., van Beers, D., Bertram, M., Fuse, K., Gordon, R. B., Gritsinin, A. et al. (2004). Multilevel cycle of anthropogenic copper. *Environmental Science and Technology, 38* (4), 1242–1252.

Greenberg, R. R., Zoller, W. H., and Gordon, G. E. (1978). Composition and size distributions of particles released in refuse incineration. *Environmental Science and Technology, 12* (5), 566–573.

Guinée, J. B. , Gorrée, M., Heijungs, R., Huppes, G., Kleijn, R., de Koning, A. et al. (2001). *Life Cycle Assessment: An Operational Guide to the ISO Standards.* The Hague, The Netherlands: Ministry of Housing, Spatial Planning and the Environment.

Guralnik, D. B., and Friend, J. H. (1968). *Webster's New World Dictionary of the American Language College Edition.* New York: World Publishing.

Güttinger, H., and Stumm, W. (1990). Ökotoxikologie am Beispiel der Rheinverschmutzung durch den Chemie-Unf all bei Sandoz in Basel. *Naturwissenschaften, 77* (6), 253–261.

Haas, G., Wetterich, F., and Köpke, U. (2001). Comparing intensive, extensified and organic grassland farming in southern Germany by process life cycle assessment. *Agriculture, Ecosystems and Environment, 83* (1–2), 43–53.

Habersatter, K. (1991). Ökobilanzen von Packstoffen—Stand 1990. *Schriftenreihe Umwelt No. 132.* Bern, Switzerland: BUWAL.

Habib, K., Schibye, P. K., Vestbø, A. P., Dall, O., and Wenzel, H. (2014). Material flow analysis of NdFeB magnets for Denmark: A comprehensive waste flow sampling and analysis approach. *Environmental Science and Technology, 48* (20), 12229–12237.

Hackl, A., and Mauschitz, G. (1997). *Emissionen aus Anlagen der österreichischen Zementindustrie II.* Vienna: Zement + Beton Handels- und Werbeges.m.b.H.

Haigh, N. (1994). The introduction of the precautionary principle into the UK. In T. O'Riordan and J. Cameron (Eds.), *Interpreting the Precautionary Principle* (Chap. 13). London: Earthscan Publications.

Hammond, A., Adriaanse, A., Rodenburg, E., Bryant, D., and Woodward, R. (1995). *Environmental Indicators: A Systematic Approach to Measuring and Reporting on Environmental Policy Performance in the Context of Sustainable Development.*

Hanley, N., and Spash, C. L. (1995). *Cost–Benefit Analysis and the Environment.* Brookfield, VT: Edward Elgar.

Hashimoto, S., Tanikawa, H., and Moriguchi, Y. (2007). Where will the large amounts of materials accumulated within the economy go? A material flow analysis of construction materials. *Waste Management, 27,* 1725–1738.

Hatayama, H., Daigo, I., Matsuno, Y., and Adachi, Y. (2012). Evolution of aluminum recycling initiated by the introduction of next-generation vehicles and scrap sorting technology. *Resources, Conservation and Recycling, 66,* 8–14.

Hedbrant, J., and Sörme, L. (2000). Data vagueness and uncertainties in urban heavy metal data collection. *Water, Air, and Soil Pollution,* Focus 1, 43–53.

Heijungs, R., Guinée, J. B., Huppes, G., Lankreijer, R. M., Udo de Haes, H. A., and Wegener-Sleeswijk, A. (1992). *Environmental Life-Cycle Assessment of Products: Guide and Backgrounds.* Leiden, The Netherlands: Centre of Environmental Science.

Heinrich, W. (1988). Carcinogenicity of cadmium—Overview of experimental and epidemiological results and their influence on recommendations for maximum concentrations in the occupational area. In M. Stoeppler and M. Piscator (Eds.), *Cadmium.* Berlin, Germany: Springer.

Hellweg, S. (2000). *Time- and Site-Dependent Life-Cycle Assessment of Thermal Waste Treatment Processes.* (Ph.D. thesis), Swiss Federal Institute of Technology, Zurich.

Hendrickson, C., Horvath, A., Joshi, S., and Lave, L. (1998). Economic input–output models for environmental life-cycle assessment. *Environmental Science and Technology, 32* (7), 184A–191A.

Henseler, G., Scheidegger, R., and Brunner, P. H. (1992). Determination of material flux through the hydrosphere of a region. *Vom Wasser, 78,* 91–116.

Hertwich, E. G., Pease, W. S., and Koshland, C. P. (1997). Evaluating the environmental impact of products and production processes: A comparison of six methods. *Science of the Total Environment, 196* (1), 13–29.

Heyde, M., and Kremer, M. (1999). *Recycling and Recovery of Plastics from Packagings in Domestic Waste.* Bayreuth, Germany: Eco-Informa Press.

Hileman, B. (1999). Prescription for a global biotechnology dialogue. *Chemical and Engineering News, 77* (29), 42.

Hinterberger, F., Luks, F., and Schmidt-Bleek, F. (1997). Material flows vs. 'natural capital': What makes an economy sustainable? *Ecological Economics, 23* (1), 1–14.

Höglmeier, K., Steubing, B., Weber-Blaschke, G., and Richter, K. (2015). LCA-based optimization of wood utilization under special consideration of a cascading use of wood. *Journal of Environmental Management, 152* (0), 158–170.

Holleman, A. F., and Wiberg, E. (1995). *Lehrbuch der Anorganischen Chemie (101st edition).* New York: Walter de Gruyter.

Huijbregts, M. A. J. (2000). Priority assessment of toxic substances in the frame of LCA: Time horizon dependency of toxicity potentials calculated with the multimedia fate, exposure and effects model USES-LCA. Retrieved from http://www.leidenuniv.nl/interfac/cml/lca2/

Huntzicker, J. J., Friedlander, S. K., and Davidson, C. I. (1975). Material balance for automobile emitted lead in Los Angeles basin. *Environmental Science and Technology, 9* (5), 448–457.

Hutterer, H., Pilz, H., Angst, G., and Musial-Mencik, M. (2000). Stoffliche Verwertung von Nichtverpackungs-Kunststoffabfällen, Kosten-Nutzen-Analyse von Massnahmen auf dem Weg zur Realisierung einer umfassenden Stoffbewirtschaftung von Kunststoffabfällen. *Monographien 124.* Vienna: Austrian Federal Environmental Agency.

Ivanova, D., Stadler, K., Steen-Olsen, K., Wood, R., Vita, G., Tukker, A., and Hertwich, E. G. (2016). Environmental impact assessment of household consumption, *Journal of Industrial Ecology, 20* (3), 526–536.

Jarupisitthorn, C., Pimtong, T., and Lothongkum, G. (2003). Investigation of kinetics of zinc leaching from electric arc furnace dust by sodium hydroxide. *Materials Chemistry and Physics, 77* (2), 531–535.

Jasinski, S. M. (1995). The materials flow of mercury in the United States. *Resources, Conservation and Recycling, 15* (3–4), 145–179.

Jelinski, L. W., Graedel, T. E., Laudise, R. A., McCall, D. W., and Patel, C. K. N. (1992). Industrial ecology: Concepts and approaches. *Proceedings of the National Academy of Sciences of the United States of America, 89* (3), 793–797.

Johansson, N., Krook, J., Eklund, M., and Berglund, B. (2013). An integrated review of concepts and initiatives for mining the technosphere: Towards a new taxonomy. *Journal of Cleaner Production, 55*, 35–44.

Johansson, P.-O. (1993). *Cost-Benefit Analysis of Environmental Change.* New York: Cambridge University Press.

Johnston, L. P. M., and Kramer, M. A. (1995). Maximum likelihood data rectification: Steady-state systems. *AIChE Journal, 41* (11), 2415–2426. doi: 10.1002/aic .690411108.

Kampel, E. (2002). *Packaging Glass and Chlorine Fluxes of Australia, Austria, and Switzerland and Their Assessment with Different Software Tools (AUDIT and POWERSIM).* (Master thesis), Technische Universität Wien, Wien.

Kelly, T. D., and Matos, G. R. (2014). Historical statistics for mineral and material commodities in the United States (2016 version): *U.S. Geological Survey Data Series 140.* Retrieved from http://minerals.usgs.gov/minerals/pubs/historical-statistics/

Kesler, S. (1994). *Mineral Resources, Economics, and the Environment.* New York: Macmillan.

Kleijn, R., Huele, R., and Van Der Voet, E. (2000). Dynamic substance flow analysis: The delaying mechanism of stocks, with the case of PVC in Sweden. *Ecological Economics, 32* (2), 241–254.

Kleijn, R., Tukker, A., and Van Der Voet, E. (1997). Chlorine in the Netherlands, Part I, an overview. *Journal of Industrial Ecology, 1* (1), 95–116.

Kleijn, R., Van Der Voet, E., and Udo de Haes, H. A. (1994). Controlling substance flows: The case of chlorine. *Environmental Management, 18* (4), 523–542.

Klinglmair, M., Zoboli, O., Laner, D., Rechberger, H., Astrup, T. F., and Scheutz, C. (2016). The effect of data structure and model choices on MFA results: A comparison of phosphorus balances for Denmark and Austria. *Resources, Conservation and Recycling, 109,* 166–175.

König, A. (1997). *The Urban Metabolism of Hong Kong: Rates, Trends, Limits and Environmental Impacts.* Paper presented at the POLMET (Pollution in the Metropolitan and Urban Environment) Conference, Hong Kong.

König, A. (2002). Personal communication: In an introduction to a MFA seminar at the University of Hong Kong, König calls MFA the acronym for "Master of Fine Arts."

Kral, U., Brunner, P. H., Chen, P.-C., and Chen, S.-R. (2014). Sinks as limited resources? A new indicator for evaluating anthropogenic material flows. *Ecological Indicators, 46,* 596–609.

Kral, U., Kellner, K., and Brunner, P. H. (2013). Sustainable resource use requires "clean cycles" and safe "final sinks." *Science of the Total Environment, 461–462* (100), 819–822.

Krauskopf, B. (1967). *Introduction to Geochemistry.* New York: McGraw-Hill.

Krauskopf, K. B. (1979). *Introduction to Geochemistry (2nd edition).* New York: McGraw-Hill.

Krolczyk, J. B., Krolczyk, G. M., Legutko, S., Napiorkowski, J., Hloch, S., Foltys, J., and Tama, E. (2015). Material flow optimization—A case study in automotive industry. *Technical Gazette, 22* (6), 1447–1456.

Krotscheck, C., and Narodoslawsky, M. (1996). The Sustainable Process Index. A new dimension in ecological evaluation. *Ecological Engineering, 6* (4), 241–258.

Krotscheck, C., König, F., and Obernberger, I. (2000). Ecological assessment of integrated bioenergy systems using the Sustainable Process Index. *Biomass and Bioenergy, 18* (4), 341–368.

Krozer, J. (1996). *Operational Indicators for Progress towards Sustainability, EU project final report, No. EV-5V-CT94-0374.*

Krozer, J., and Vis, J. C. (1998). How to get LCA in the right direction? *Journal of Cleaner Production, 6* (1), 53–61.

Kytzia, S. (1998). Wie kann man Stoffhaushaltssysteme mit ökonomischen Daten verknüpfen? In T. Lichtensteiger (Ed.), *Ressourcen im Bauwesen; Aspekte einer nachhaltigen Ressourcenbewirtschaftung im Bauwesen* (pp. 69–80). Zürich: Vdf Hochschulverlag an der ETH Zürich.

Kytzia, S., and Faist, M. (2002). *Joining economic and engineering perspectives in an integrated assessment of economic systems.* Paper presented at the International Conference on Input-Output Techniques, Montreal.

Kuriki, S., Daigo, I., Matsuno, Y., and Adachi, Y. (2010). Recycling potential of platinum Group Metals in Japan. *Journal of the Japan Institute of Metals and Materials, 74* (12), 801–805.

Lahner, T. (1994). Steine und Erden—Für die ökologische Bewertung des Bauwesens ist eine Stoff- und Güterbilanz notwendig. *Müll Magazin, 7* (1), 9–10.

Landfield, A. H., and Karra, V. (2000). Life cycle assessment of a rock crusher. *Resources, Conservation and Recycling, 28* (3–4), 207–217.

Landner, L., and Lindeström, L. (1998). *Zinc in society and in the environment*. Kil, Sweden: Swedish Environmental Research Group (MFG), serg@mfg.se.

Landner, L., and Lindeström, L. (1999). *Copper in Society and in the Environment (2nd Edition)*. Vaesteras, Sweden: Swedish Environmental Research Group.

Laner, D., and Rechberger, H. (2007). Treatment of cooling appliances: Interrelations between environmental protection, resource conservation, and recovery rates. *Resources, Conservation and Recycling, 52* (1), 136–155.

Laner, D., and Rechberger, H. (2016). Material flow analysis. In M. Finkbeiner (Ed.), *Special Types of Life Cycle Assessment*. Dordrecht: Springer.

Laner, D., Rechberger, H., and Astrup, T. (2014). Systematic evaluation of uncertainty in material flow analysis. *Journal of Industrial Ecology, 18* (6), 859–870. doi: 10.1111/jiec.12143.

Lankey, R. L., Davidson, C. I., and McMichael, F. C. (1998). Mass balance for lead in the California South Coast Air Basin: An update. *Environmental Research, 78* (2), 86–93.

Lantzy, R. J., and Mackenzie, F. T. (1979). Atmospheric trace metals: Global cycles and assessment of manis impact. *Geochimica et Cosmochimica Acta, 43*, 511.

Larsen, T. A., Peters, I., Alder, A., Eggen, R., Maurer, M., and Muncke, J. (2001). Re-engineering the toilet for sustainable wastewater management. *Environmental Science and Technology, 35* (9), 192A–197A.

Lassen, C., and Løkke, S. (1999). *Brominated Flame Retardants—Substance Flow Analysis and Assessment of Alternatives*. Copenhagen, Denmark: The Danish Environmental Protection Agency.

Lederer, J., and Rechberger, H. (2010). Comparative goal-oriented assessment of conventional and alternative sewage sludge treatment options. *Waste Management, 30* (6): 1043–1056.

Lederer, J., Laner, D., and Fellner, J. (2014). A framework for the evaluation of anthropogenic resources: The case study of phosphorus stocks in Austria. *Journal of Cleaner Production, 84*, 368–381.

Legarth, J. B., Alting, L., Danzer, B., Tartler, D., Brodersen, K., Scheller, H., and Feldmann, K. F. (1995). A new strategy in the recycling of printed circuit boards. *Circuit World, 21* (3), 10–15.

Leisewitz, A., and Schwarz, W. (2000). Erarbeitung von Bewertungsgrundlagen zur Substitution umweltrelevanter Flammschutzmittel Band II: Flammhemmende Ausrüstung ausgewählter Produkte—anwendungsbezogene Betrachtung: Stand der Technik, Trend, Alternativen. Dessau, Germany: Umweltbundesamt.

Lentner, C. (1981). *Geigy Scientific Tables (8th Edition*, Vol. 1). Basle, Switzerland: Ciba-Geigy Ltd.

Lenzen, M., and Munksgaard, J. (2002). Energy and CO_2 life-cycle analyses of wind turbines-review and applications. *Renewable Energy, 26* (3), 339–362.

Leontief, W. (1977). *The Structure of the American Economy, 1919–1929*. White Plains, NY: repr. International Arts and Sciences Press (now M.E. Sharpe).

Leontief, W. W. (1966). *Input–Output Economics*. New York: Oxford University Press.

Lifset, R., and Graedel T. E. (2002). Industrial ecology: Goals and definition. In R. U. Ayres and L. W. Ayres (Eds.), *A Handbook of Industrial Ecology*. Northampton, MA: Edward Elgar.

Liu, G., Bangs, C. E., and Müller, D. B. (2012). Stock dynamics and emission pathways of the global aluminium cycle. *Nature Climate Change, 3* (4), 338–342.

Llewellyn, T. O. (1994). *Cadmium Materials Flow, Information Circular 9380*.

Lohm, U., Anderberg, S., and Bergbäck, B. (1994). Industrial metabolism at the national level: A case-study on chromium and lead pollution in Sweden, 1880–1980. In R. U. Ayres and U. E. Simonis (Eds.), *Industrial Metabolism—Restructuring for Sustainable Development*. Tokyo: The United Nations University.

Løvik, A. N., Modaresi, R., and Müller, D. B. (2014). Long-term strategies for increased recycling of automotive aluminum and its alloying elements. *Environmental Science and Technology, 48* (8), 4257–4265.

Luo, Y., Luo, X. J., Lin, Z., Chen, S. J., Liu, J., Mai, B. X., and Yang, Z. Y. (2009). Polybrominated diphenyl ethers in road and farmland soils from an e-waste recycling region in southern China: Concentrations, source profiles, and potential dispersion and deposition. *Science of the Total Environment, 407* (3), 1105–1113.

Madron, F. (1992). *Process Plant Performance: Measurement and Data Processing for Optimization and Retrofits*. Chichester, West Sussex, England: Ellis Horwood Limited Co.

Major, R. H. (1938). Santorio Santorio. *Annals of Medical History, 10*, 369.

Månsson, N., Bergbäck, B., Hjortenkrans, D., Jamtrot, A., and Sörme, L. (2009). Utility of substance stock and flow studies—The Stockholm example. *Journal of Industrial Ecology, 13* (5), 674–686.

Martin, P. H. (1997). If you don't know how to fix it, please stop breaking it! The precautionary principle and climate change. *Foundations of Science, 2* (2), 263–292.

Matthews, E., Amann, C., Bringezu, S., Fischer-Kowalski, M., Huttler, W., Kleijn, R. et al. (2000). *The Weight of Nations, Material Outflows from Industrial Economies*. Washington, DC: World Resources Institute.

Maystre, L. Y., and Viret, F. (1995). A goal-oriented characterization of urban waste. *Waste Management and Research, 13* (3), 207–218.

Meadows, D. H., Meadows, D. L., Randers, J., and Behrens III, W. W. (1972). *The Limits to Growth*. London, GB: Earth Island.

Mee, L. D. (1992). The Black Sea in crisis: A need for concerted international action. *Ambio, 21* (4), 278–286.

Melo, M. T. (1999). Statistical analysis of metal scrap generation: The case of aluminium in Germany. *Resources, Conservation and Recycling, 26* (2), 91–113.

Metallgesellschaft Aktiengesellschaft. (1993). *Metallstatistik (Metal Statistics)*. Frankfurt am Main, Germany: Metallgesellschaft Aktiengesellschaft.

Michaelis, P., and Jackson, T. (2000a). Material and energy flow through the UK iron and steel sector. Part 1: 1954–1994. *Resources, Conservation and Recycling, 29* (1–2), 131–156.

Michaelis, P., and Jackson, T. (2000b). Material and energy flow through the UK iron and steel sector. Part 2: 1994–2019. *Resources, Conservation and Recycling, 29* (3), 209–230.

Miller, R., and Blair, P. (1985). *Input-output analysis: Foundations and extentions*. Englewood Cliffs, NJ: Prentice-Hall.

Modaresi, R., and Müller, D. B. (2012). The role of automobiles for the future of aluminum recycling. *Environmental Science and Technology, 46* (16), 8587–8594.

Morf, L. S., and Brunner, P. H. (1998). The MSW incinerator as a monitoring tool for waste management. *Environmental Science and Technology, 32* (12), 1825–1831.

Morf, L. S., Brunner, P. H., and Spaun, S. (2000). Effect of operating conditions and input variations on the partitioning of metals in a municipal solid waste incinerator. *Waste Management and Research, 18* (1), 4–15.

Morf, L., Buser, A., Taverna, R., Bader, H.-P., and Scheidegger, R. (2008). Dynamic substance flow analysis as a valuable risk evaluation tool—A case study for brominated flame retardants as an example of potential endocrine disrupters. *CHIMIA, 62* (5), 424–431.

Morf, L. S., Ritter, E., and Brunner, P. H. (1997). *Güter- und Stoffbilanz der MVA Wels:* Institut für Wassergüte und Abfallwirtschaft, TU Wien.

Morf, Ritter, and Brunner, 1997.

Morf, L. S., Taverna, R., Daxbeck, H., and Smutny, R. (2003). *Selected polybrominated flame retardants, PBDEs and TBBPA, substance flow analysis.* Bern, Switzerland: Swiss Federal Office for the Environment.

Morf, L., Tremp, J., Gloor, R., Huber, Y., Stengele, M., and Zennegg, M. (2005). Brominated flame retardants in waste electrical and electronic equipment: Substance flows in a recycling plant. *Environmental Science and Technology, 39* (22), 8691–8699.

Morgan, M. G., Henrion, M., and Small, M. (1990). *Uncertainty—A Guide to Dealing with Uncertainty in Quantitative Risk and Policy Analysis.* Cambridge, U.K.: Cambridge University Press.

Morris, W. (1982). *American Heritage Dictionary (2nd College Edition).* Boston: Houghton Mifflin.

Müller, D., Bader, H.-P., and Baccini, P. (2004). Physical characterization of regional timber management for a long-term scale. *Journal of Industrial Ecology, 8* (3), 65–88.

Müller, D. B. (2006). Stock dynamics for forecasting material flows—Case study for housing in The Netherlands. *Ecological Economics, 59* (1), 142–156.

Müller, E., Hilty, L. M., Widmer, R., Schluep, M., and Faulstich, M. (2014). Modeling metal stocks and flows: A review of dynamic material flow analysis methods. *Environmental Science and Technology, 48* (4), 2102–2113.

Müller, M. (2013). *Developing a method for economic and ecologic optimization of manufacturing processes based on material flow analysis (MFA)* (Doctoral dissertation). Vienna, Austria: Technische Universität Wien.

Müller-Wenk, R. (1978). *Die ökologische Buchhaltung: ein Informations- und Steuerungsinstrument für umweltkonforme Unternehmenspolitik.* Frankfurt: Campus Verlag.

MUNLV. (2000). *Die Stofflussanalyse im immissionsschutzrechtlichen Genehmigungsverfahren (Erlass).* Düsseldorf, Germany: Ministerium für Umwelt und Naturschutz, Landwirtschaft und Verbraucherschutz des Landes N. W.

Myer, A., and Chaffee, C. (1997). Life-cycle analysis for design of the Sydney Olympic Stadium. *Renewable Energy, 10* (2–3), 169–172.

Narasimhan, S., and Jordache, C. (2000). *Data Reconciliation and Gross Error Detection—An Intelligent Use of Process Data.* Houston, Texas: Gulf Publishing.

Narodoslawsky, M., and Krotscheck, C. (1995). The sustainable process index (SPI): Evaluating processes according to environmental compatibility. *Journal of Hazardous Materials, 41* (2–3), 383–397.

Nassar, N. T., Barr, R., Browning, M., Diao, Z., Friedlander, E., Harper, E. M., ... Graedel, T. E. (2011). Criticality of the geological copper family. *Environmental Science and Technology, 46* (2), 1071–1078.

Newcombe, K., Kalma, I. D., and Aston, A. R. (1978). The metabolism of a city: The case of Hong Kong. *Ambio, 7* (3), 3–15.

Nriagu, J. O. (1994). *Arsenic in the Environment*. West Sussex, U.K.: Wiley.

Obernberger, I., Biedermann, F., Widmann, W., and Riedl, R. (1997). Concentrations of inorganic elements in biomass fuels and recovery in the different ash fractions. *Biomass and Bioenergy, 12* (3), 211–224.

Obernosterer, R. W. (1994). *Flüchtige Halogenkohlenwasserstoffe FCKW, CKW, Halone Stoffflußanalyse Österreich*. (Diplomarbeit), TU Wien, Wien.

Obernosterer, R., and Brunner, P. H. (2001). Urban management: The example of lead. *Water, Air and Soil Pollution*, (Focus 1), 241–253.

Obernosterer, R., Brunner, P. H., Daxbeck, H., Gagan, T., Glenck, E., Hendriks, C. et al. (1998). *Materials Accounting as a Tool for Decision Making in Environmental Policy—Mac TEmPo Case Study Report—Urban Metabolism, The City of Vienna*.

Obrist, J., Steiger, B., Schulin, R., Schärer, F., and Baccini, P. (1993). Regionale Früherkennung der Schwermetall- und Phosphorbelastung von Landwirtschaftsböden mit der Stoffbuchhaltung "Proterra." *Landwirtsch. Schweiz*, (6), 513.

OECD. (2008). *Measuring Material Flows and Resource Productivity. The OECD Guide Volume I*. Paris, France: OECD Publishing.

Oliaei, F., Weber, R., and Watson, A. (2010). PBDE contamination in minnesota landfills, waste water treatment plants and sediments as PBDE sources and reservoirs. *Organohalogen Compounds, 72*, 1346–1349.

Organization for Economic Cooperation and Development (OECD). (1994). *Environmental Indicators*. Paris, France: OECD Publications.

O'Rourke, D., Connelly, L., and Koshland, C. P. (1996). Industrial ecology: A critical review. *International Journal of Environment and Pollution, 6* (2–3), 89–112.

Oswald, F., and Baccini, P. (2003). *Netzstadt—Designing the Urban*. Basel, Switzerland: Birkhäuser.

Ott, C., and Rechberger, H. (2012). The European phosphorus balance. *Resources, Conservation and Recycling, 60* (0), 159–172.

Patten, B. C. (1971). A primer for ecological modeling and simulation with analog and digital computers. In B. C. Patten (Ed.), *Systems Analysis and Simulation in Ecology* (Chap. 1). New York: Academic Press.

Pauliuk, S., Wang, T., and Müller, D. B. (2013). Steel all over the world: Estimating in-use stocks of iron for 200 countries. *Resources, Conservation and Recycling, 71* (0), 22–30.

Paumann, R., Obernosterer, R., and Brunner, P. H. (1997). *Wechselwirkungen zwischen anthropogenem und natürlichem Stoffhaushalt der Stadt Wien am Beispiel von Kohlenstoff, Stickstoff und Blei*.

Penman, H. L. (1948). Natural evaporation from open water, bare soil and grass. *Proceedings of the Royal Society of London A, 193*, 120–145.

Pesendorfer, D. (2002). Integrierte Stoff- und Ressourcenpolitik—Beitrag und Grenzen von Material Flow Analysis zu einem inputorientierten Ansatz. *Europäische Hochschulschriften, Band Politikwissenschaft, 31* (445), ISBN 978-3-631-38807-5.

Pitard, F. F. (1989). *Pierre Gy's Sampling Theory and Sampling Practice* (Vol. II). Boca Raton, FL: CRC Press.

Potting, J., and Blok, K. (1995). Life-cycle assessment of four types of floor covering. *Journal of Cleaner Production, 3* (4), 201–213.

Primault, B. (1962). Du calcul de l'évapotranspiration. *Archiv für Meteorologie, Geophysik und Bioklimatologie Serie B, 12* (1), 124–150.

Raghu, N. K., and James, F. L. (2007). Trapezoidal and triangular distributions for type B evaluation of standard uncertainty. *Metrologia, 44* (2), 117.

Rant, Z. (1956). Exergie, ein neues Wort für technische Arbeitsfähigkeit. *Forschung im Ingenieurwesen, 22* (1), 36.

Rauhut, A., and Balzer, D. (1976). Verbrauch und Verbleib von Cadmium in der Bundesrepublik Deutschland im Jahr 1973. *Metall, 30,* 269.

Rechberger, H. (1999). *Entwicklung einer Methode zur Bewertung von Stoffbilanzen in der Abfallwirtschaft. (Doctoral Thesis).* Vienna, Austria: Technische Universität Wien.

Rechberger, H., and Brunner, P. H. (2002a). Die Methode der Stoffkonzentrierungseffizienz (SKE) zur Bewertung von Stoffbilanzen in der Abfallwirtschaft. In G. Hösel, B. Bilitewski, W. Schenkel, and H. Schnurer (Eds.), *Müllhandbuch* (pp. 1–18). Berlin: Erich Schmidt Verlag.

Rechberger, H., and Brunner, P. H. (2002b). A new, entropy based method to support waste and resource management decisions. *Environmental Science and Technology, 36* (4), 809–816.

Rechberger, H., and Graedel, T. E. (2002). The contemporary European copper cycle: Statistical entropy analysis. *Ecological Economics, 42* (1–2), 59–72.

Reijnders, L. (1998). The factor X debate: Setting targets for eco-efficiency. *Journal of Industrial Ecology, 2* (1), 13–22.

Reimann, D. O. (1989). Heavy metals in domestic refuse and their distribution in incinerator residues. *Waste Management and Research, 7* (1), 57–62.

Resource Conservation and Recovery Act. (1976). U.S. Code Collection, Title 42, Chap. 82, Solid Waste Disposal, Subchap. I, General Provisions, Sec. 6902. Retrieved from http://www4.law.cornell.edu/uscode/42/6902.html

Ritthoff, M., Rohn, H., and Liedtke, C. (2002). MIPS berechnen—Ressourcenproduktivität von Produkten und Dienstleistungen. *Wuppertal Spezial 27.* Wuppertal, Germany: Wuppertal Institute.

Romagnoli, J. A., and Sánchez, M. C. (2000). *Data processing and reconciliation for chemical process operations* (Vol. 2). San Diego, CA: Academic.

Rosen, M. A., and Dincer, I. (2001). Exergy as the confluence of energy, environment and sustainable development. *Exergy, An International Journal, 1* (1), 3–13.

Ruth, M. (1995). Thermodynamic implications for natural resource extraction and technical change in U.S. copper mining. *Environmental and Resource Economics, 6* (2), 187–206.

Rydh, C. J., and Karlström, M. (2002). Life cycle inventory of recycling portable nickel–cadmium batteries. *Resources, Conservation and Recycling, 34* (4), 289–309.

Saltelli, A., and Annoni, P. (2010). How to avoid a perfunctory sensitivity analysis. *Environmental Modelling and Software, 25* (12), 1508–1517. doi: http://dx.doi.org/10.1016/j.envsoft.2010.04.012

Saltelli, A., Ratto, M., Andres, T., Campolongo, F., Cariboni, J., Gatelli, D.,... and Tarantola, S. (2008). *Global Sensitivity Analysis. The Primer.* Chichester, England: John Wiley & Sons, Ltd.

Sand, P. H. (2000). The precautionary principle: A European perspective. *Human and Ecological Risk Assessment (HERA), 6* (3), 445–458.

Sandin, P. (1999). Dimensions of the precautionary principle. *Human and Ecological Risk Assessment (HERA), 5* (5), 889–907.

Sax, N. I., and Lewis, R. J. (1987). *Hawley's Condensed Chemical Dictionary (11th Edition)*. New York: Van Nostrand Reinhold.

Schachermayer, E., Bauer, G., Ritter, E., and Brunner, P. H. (1995). *Messung der Güter- und Stoffbilanz einer Müllverbrennungsanlage (Project MAPE)*. Vienna, Austria: Umweltbundesamtes Wien GmbH.

Schachermayer, E., Lahner, T., and Brunner, P. H. (2000). Assessment of two different separation techniques for building wastes. *Waste Management and Research, 18* (1), 16–24.

Schachermayer, E., Rechberger, H., Maderner, W., and Brunner, P. H. (1995). *Systemanalyse und Stoffbilanz eines kalorischen Kraftwerkes (SYSTOK)*. Vienna, Austria: Umweltbundesamt Wien GmbH.

Scheffer, F. (1989). *Lehrbuch der Bodenkunde (12th Edition)*. Stuttgart, Germany: Ferdinand Enke.

Schleisner, L. (2000). Life cycle assessment of a wind farm and related externalities. *Renewable Energy, 20* (3), 279–288.

Schlummer, M., Gruber, L., Mäurer, A., Wolz, G., and Van Eldik, R. (2007). Characterisation of polymer fractions from waste electrical and electronic equipment (WEEE) and implications for waste management. *Chemosphere, 67*, 1866–1876.

Schmidt, M., and Häuslein, A. (1997). *Ökobilanzierung mit Computerunterstützung—Produktbilanzen und betriebliche Bilanzen mit dem Programm Umberto*. Berlin, Germany: Springer Verlag.

Schmidt-Bleek, F. (1994). *Wieviel Umwelt braucht der Mensch? MIPS—Das Mass für ökologisches Wirtschaften*. Berlin: Birkhäuser.

Schmidt-Bleek, F. (1997). *Wieviel Umwelt braucht der Mensch? Faktor 10—das Maß für ökologisches Wirtschaften*. Munich: DTV.

Schmidt-Bleek, F. (1998). *Das MIPS-Konzept*. Munich: Droemer Knaur.

Schmidt-Bleek, F., Bringezu, S., Hinterberger, F., Liedtke, C., Spangenberg, J., Stiller, H., and Welfen, M. J. (1998). *MAIA-Einführung in die Material-Intensitäts-Analyse nach dem MIPS-Konzept*. Berlin: Birkhäuser.

Schönbäck, W., Kosz, M., and Madreiter, T. (1997). *Nationalpark Donauauen: Kosten-Nutzen-Analyse*. New York: Springer.

Schwab, O., Laner, D., and Rechberger, H. (in press). Quantitative evaluation of data quality in regional material flow analysis. *Journal of Industrial Ecology*.

Schwab, O., Zoboli, O., and Rechberger, H. (2016). Data characterization framework for material flow analysis. *Journal of Industrial Ecology, 20*, 1–10.

Science Communication Unit. (2013). *Policy In-depth Report: Resource Efficiency Indicators*. University of the West of England, Ed.; Science for Environment Report Produced for the European Commission DG Environment. Bristol, UK: University of the West of England.

Seppälä, J., Koskela, S., Melanen, M., and Palperi, M. (2002). The Finnish metals industry and the environment. *Resources, Conservation and Recycling, 35* (1–2), 61–76.

Seppälä, J., Melanen, M., Jouttijärvi, T., Kauppi, L., and Leikola, N. (1998). Forest industry and the environment: A life cycle assessment study from Finland. *Resources, Conservation and Recycling, 23* (1–2), 87–105.

Seppelt, R., Manceur, A. M., Liu, J., Fenichel, E. P., and Klotz, S. (2014). Synchronized peak-rate years of global resources use. *Ecology and Society, 19* (4), 50.

Settle, D. M., and Patterson, C. C. (1980). Lead in albacore: Guide to lead pollution in Americans. *Science, 207* (4436), 1167–1176.

Shannon, C. E. (1948). A mathematical theory of communications. *Bell System Technical Journal (formerly AT&T Tech. J.), (27),* 379.

Sindiku, O., Babayemi, J., Osibanjo, O., Schlummer, M., Schluep, M., Watson, A., and Weber, R. (2014). Polybrominated diphenyl ethers listed as Stockholm Convention POPs, other brominated flame retardants and heavy metals in e-waste polymers in Nigeria. *Environmental Science and Pollution Research International, 22* (19), 14489–14501.

Sobańtka, A., and Rechberger, H. (2013). Extended statistical entropy analysis (eSEA) for improving the evaluation of Austrian wastewater treatment plants. *Water Science and Technology, 67* (5), 1051–1057.

Sobańtka, A. P., Zessner, M., and Rechberger, H. (2012). The extension of statistical entropy analysis to chemical compounds. *Entropy, 14,* 2413–2426.

Sobańtka, A. P., Pons, M.-N., Zessner, M., and Rechberger, H. (2013a). Implementation of extended statistical entropy analysis to the effluent quality index of the benchmarking simulation model no. 2, *Water, 6,* 86–103.

Sobańtka, A. P., Thaler, S., Zessner, M., and Rechberger, H. (2013b). Extended statistical entropy analysis for the evaluation of nitrogen budgets in Austria. *International Journal of Environmental Science and Technology, 11* (7), 1947–1958.

Somlyódy, L., Brunner, P. H., and Kroiß, H. (1999). Nutrient balances for Danube countries: A strategic analysis. *Water Science and Technology, 40* (10), 9–16.

Somlyódy, L., Brunner, P. H., Fenz, R., Kroiß, H., Lampert, C., and Zessner, M. (1997). *Nutrient Balances for Danube Countries.* Final Report Project EU/AR/102A/91, Service Contract 95–0614.00, PHARE Environmental Program for the Danube River Basin ZZ 9111/0102. Vienna, Austria: Consortium TU Vienna Institute for Water Quality and Waste Management, and TU Budapest Department of Water and Waste Water Engineering.

Song, H.-S., and Hyun, J. C. (1999). A study on the comparison of the various waste management scenarios for PET bottles using the life-cycle assessment (LCA) methodology. *Resources, Conservation and Recycling, 27* (3), 267–284.

Spatari, S., Bertram, M., Fuse, K., Graedel, T. E., and Rechberger, H. (2002). The contemporary European copper cycle: 1 year stocks and flows. *Ecological Economics, 42* (1–2), 27–42.

Stahel, U., Fecker, I., Förster, R., Maillefer, C., and Reusser, L. (1998). Bewertung von Ökoinventaren für Verpackungen. *Schriftenreihe Umwelt No. 300.* Bern, Switzerland: BUWAL.

STAN. (2016). STAN software for substance flow analysis [Software]. Retrieved from http://www.stan2web.net/

Steen, I. (1998). Phosphorous availability in the 21st century—Management of a non-renewable resource. *Phosphorus and Potassium, 217,* 25–31.

Stegemann, J. A., Caldwell, R. J., and Shi, C. (1997). Variability of field solidified waste. *Journal of Hazardous Materials, 52* (2–3), 335–348.

Steindl-Rast, D., and Lebell, S. (2002). *Music of Silence.* Berkeley, CA: Seastone.

Steinmüller, H., and Krotscheck, C. (1997). *Regional assessment with the sustainable process index (SPI) concept.* Paper presented at the 7th International Conference on Environ-Metrics, Innsbruck.

Stigliani, W. M., Jaffé, P. R., and Anderberg, S. (1993). Heavy metal pollution in the Rhine Basin. *Environmental Science and Technology, 27* (5), 786–793.

Stockholm Convention. (2004). The Stockholm Convention on Persistent Organic Pollutants. Retrieved from http://chm.pops.int/TheConvention/Overview/tabid /3351/Default.aspx

Stumm, W., and Davis, J. (1974). Kann Recycling die Umweltbeeinträchtigung vermindern? *Recycling: Lösung der Umweltkrise?* Zurich: Gottlieb Duttweiler-Institut für wirtschaftliche und soziale Studien.

Szargut, J., Morris, D., and Steward, F. (1988). *Exergy Analysis of Thermal, Chemical, and Metallurgical Processes.* New York: Hemisphere Publishing.

Tanikawa, H., Fishman, T., Okuoka, K., and Sugimoto, K. (2015). The weight of society over time and space: A comprehensive account of the construction material stock of Japan, 1945–2010. *Journal of Industrial Ecology, 19* (5), 778–791.

Tasaki, T., Takasuga, T., Osako, M., and Sakai, S. (2004). Substance flow analysis of brominated flame retardants and related compounds in waste TV sets in Japan. *Waste Management, 24* (6), 571–580.

Tauber, C. (1988). *Spurenelemente in Flugaschen.* Köln: Verlag TÜV Rheinland GmbH.

Thomas, V., and Spiro, T. (1994). *Emissions and Exposures to Metals: Cadmium and Lead, in Industrial Ecology and Global Change.* Cambridge, U.K.: Cambridge University Press.

Tonini, D., Martinez-Sanchez, V., and Astrup, T. F. (2013). Material resources, energy, and nutrient recovery from waste: Are waste refineries the solution for the future? *Environmental Science and Technology, 47* (15), 8962–8969.

Trinkel, V., Kieberger, N., Rechberger, H., and Fellner, J. (2015). Material flow accounting at plant level—Case study heavy metal flows in blast furnace processes. In J. Lederer, D. Laner, H. Rechberger, and J. Fellner (Eds.), *International Workshop Mining the Technosphere: Drivers and Barriers, Challenges and Opportunities*, Vienna, Austria: Technische Universität Wien.

Truttmann, N., and Rechberger, H. (2006). Contribution to resource conservation by reuse of electrical and electronic household appliances. *Resources, Conservation and Recycling, 48* (3), 249–262.

Tukker, A., and Kleijn, R. (1997). *Using SFA and LCA in a precautionary approach: The case of chlorine and PVC.* In ConAccount workshop, edited by S. Bringezu et al. Leiden, The Netherlands: Wuppertal Institute for Climate, Environment and Energy.

U.S. Bureau of Mines. (1933–1996). *Minerals Yearbook 1932–1994.* Washington, DC: U.S. Department of the Interior (formerly U.S. Bureau of Mines).

U.S. EPA. (2002). *Municipal solid waste in the United States: 2000 facts and figures EPA 530-R-02-001.* Washington, DC: Office of Solid Waste and Emergency Response.

U.S. Geological Survey. (2001a). Mineral Commodity Summaries, Cadmium. Retrieved from http://minerals.usgs.gov/minerals/pubs/commodity/cadmium/

U.S. Geological Survey. (2001b). Minerals Yearbook, Cadmium. Retrieved from http://minerals.usgs.gov/minerals/pubs/commodity/cadmium/

Udluft, P. (1981). Bilanzierung des Niederschlagseintrags und Grundwasseraustrags von anorganischen Spurenstoffen. *Hydrochemisch-hydrogeologische Mitteilungen, 4*, 61–80.

Udo de Haes, H. A. (1996). *Towards a Methodology for Life Cycle Assessment.* Brussels, Belgium: SETAC Europe.

Udo de Haes, H. A., Heijungs, R., Huppes, G., Van der Voet, E., and Hettelingh, J.-P. (2000). Full mode and attribution mode in environmental analysis. *Journal of Industrial Ecology, 4* (1), 45–56.

UNEP. (2015a). Draft guidance for the inventory of polybrominated diphenyl ethers (PBDEs) listed under the Stockholm Convention on Persistent Organic Pollutants. Retrieved from http://chm.pops.int/Implementation/NIPs/Guidance/Guidance fortheinventoryofPBDEs/tabid/3171/Default.aspx

UNEP. (2015b). Guidance on best available techniques and best environmental practices for the recycling and disposal of wastes containing polybrominated diphenyl ethers (PBDEs) listed under the Stockholm Convention on Persistent Organic Pollutants. United Nations Environment Programme. Retrieved from http://chm.pops.int/Portals/0/download.aspx?d=UNEP-POPS-COP.7-INF -22.English.pdf

United Nations Conference on Environment and Development. (1992). *Report of the Conference, Rio de Janeiro, A/CONF.151/26/Rev.1.1*, New York.

Vadenbo, C., Hellweg, S., and Guillén-Gosálbez, G. (2014a). Multi-objective optimization of waste and resource management in industrial networks—Part I: Model description. *Resources, Conservation and Recycling, 89* (0), 52–63.

Vadenbo, C., Guillén-Gosálbez, G., Saner, D., and Hellweg, S. (2014b). Multi-objective optimization of waste and resource management in industrial networks—Part II: Model application to the treatment of sewage sludge. *Resources, Conservation and Recycling, 89* (0), 41–51.

Valsaraj, K. T. (2000). *Elements of Environmental Engineering*. Boca Raton, FL: CRC Press.

van der Voet, E. (1996). *Substances from Cradle to Grave: Development of a Methodology for the Analysis of Substance Flows through the Economy and the Environment of a Region*. Amsterdam: Optima Druck.

van der Voet, E. (2002). Substance flow analysis methodology. In: R. U. Ayres and L. W. Ayres (Eds.), *A Handbook of Industrial Ecology*. Cheltenham, UK: Edward Elgar.

van der Voet, E., Guinée, J. B., and Udo de Haes, H. A. (2000). *Heavy Metals: A Problem Solved?* Dordrecht, The Netherlands: Kluwer Academic Publishers.

van der Voet, E., van Egmond, L., Kleijn, R., and Huppes, G. (1994). Cadmium in the European Community: A policy-oriented analysis. *Waste Management and Research, 12* (6), 507–526.

Vehlow, J., and Mark, E. (1997). *Electrical and Electronic Plastics Waste Co-combustion with Municipal Solid Waste for Energy Recovery*. Brussels, Belgium: Association of Plastics Manufacturers in Europe (APME). Retrieved from http://www.seas .columbia.edu/earth/wtert/sofos/nawtec/nawtec05/nawtec05-19.pdf

Vehlow, J. (1993). *Heavy Metals in Waste Incineration*. Paper presented at the DAKOFA Conference, Copenhagen, Denmark.

Verougstraete, V., Lison, D., and Hotz, P. (2002). A systematic review of cytogenetic studies conducted in human populations exposed to cadmium compounds. *Mutation Research—Reviews in Mutation Research, 511* (1), 15–43.

Vidal, B. (1985). *Histoire de la Chemie*. Paris: Presses Universitaire de France.

Vogel, F. (1996). *Beschreibende und schließende Statistik*. Wien: R. Oldenbourg.

Von Steiger, B., and Baccini, P. (1990). *Regionale Stoffbilanzierung landwirtschaftlicher Böden*. Bern-Liebefeld, Schweiz: Nationales Forschungsprogramm Boden NFP 22.

von Weizsäcker, E. U., Lovins, A. B., and Lovins, L. H. (1997). *Faktor vier. Doppelter Wohlstand—halbierter Verbrauch*. Munich: Droemer Knaur.

Vyzinkarova, D., and Brunner, P. H. (2013). Substance flow analysis of wastes containing polybrominated diphenyl ethers: The need for more information and for final sinks. *Journal of Industrial Ecology, 17* (6), 900–911.

Waalkes, M. P. (2000). Cadmium carcinogenesis in review. *Journal of Inorganic Biochemistry, 79* (1–4), 241–244.

Wackernagel, M., and Rees, W. (1996). *Our Ecological Footprint—Reducing Human Impact on the Earth.* Philadelphia: New Society Publishers.

Wackernagel, M., Monfreda, C., and Deumling, D. (2002). *Ecological Footprint of Nations* (November 2002 update), Redefining Progress. Retrieved from http://www.redefiningprogress.org/publications/ef1999.pdf

Wäger, P. A., Schluep, M., Müller, E., and Gloor, R. (2011). RoHS regulated substances in mixed plastics from waste electrical and electronic equipment. *Environmental Science and Technology, 46* (2), 28–635.

Wall, G. (1977). *Exergy—A Useful Concept within Resource Accounting.* Göteborg, Sweden: Institute of Theoretical Physics, Chalmers University of Technology and University of Göteborg.

Wall, G. (1993). *Exergy, ecology and democracy—Concepts of a vital society.* Paper presented at the ENSEC '93 International Conference on Energy Systems and Ecology, Krakow, Poland.

Wall, G., Sciubba, E., and Naso, V. (1994). Exergy use in the Italian society. *Energy, 19,* 1267–1274.

Widman, J. (1998). Environmental impact assessment of steel bridges. *Journal of Constructional Steel Research, 46* (1–3), 291–293.

Wiedmann, T., Schandl, H., Lenzen, M., Moran, D., Su, S., West, J., and Kanemoto, K. (2013). The material footprint of nations. *PNAS, 112* (20), 6271–6276.

Wilson, D. (1990). Recycling of demolition wastes. In F. Moavenzadeh (Ed.), *Concise Encyclopedia of Building and Construction Materials* (pp. 517–518). Oxford: Pergamon/Elsevier.

Winterstetter, A., Laner, D., Rechberger, H., and Fellner, J. (2015). Framework for the evaluation of anthropogenic resources: A landfill mining case study—Resource or reserve? *Resources, Conservation and Recycling, 96* (0), 19–30.

Wrisberg, N., Udo de Haes, H. A., Triebswetter, U., Eder, P., and Clift, R. (2002). *Analytical Tools for Environmental Design and Management in a Systems Perspective.* Dordrecht, The Netherlands: Kluwer Academic Publishers.

Xia, D. K., and Pickles, C. A. (2000). Microwave caustic leaching of electric arc furnace dust. *Minerals Engineering, 13* (1), 79–94.

Yamasue, E., Matsubae, K., Nakajima, K., Hashimoto, S., and Nagasaka, T. (2013). Using total material requirement to evaluate the potential for recyclability of phosphorous in steelmaking dephosphorization slag. *Journal of Industrial Ecology, 17* (5), 722–730.

Youcai, Z., and Stanforth, R. (2000). Integrated hydrometallurgical process for production of zinc from electric arc furnace dust in alkaline medium. *Journal of Hazardous Materials, 80* (1–3), 223–240.

Yu, C. C., and Maclaren, V. (1995). A comparison of two waste stream quantification and characterization methodologies. *Waste Management and Research, 13* (4), 343–361.

Yue, Q., Lu, Z. W., and Zhi, S. K. (2009). Copper cycle in China and its entropy analysis. *Resources, Conservation and Recycling, 53* (12), 680–687.

Zadeh, L. A. (1965). Fuzzy sets. *Information and Control, 8,* 338–353.

Zeltner, C., Bader, H. P., Scheidegger, R., and Baccini, P. (1999). Sustainable metal management exemplified by copper in the USA. *Regional Environmental Change, 1* (1), 31–46.

Zessner, M., Fenz, R., and Kroiss, H. (1998). Waste water management in the Danube Basin. *Water Science and Technology, 38,* 41–49.

Zoboli, H. (2016). Novel approaches to enhance regional nutrients management and monitoring applied to the Austrian phosphorous case study (PhD Thesis). Vienna: Technische Universität Wien.

Zoboli, O., Laner, D., Zessner, M., and Rechberger, H. (2015). Added values of time series in material flow analysis: The Austrian phosphorus budget from 1990 to 2011. *Journal of Industrial Ecology.* doi: 10.1111/jiec.1238.

Zoboli, O., Zessner, M., and Rechberger, H. (2016). Supporting phosphorus management in Austria: Potential, priorities and limitations. *Science of the Total Environment, 565,* 313–323.

Zuser, A., and Rechberger, H. (2011). Considerations of resource availability in technology development strategies: The case study of photovoltaics. *Resources, Conservation and Recycling, 56,* 56–65.

Index